STRENGTH OF MATERIALS

Lab Testing • Theory & Problems

H M Raghunath

BE, MSc., (Engg.)

Former, Professor in Civil Engineering
Manipal Institute of Technology
Manipal, Karnataka
Now, Advisor in Water Resources Evaluation

NEW AGE INTERNATIONAL (P) LIMITED, PUBLISHERS

LONDON • NEW DELHI • NAIROBI

Bangalore • Chennai • Cochin • Guwahati • Hyderabad • Kolkata • Lucknow • Mumbai

Visit us at **www.newagepublishers.com**

Published by New Age International (P) Ltd., Publishers
First Edition: 2010
Reprint: 2018

GLOBAL OFFICES

- **New Delhi** **NEW AGE INTERNATIONAL (P) LIMITED, PUBLISHERS**
7/30 A, Daryaganj, New Delhi-110002, (INDIA)
Tel.: (011) 23253771, 23253472, **Telefax:** 23267437, 43551305
E-mail: contactus@newagepublishers.com • Visit us at www.newagepublishers.com
- **London** **NEW AGE INTERNATIONAL (UK) LTD.**
27 Old Gloucester Street, London, WC1N 3AX, UK
E-mail: info@newacademicscience.co.uk • Visit us at www.newacademicscience.co.uk
- **Nairobi** **NEW AGE GOLDEN (EAST AFRICA) LTD.**
Ground Floor, Westlands Arcade, Chiromo Road (Next to Naivas Supermarket)
Westlands, Nairobi, KENYA, **Tel.:** 00-254-713848772, 00-254-725700286
E-mail: kenya@newagepublishers.com

BRANCHES

- **Bangalore** 37/10, 8th Cross (Near Hanuman Temple), Azad Nagar, Chamarajpet, Bangalore- 560 018
Tel.: (080) 26756823, **Telefax:** 26756820, **E-mail: bangalore@newagepublishers.com**
- **Chennai** 26, Damodaran Street, T. Nagar, Chennai-600 017, **Tel.:** (044) 24353401, **Telefax:** 24351463
E-mail: chennai@newagepublishers.com
- **Cochin** CC-39/1016, Carrier Station Road, Ernakulam South, Cochin-682 016, **Tel.:** (0484) 2377303, **Telefax:** 4051304
E-mail: cochin@newagepublishers.com
- **Guwahati** Hemsen Complex, Mohd. Shah Road, Paltan Bazar, Near Starline Hotel, Guwahati-781 008,
Tel.: (0361) 2513881, **Telefax:** 2543669, **E-mail: guwahati@newagepublishers.com**
- **Hyderabad** 105, 1st Floor, Madhiray Kaveri Tower, 3-2-19, Azam Jahi Road, Near Kumar Theater, Nimboliadda Kachiguda, Hyderabad-500 027, **Tel.:** (040) 24652456, **Telefax:** 24652457
E-mail: hyderabad@newagepublishers.com
- **Kolkata** RDB Chambers (Formerly Lotus Cinema) 106A, 1st Floor, S N Banerjee Road, Kolkata-700 014
Tel.: (033) 22273773, **Telefax:** 22275247, **E-mail: kolkata@newagepublishers.com**
- **Lucknow** 16-A, Jopling Road, Lucknow-226 001, **Tel.:** (0522) 2209578, 4045297, **Telefax:** 2204098
E-mail: lucknow@newagepublishers.com
- **Mumbai** 142C, Victor House, Ground Floor, N.M. Joshi Marg, Lower Parel, Mumbai-400 013
Tel.: (022) 24927869, **Telefax:** 24915415, **E-mail: mumbai@newagepublishers.com**
- **New Delhi** 22, Golden House, Daryaganj, New Delhi-110 002, **Tel.:** (011) 23262368, 23262370, **Telefax:** 43551305
E-mail: sales@newagepublishers.com

ISBN: 97881-224-2761-5

C-17-06-10539

Printed in India at Ajit Printing Press, Delhi.
Typeset at Goswami Associates, Delhi.

NEW AGE INTERNATIONAL (P) LIMITED, PUBLISHERS
7/30 A, Daryaganj, New Delhi-110002
Visit us at **www.newagepublishers.com**
(CIN: U74899DL1966PTC004618)

Preface

It is necessary to check the proper heat treatment and quality of raw stock by mechanical testing into destruction. Surfaces of some polished and finished products have to be tested for hardness without leaving a severe mark. New materials have to be developed to meet the continuous demands of modern industry and standard mechanical tests are of great utility in research and development of new products.

The products of each batch have to be tested for quality control and to avoid failures in service. The finished products have to be mechanically tested to check whether they satisfy the standard specifications for commercial acceptance.

Sometimes it becomes necessary to test a structure or component without destruction or disruption as in rail-flaw detection when **non-destructive testing techniques** have to be employed.

It also becomes necessary to measure the static and dynamic strains at a point in a structure or machine component by mounting **strain gauges or strain gauge rosette** measurements.

When the stresses developed are to be known at a number of points, or stress pattern over the whole field is required **photoelastic method** is employed.

An introduction to all the above testing techniques with suitable testing equipment, test procedures, observation, graphs and analysis of test results and reporting for commercial acceptance are dealt in a logical sequence employing the laws (equations) of the mechanics (strength) of materials and the relevant standard specifications, in Parts—A & B.

In Part—C, Simple Theory of **Strength of Materials** with graded worked examples and classified problems for Assignment are given. Intelligence Questions (Objective and Viva) are given at the end.

The book is written in a concise and lucid style and profusely illustrated with schematic diagrams. It covers a complete course of Strength of Materials—Theory and Lab.

The subject is interdisciplinary and the book is of utility to Degree, Post-Degree, Diploma, AMIE & UPSC students, training and test houses, professionals in materials manufacture, research and product development.

Suggestions for improvement are welcome and will be incorporated to the extent possible in the next edition.

H.M. Raghunath

Preface

Contents

QUESTIONS AND PROBLEMS

PART A

MECHANICAL TESTING OF METALS AND TIMBER

1 Mechanical Testing of Metals and Timber

Why Mechanical Testing?

(*i*) A manufacturer producing a metal from the raw material into finished bars and sections has to check whether the mechanical properties are up to the Standard Specifications (I.S., B.S., A.S.T.M., etc.).

(*ii*) Specific service situations call for a knowledge of the hardness, Young's modulus, and ductility of the material; for instance ductility is relevent when the method of forming is considered.

(*iii*) To avoid failure in service from the use of materials with inadequate properties.

(*iv*) Production of new materials to meet the demands of modern industry by Research and Development (*R* & *D*) making use of standard mechanical tests.

(*v*) Quality control in manufacture by conducting mechanical tests on the finished product of each batch to satisfy the customers requirements. For example, the tensile strength of a casting can be altered by small changes in composition and heat treatment.

Test Specimens. Mechanical testing is usually carried out on test specimens cut from a piece of material being used. Considerable care must be exercised in selecting the representative sample, the method of preparation of the test-piece, shape and dimensions, all conforming to the standard specifications (I.S., B.S., A.S.T.M., etc.) so that the results obtained by different **test houses** are comparable.

Type of Tests. The external forces acting on a structural component may be such as to cause direct axial stress (tension or compression), direct shear stress (tangential to the surface), flexure stress (normal stress on a cross-section due to bending or bending moment), torsional shear stress (due to twisting), hardness (indentation), impact (shock) or fatigue (cyclic reversal of stresses), in the material. The relevent mechanical tests are conducted to check the strength of the material against the types of stresses that are going to be caused during service and estimate the factor of safety.

The Indian Standard Institution has prepared specifications (I.S.S. Booklets) with I.S. No. and Year and standardised the test procedures so that comparisons can easily be made. Other countries also have their own specifications like B.S. in London and A.S.T.M. in U.S.A. The relevent I.S. Specifications Booklets for each test is listed at the end of this book.

Methods of Application of Loads

(*a*) **By Adding Weights.** Though it is simple, handling of larger weights is tedious and costly.

(*b*) **Weights and Levers.** This method is specially desirable when a constant load has to be applied for a long period of time as for example in **Creep Tests.** For application of larger loads, a compound system of levers can be employed.

(*c*) **Screw-gear Mechanism.** The load may be applied by mechanical means either by turning by hand or by an electric motor.

(*d*) **Hydraulic System.** This system is often employed to move the cross-head of a testing machine. This hydraulic loading system depends on the movement of a piston or ram in a cylinder by means of oil pressure. Oil is pumped into the cylinder by a motor driven pump and valves are used to regulate the rate of application of load as indicated by a pointer on the dial gauge.

(*e*) **Electronic System.** Electronically operated Universal Testing Machines (UTM) are available.

Strain Measurements. The deformations (or strains) caused due to application of loads are usually measured by extensometers or strain gauges. Different types of strain gauges are available such as mechanical, optical, electrical, though mechanical strain gauges (in which the deformations are magnified by the spring-lever mechanism) are commonly used.

Testing Machines. Testing machines for conducting different types of tests are available. It is now usual to employ testing machines which are readily adoptable for conducting several tests like tension, compression, shear and bending. In some cases, they are provided with attachments for making torsion and hardness tests also. Such machines are called Universal Testing Machines (UTM) and are used in laboratories where owing to cost and space restrictions, it is not possible to have a number of separate machines.

Universal Testing Machine (UTM). The UTM consists of a testing (straining) unit and a control (measuring or recording) unit, connected to each other by means of three hydraulic pipes and electric cable, see Figure 1.1.

The testing unit consists of a hydraulic piston and cylinder. To this piston two cross-heads—one at bottom (B.C.) and one at top (T.C.), are rigidly fixed on two vertical columns. Another adjustable cross-head (A.C.) is supported on two screwed spindles and can be raised or lowered by rotation of the screwed spindles by an electric motor (housed in the testing unit through a chain-gear system).

The control unit has an oil tank from which oil under pressure is delivered to the cylinder in the testing unit by means of a hydraulic oil pump. The oil pump (Duplex) is driven by an electric motor housed in the control unit. The rate at which oil is pumped can be controlled by adjusting valves, thus controlling the rate of application of load.

When oil under pressure enters the cylinder in the testing unit, it lifts up the piston which actuates the bottom and top cross-heads (B.C. and T.C.) also to move up. Since the adjustable cross-head (A.C.) at the middle will be stationary during the test, a specimen held in-between this and the top cross-head will be subjected to tension.

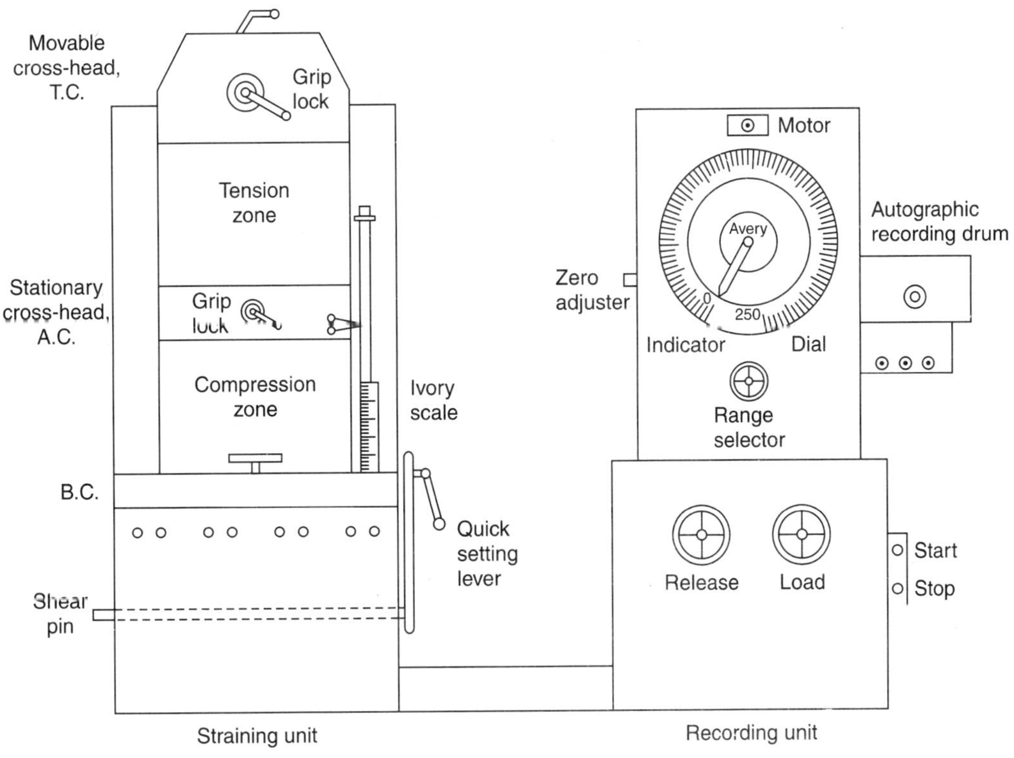

Figure 1.1. Universal testing machine

Specimens to be tested for compression, shear and bending will be placed between the adjustable cross-head and the bottom cross-head.

Universal testing machines of different capacities are available. A small UTM can be operated in three load ranges : 0–40, 0–200, and 0–400 kN and a bigger UTM in 0–100, 0–400 and 0–1000 kN. The required load range for any particular test is obtained by operating the lever provided in the control unit for this purpose.

St. Venant's Principle. If the forces acting on a small area of the body are replaced by a statically equivalent system of forces acting on the same area, there will be considerable changes in the local stress distribution, but the effect on the stress at great distances (compared with the area on which the forces act) will be negligible. For example in a tension test, the stress distribution on the gripped ends of the rod will vary according to whether gripping is by wedge jaws, screw thread or bottom head. But at distances of more than three diameters from the ends, the stress distribution is quite uniform across the cross-section.

2 Tension Test

The tension test involves clamping one end of the test specimen and applying a tensile force (load) gradually at the other end. The load applied is recorded on the dial of the measuring unit. The amount of stretching (or extension x) is noted at regular load increments. The load is gradually increased until fracture (breaking) occurs and a load-extension curve is plotted.

Test specimens may be flat or round with reduced section at the central portion and machined with parallel sides, see Figure 2.1. Flat strips can be cut directly from sheet or plate material. Round test specimens are machined from thick plates, bar stock castings, moulded components, etc.

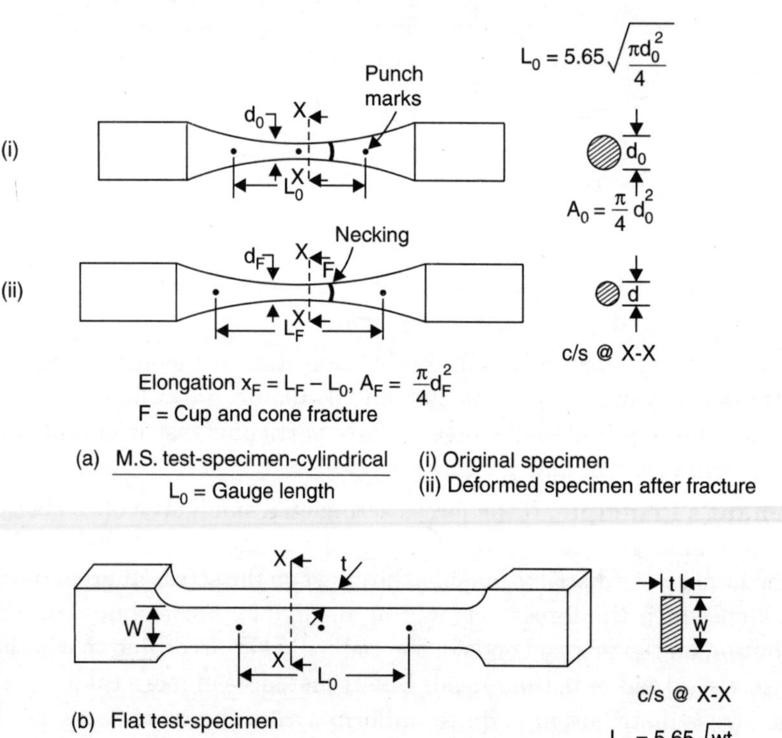

Figure 2.1. Specimens for tension test

A gauge length is marked in the central portion and is used to measure extension. It is important that the fracture occurs within the gauge length, and therefore the gauge length L_0, is considered in relation to the volume of the material being deformed or cross-sectional area A_0, of the specimen

$$L_0 = 5.65\sqrt{A_0} = 5\, d_0 \text{ (for rods)} \qquad ...(2.1)$$

Two punch marks at a length $\frac{L_0}{2}$ on either side from the centre of the gauge length are made. The **extensometer** is mounted on the specimen, lower ends of the screws (or knife edges) bearing against the punch marks, see Figure 2.2. The readings on the dial of the extensometer and the ivory scale are adjusted to zero.

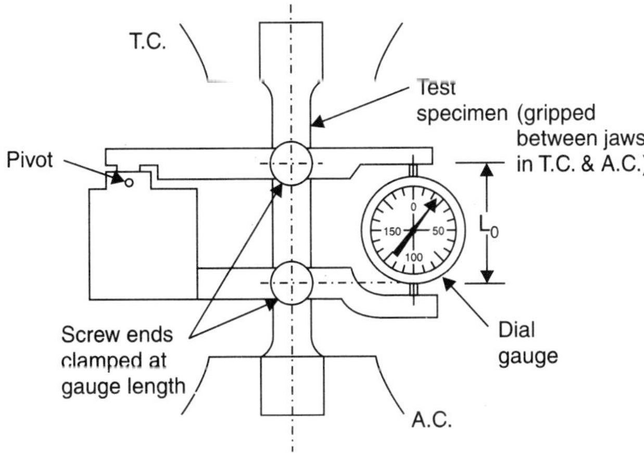

Figure 2.2. Extensometer mounted on the test-specimen

The hydraulic pump is now put into operation and the rate of loading is adjusted not to exceed 10 N/mm^2 per sec. upto the yield point. The extensometer readings are noted at regular load increments (at least 8 readings before the yield point).

As the yield point is approached, the pointer of the load measuring gauge remains stationary and the pointer of the extensometer moves very rapidly indicating a plastic flow of the material. The valve of the load at the yield point (YP) is recorded and the extensometer is removed to avoid damage.

Now in the plastic range, the rate of extension per unit load is more than that in the elastic range. When the maximum load is reached, the dummy pointer stays and the load pointer starts moving backwards because of the **neck formation**. Although the load goes on decreasing, the length of the specimen goes on increasing (see Figure 2.3), because the **true stress** (load/actual area at the neck) at the neck goes on increasing, see Figure 2.10. The specimen finally breaks at the neck with a thundering noise. The maximum (ultimate) load, and the breaking (crushing, rupture or collapse) load are recorded.

The two pieces of the broken specimen are taken out and it can be observed that the specimen has heated up indicating that in the plastic range, part of the energy is dissipated into **heat and sound** while in the elastic range, all the work done on the specimen is stored in the form of **strain energy.** The two broken pieces are put together, the length between the gauge points L_F, and the diameter at the neck d_F, are noted, see Figure 2.1(*a-ii*).

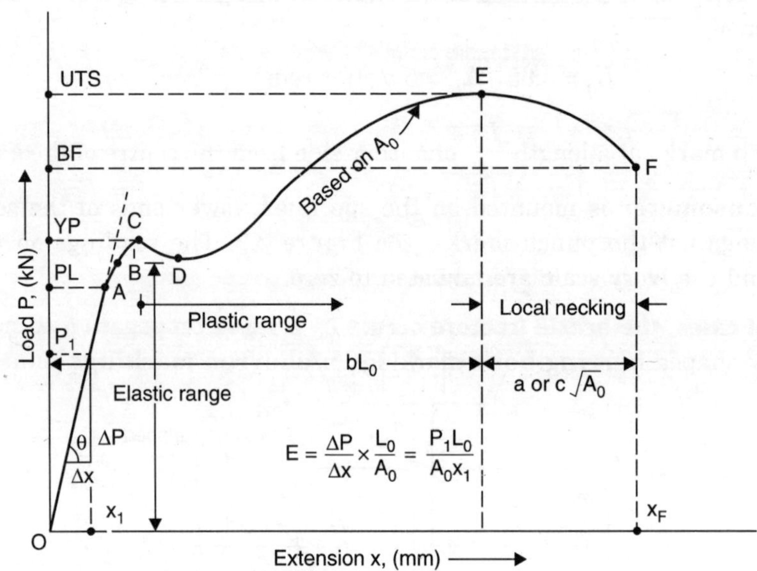

Figure 2.3. Load-extension curve for low-carbon steel [Static tension test]

Total elongation of gauge length (upto fracture), see Figure 2.3.

$$x_F = L_F - L_0 \quad \text{...(2.2)}$$

$$x_F = a + bL_0 \qquad \text{(Barba)} \quad \text{...(2.3)}$$

$$\left.\begin{aligned} x_F &= c\sqrt{A_0} + bL_0 \\ a &= c\sqrt{A_0} = \text{neck elongation} \end{aligned}\right\} \text{Unwin} \quad \text{...(2.3}a\text{)}$$

where *a, b* and *c* are constants and *b* and *c* are called **Unwin's constants.** The constants considerably vary from one metal to another depending on the grain size and microstructure.

Fracture. When the test specimen breaks under load, the fracture is studied with respect to the shape and structure of the material. The type of fracture can be classified as **ductile** or **brittle.**

Ductile fracture takes place after considerable plastic deformation and the fracture surface appears as dull, grey and fibrous. For a mild steel (M.S.) cylindrical tension specimen, the fracture is of the **cup and cone** type, which is typical of a ductile fracture, see Figure 2.4(*i*). It has a silky shining appearance and the plane of failure (shear plane) is at 45° to the axis of the specimen. This type of fracture also occurs in tensile test specimens of copper and aluminium.

In brittle materials like cast iron (C.I.), the fracture is almost flat (no necking) and the fractured surface has a bright granular appearance, see Figure 2.4 (*ii*). When rotated, the brittle failure surface depicts bright star like reflections. The brittle fracture is irregular due to shear.

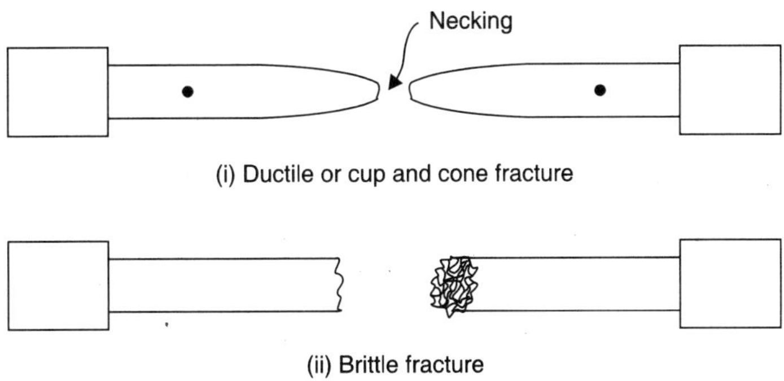

Figure 2.4. Fracture in cylindrical tensile specimens

In actual cases, the brittle fracture occurs by a rapid propagation of cracks. In plates, characteristic V-shaped **herring-bone** markings or **chevron markings** point to the origin of crack, see Figure 2.5.

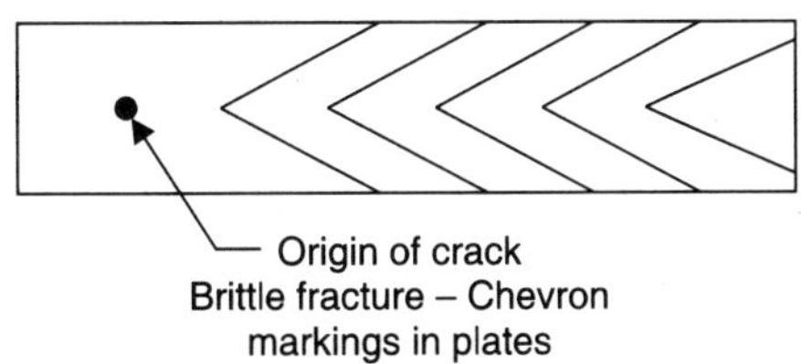

Figure 2.5. Types of fracture

Certain Definitions and Salient Points on the Load-Extension Curve

1. Stress (σ). When a structural member (or test specimen) is subjected to an external force (load P), it develops resistance i.e., internal forces to resist the external load. Stress may be defined as the force (or load per unit area). Usually, the original area of cross-section (A_0) of the test specimen is taken.

$$\sigma = \frac{P}{A_0}\text{, N/mm}^2 \qquad ...(2.4)$$

2. Strain (ε). When a load is applied, the specimen (or member) will deform (i.e., elongation or compression x, according to the axial load being tensile or compressive). Strain is deformation x, per unit length (original length (L_0) is always taken) and has no dimensions.

$$\varepsilon = \frac{x}{L_0}\text{, a decimal} \qquad ...(2.5)$$

$$\varepsilon = \frac{x}{L_0} \times 100\ \% \qquad ...(2.5a)$$

It can be seen from Eqns. (2.4) and (2.5), that the **load-extension curve** is similar to the **stress-strain curve** but only for a different scale.

3. Hooke's Law. It states that the strain is proportional to the stress producing it, within the elastic limit (EL) rather proportional limit (PL); the two points (EL) and (PL) cannot be distinguished in many curves. This constant of proportionality i.e., the ratio of normal stress to axial strain, is termed as Young's modulus of elasticity E. It is the slope of the initial straight portion of the stress-strain curve (see Figure 2.10), and can be calculated from a point in the initial straight portion of the load-extension curve, see Figure 2.3, as

$$E = \frac{\sigma}{\varepsilon} = \frac{P/A_0}{x/L_0} = \frac{PL_0}{A_0 x} \text{ N/mm}^2 \qquad \text{...(2.6)}$$

$$E = \text{Slope of the load-extn. plot} \times \frac{L_0}{A_0} \text{ N/mm}^2 \qquad \text{...(2.7)}$$

Note. Slope of the load-extn. plot = $\frac{P_1}{x_1}$, (see Figure 2.3)

4. Proportional Limit (PL). This is the point beyond which Hooke's law ceases to be obeyed (i.e., the end of the straight line plot, A) although the material may still be in the elastic state, in the sense if the load is removed, the strain becomes zero.

5. Elastic Limit (EL). This is the point B on the curve at which the elasticity ceases.

6. Permanent Set. If the material is stressed beyond the elastic limit B, on removal of stress (load), it will not regain its original length i.e., there is some residual strain. This permanent change in the length of the body (expressed as percent strain) is called **permanent set**.

7. Yield Point (YP). At this point (C) an appreciable elongation (strain) of the test specimen occurs even at constant load i.e., there is a plastic flow of the material and the material enters the plastic range. During a careful test on a M.S. specimen, a drop in load occurs immediately the yielding commences, so that there are two values known as **upper- and lower-yield points** (*Yu*, *Yl* corresponding to the points C and D).

8. Proof Stress. Materials like some alloy steels and light alloys of aluminium and magnesium do not show clearly a definite limit of proportionality or yield point in a tensile test, see Figure 2.6. For such materials showing no definite yield, a 'proof stress' is used to determine the onset of plastic range.

Proof stress is the stress at which if the material is unloaded, there will be a specified percentage of strain permanently left in it. (i.e., permanent set = 0.1% or 0.2% usually)

To determine, say 0.2% proof stress (see Figure 2.6), a tangent to the curve at the origin OT is drawn which is the initial straight portion of the curve. A 0.2% **offset** OP (= x_1) is measured such as

$$\varepsilon = 0.2\% = 0.002 = \frac{x_1}{L_0}$$

or

$$OP = x_1 = 0.002 \times L_0 \qquad \text{...(2.8)}$$

where L_0 is the original gauge length of the test specimen. Now, a line $PQ // OT$ is drawn intersecting the curve at Q. Then the stress at Q is called the **0.2% proof stress** and is used in place of yield stress.

$$0.2\% \text{ proof stress} = \frac{0.2\% \text{ proof load } (P_1)}{A_0} \qquad \text{...(2.9)}$$

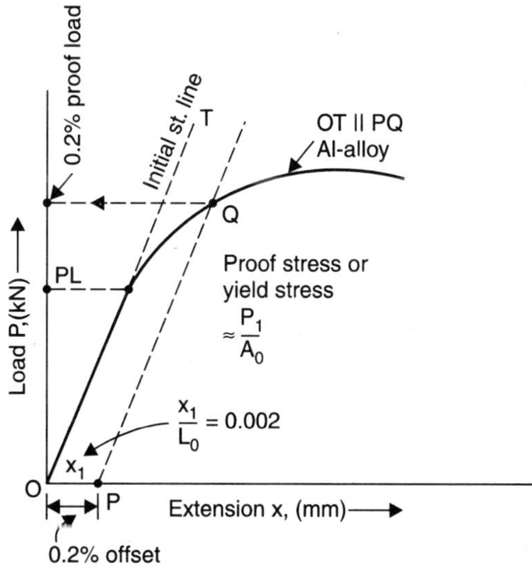

Figure 2.6. Proof stress by offset method

9. Secant Modulus. In the case of materials like C.I., wood, concrete, plastics, etc. there is no clearly defined straight part i.e., the load-extension (or stress-strain) curve is curved from the beginning (origin) itself, see Figure 2.7. In such cases, it is difficult to compute the elastic modulus and an arbitrary value is adopted known as the **secant modulus**.

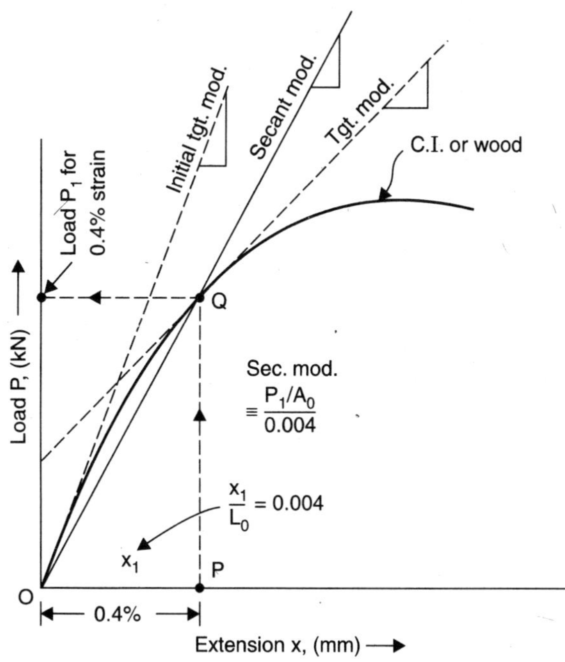

Figure 2.7. Determination of secant modulus

A point Q on the stress-strain curve (or load-extn. curve) at which a specified strain (0.1, 0.2, or 0.4%) is produced during the test is taken. The slope of the line joining this point to the origin is the **secant modulus** for the material and is obtained in Figure 2.7.

The slope of the tangent at any specified point (say, Q) on the stress-strain curve is called the **tangent modulus.**

The tangent to the curve at the origin is called the **initial tangent.**

10. Ductility. The capacity for being drawn out plastically before breaking (fracture) is called the ductility of the material. In a tensile test, this is measured by the following two quantities.

$$\% \text{ Elongation} = \frac{L_F - L_0}{L_0} \times 100 \quad \text{...(2.10)}$$

Typical values are 25-30% for ductile materials like **pure copper** and **low carbon steel** while for very high tensile strength steels and C.I. as little as 5%.

$$\% \text{ Reduction in area or contraction} = \frac{A_0 - A_F}{A_0} \times 100 \quad \text{...(2.11)}$$

Typical values are 40-60% for ductile materials and 5-10% for brittle materials.

% Contraction is considered to be a better measure of ductility being independent of the gauge length.

11. Tensile Strength (UTS). The maximum load at E, (see Figure 2.3), reached in a tensile test is called **the ultimate load.** The maximum stress or the ultimate tensile stress (UTS) is called **the tensile strength**

$$\text{UTS} = \frac{\text{Ultimate load}}{A_0} \text{ N/mm}^2 \quad \text{...(2.12)}$$

12. Breaking Stress (BF). The stress when the specimen fails (breaks or fractures) is called rupture stress (or breaking stress BF)

$$\text{Rupture stress} = \frac{\text{Breaking or collapse load}}{A_0} \quad \text{...(2.13)}$$

Calculated in this way BF < UTS, i.e., rupture stress is always lower than the tensile strength, while the true stress on the fractured area A_F, is significantly higher, see Figure 2.10.

13. Working Stress (σ). It is the safe maximum stress to which the material will be actually subjected to in service.

$$\text{Factor of safety (F.S.)} = \frac{\text{UTS}}{\text{Working stress } (\sigma)} \quad \text{...(2.14)}$$

F.S. = 3, for dead loads accurately known,

= 12, for shock loads of indefinite magnitude.

The value of the factor of safety depends on many considerations as the importance of the structure, types of loading, etc. and usually varies from 2 to 4.

$$\begin{pmatrix}\text{Working load}\\ \text{or design load}\end{pmatrix} = \begin{pmatrix}\text{Working}\\ \text{stress } (\sigma)\end{pmatrix} \times \begin{pmatrix}\text{Area of}\\ \text{cross-section}\end{pmatrix} \quad ...(2.15)$$

14. Load Factor (L.F.). Ductile materials and struts work with a load factor

$$\text{Load factor} = \frac{\text{Load at fracture (collapse)}}{\text{Working load}} \quad ...(2.16)$$

where rigidity (rather stiffness) is the main criterion, design may be based on a limiting deflection when subjected to working load.

15. Strain Energy, Resilience. For gradually applied or static load P, the work done or the **strain energy (S.E.)** in stretching the bar upto the yield point

$$\text{S.E.} = \frac{1}{2} P \times x = \frac{1}{2} (\sigma \times A) \left(\frac{\sigma}{E} \times L\right) = \left(\frac{\sigma^2}{2E}\right) AL$$

$$\text{S.E.} = \frac{\sigma}{2E} \times \text{Volume of the bar} \quad ...(2.17)$$

and is expressed in N·m or J.

The strain energy per unit volume is called **resilience (U),** and in simple tension or compression,

$$U = \frac{\sigma^2}{2E} \quad \text{N·m/m}^3 \text{ or J/m}^3 \quad ...(2.18)$$

$$= \text{area OAB in Figure 2.8}$$

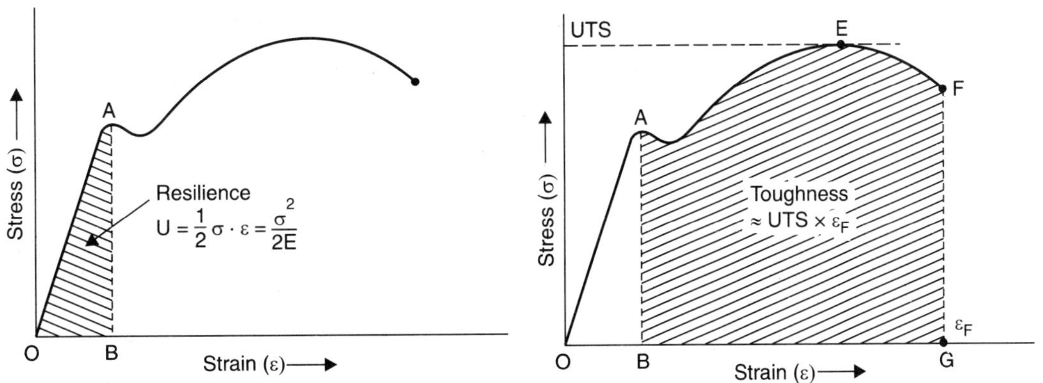

Figure 2.8. Resilience

Figure 2.9. Toughness

16. Toughness. It is the ability of the material to absorb energy in the permanent or plastic deformation range.

Toughness represents the work done or energy absorbed till fracture, per unit volume.

$$\text{Toughness} = \text{Area BAFG, (see Figure 2.9)}$$

$$\approx \text{UTS} \times \text{Strain at fracture (for ductile materials)} \quad ...(2.19)$$

$$\approx \frac{2}{3}\text{ (UTS)} \times \text{Strain at fracture (for brittle materials)} \quad ...(2.20)$$

Alternatively, toughness can also be determined from 'impact tests' described later. High toughness is particularly desirable in machine parts such as couplings, gears, chains and crane hooks.

Typical stress-strain curves for alloy steels, medium and high carbon steels, cast iron, aluminium, etc. are shown in Figure 2.10.

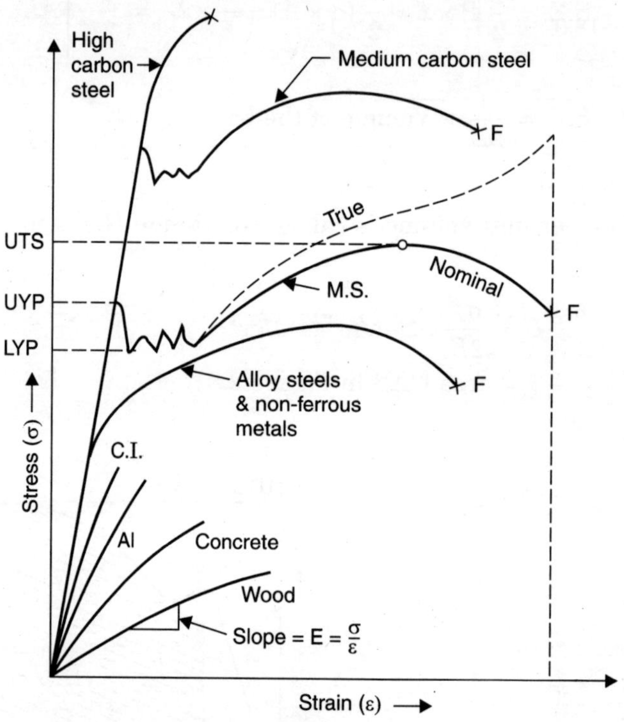

Figure 2.10. Typical stress-strain curves

Regular Tension Test. Tension test data on a steel specimen are given below. Report the results qualitatively.

Original gauge length = 50 mm

Original diameter = 12 mm

Final dia. at neck = 9.7 mm

Final length between gauge points = 62.5 mm

Sl. No.	*Load (kN)*	*Extensometer reading (L.C. = 0.0025 mm)*	*Extn. computed (mm)*
0	0	0	0
1	2.7	2	0.005
2	5.4	4	0.01
3	8.1	6	0.015
4	10.8	8	0.02
5	13.5	10	0.025
6	16.2	12	0.03
7	18.9	14	0.035
8	21.6	16	0.04
9	24.3	18	0.045
10	27.0	20	0.05
11	29.5	24	0.06
12	31.8	26	0.065
13	33.6	32	0.08
14	36.3	38	0.095
15	37.6	56	0.14
16	38.7	68	0.17
17	38.2	80 ←Extensometer removed	0.20
		Extn. noted on autographic recording (mm)	
18	37.3	0.25	
19	38.2	0.75	
20	42.7	1.20	
21	52.0	2.50	
22	56.0	3.80	
23	58.0	5.0	
24	61.4	7.5	
25	62.3	9.5	
26	60.0	11.2	
27	52.3	12.5	

TEST RESULTS

1. The load-extension diagram is drawn, see Figure 2.11.

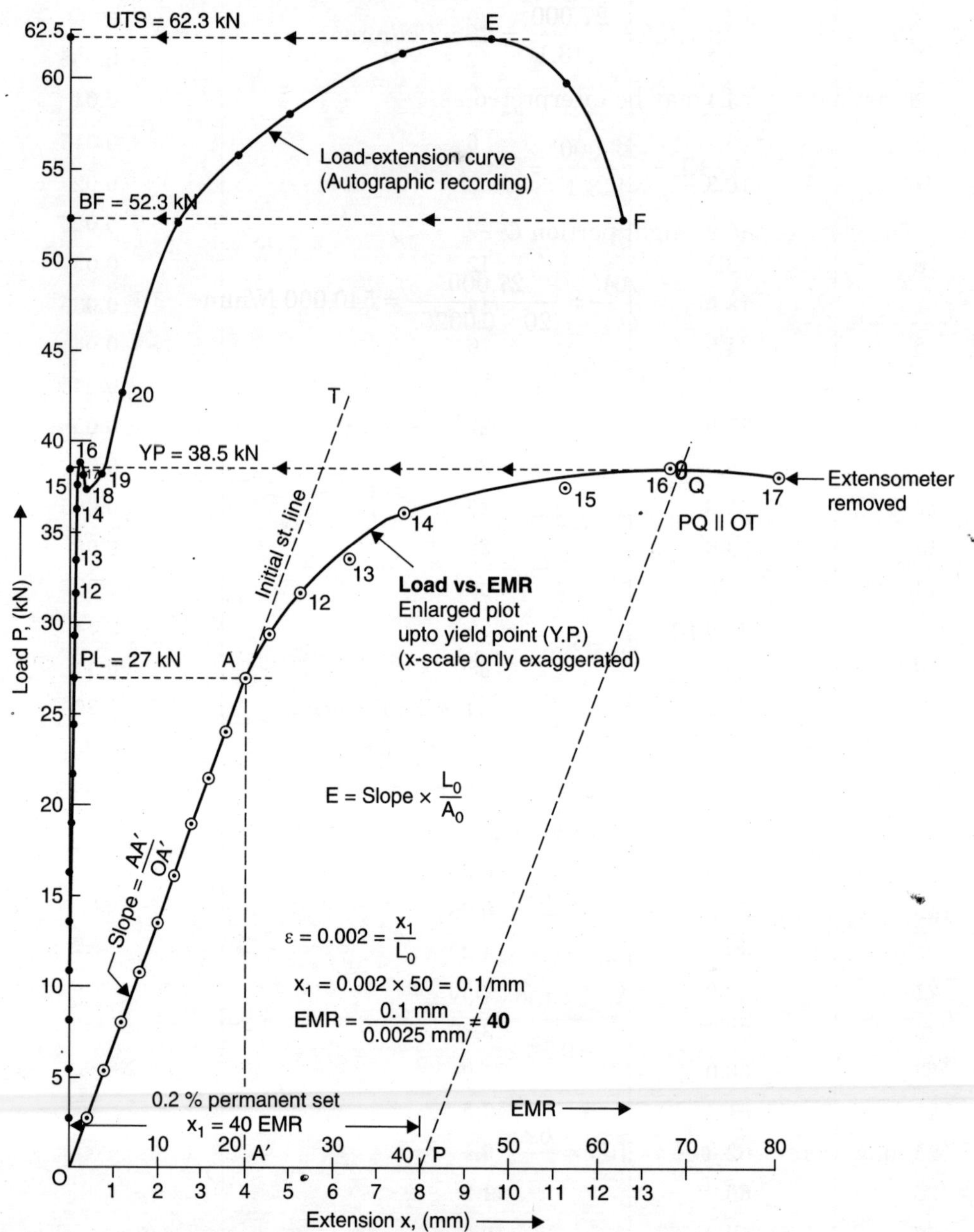

Figure 2.11. Load-extension curve for steel specimen

2. To find the yield point and the modulus of elasticity E, several points are clustered in the first part of the graph (near the yield point). To distinguish these points, the first part of the graph with the extensometer readings (EMR) is drawn to an enlarged scale i.e., Load P vs EMR.

3. The initial portion of the enlarged plot is a straight line which terminates at *A*.

$$A_0 = \frac{\pi}{4}(12^2) = 113.1 \text{ mm}^2$$

$$\text{Proportional limit (PL)} = \frac{27{,}000}{113.1} = \mathbf{238.7\ N/mm^2}$$

Elastic limit (EL) may be interpreted as

$$\text{EL} = \frac{29{,}000}{113.1} = \mathbf{256.4\ N/mm^2}$$

The slope of the straight portion *OA*

$$= \frac{AA'}{OA'} = \frac{27{,}000}{(20 \times 0.0025)} = 540{,}000 \text{ N/mm}$$

Young's modulus of elasticity

$$E = \text{Slope} \times \frac{L_0}{A_0} = 540{,}000 \frac{\text{N}}{\text{mm}} \times \frac{50}{113.1} \frac{\text{mm}}{\text{mm}^2} = \mathbf{238{,}732\ N/mm^2}$$

4. Since several points are clustered near the yield point, 0.2% proof stress is taken.

$$\text{Offset } OP = 0.002 \times L_0 = 0.002 \times 50 = 0.1 \text{ mm}$$

$$\text{EMR} - \frac{0.1}{0.0025} = 40 = x_1 \quad (PQ \parallel OT \text{ is drawn.})$$

$$\text{Yield stress} \approx \text{stress at } Q = \frac{38.5 \times 1000}{113.1} = \mathbf{340\ N/mm^2}$$

5. $\text{UTS} = \dfrac{\text{Max. load at } E}{A_0} = \dfrac{62.3 \times 1000}{113.1} = \mathbf{551\ N/mm^2}$

6. $\text{Breaking stress} = \dfrac{\text{Load at fracture } (F)}{A_0} = \dfrac{52.3 \times 1000}{113.1} = \mathbf{462\ N/mm^2}$

7. $\text{True stress (at fracture)} = \dfrac{52.3 \times 1000}{\frac{\pi}{4}(9.7^2)} = \mathbf{707.73\ N/mm^2}$

8. $\%\ \text{Elongation} = \dfrac{L_F - L_0}{L_0} \times 100 = \dfrac{62.5 - 50}{50} \times 100 = \mathbf{25\%}$

9. $\%\ \text{Reduction in area} = \dfrac{A_0 - A_F}{A_0} \times 100 = \dfrac{12^2 - 9.7^2}{12^2} \times 100 = \mathbf{34.7\%}$

10. $\text{Working stress } \sigma_t = \dfrac{\text{UTS}}{\text{F.S.}} = \dfrac{551}{3} = \mathbf{183.6\ N/mm^2}$ [$\because$ F.S. = 3]

11. Resilience (U) = $\dfrac{\sigma^2}{2E}$ per unit volume = $\dfrac{340^2}{2 \times 238{,}732}$ = **0.242 N/mm² or N·mm/mm³**

12. Toughness = UTS × ε_F, from Eq. (2.19) = 551 × $\dfrac{12.5}{50}$ = **137.75 N/mm²**

The steel bar has a good yield strength, tensile strength, ductility and toughness and passes the test for acceptance.

Commercial Tension Test. In a routine commercial test, the specimen or the rod is subjected to gradually increasing load upto failure, see Figure 2.12. No special preparation of the specimen is required, no extensometer is mounted and no graph need to be plotted. Only the yield stress, tensile strength and elongation are required.

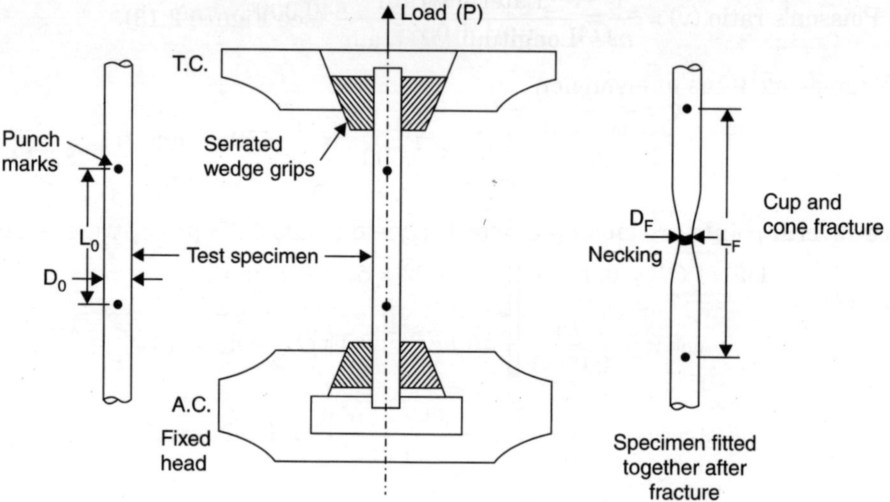

Figure 2.12. Commercial tension test

For illustration, the results obtained in a commercial test on a M.S. specimen are given below :

Original gauge length	=	200 mm
Original diameter	=	12 mm
Yield load	=	44 kN
Max. load reached	=	65 kN
Load at failure	=	58 kN
Final gauge length (when the two broken pieces were fitted together)	=	237 mm
Final diameter at neck	=	8.5 mm

Test Results : YP = 382 N/mm², UTS = 564 N/mm², BF = 503 N/mm², Elgn. = 18.5%, Contn. = 49.8%, passes the acceptance test.

Determination of the Three Moduli of Elasticity

Example 2.1. *During a tensile test on a M.S. bar of gauge length 400 mm and diameter 20 mm, the elongation under a load of 32 kN was measured as 0.2 mm, and the lateral contraction 0.003 mm. If the load was well within the elastic limit, compute the three moduli of elasticity.*

Solution. The relation between the three moduli of elasticity E, G and K $\left(\text{Young's modulus } E = \frac{\sigma}{\varepsilon}, \text{ Rigidity modulus } G = \frac{\tau}{\gamma}, \text{ Bulk modulus } K = \frac{\sigma}{\Delta V/V}\right)$ is given by

$$\boldsymbol{E = 2G(1 + \nu) = 3K(1 - 2\nu)}$$

$$\text{Poisson's ratio } (\nu) = \frac{1}{m} = \frac{\text{Lateral strain}}{\text{Longitudinal strain}}, \text{ (see Figure 2.13)}$$

$$\nu = \frac{0.003/20}{0.2/400} = 0.3$$

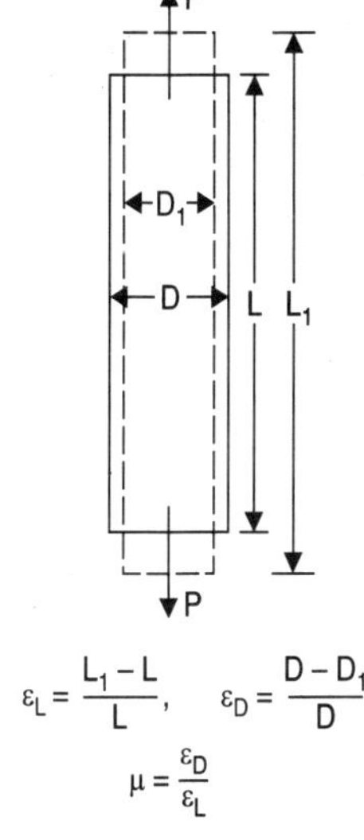

Figure 2.13. Poisson's ratio

For most metals μ lies in the range of 0.25 to 0.35.

$$(i)\ E = \frac{\sigma}{\varepsilon} = \frac{32{,}000/\pi/4\,(20^2)}{0.2/400} = \mathbf{203{,}718\ N/mm^2}$$

$$(ii)\ G = \frac{E}{2(1+\nu)} = \frac{203{,}718}{2(1+0.3)} = \mathbf{78{,}353\ N/mm^2}$$

$$(iii)\ K = \frac{E}{3(1-2\nu)} = \frac{203{,}718}{3(1-2\times 0.3)} = \mathbf{169{,}765\ N/mm^2}$$

EXERCISE 2

1. The following results were obtained in a regular tension test on a M.S. specimen of gauge length 50 mm and diameter 14 mm:

Load (kN)	0	1.5	7	10	14	18.5	22	25	28.5	34	35.5
EMR	0	2	4	6	8	10	12	14	16	18	20

L.C. of extensometer = 0.0025 mm
Yield load = 35.5 kN
Ultimate load = 58 kN
Breaking load = 46 kN
Final gauge length = 73 mm
Final gauge dia at neck = 8.4 mm
Report the results for acceptance.
(**Ans.** PL = 233, UTS = 381, BF = 302, E = 2.33 × 10^5 N/mm^2, Elgn. = 46%, Contn. = 63.5%)

3 Compression Test on Metals

Desirability of a tension or a compression test on a material depends on the following factors:

(*i*) Suitability of the material to perform under a given type of load.

(*ii*) Difference in the properties of the material under tensile and compression loading.

(*iii*) Relative difficulties and complications involved in gripping (tension test) or end bearing (compression test) on the test piece.

(*iv*) Type of service to which the material is subjected in actual practice.

The mechanical properties of a ductile material are generally obtained from a tension test. The compression behaviour is of interest in the metal forming industry involving rolling, forging, etc. and also of the forming equipment (rolling mill).

The elastic range comprising elastic modulus, proportional limit and yield point or proof stress has closely corresponding values both in compression and tension. The problem arises in the compression test when the metal enters the plastic range. The compression testing machine is shown in Figure 3.1. The length of the test piece should be equal to or less than twice the least lateral dimension to avoid the possibility of instability or buckling. The axial compression is accompanied by lateral expansion and this is restrained at the ends of the specimen due to friction between the machine platens and the end faces resulting in **barrelling.**

Figure 3.2(*i*) shows less deformed end cones. Due to this barrel effect, only an average stress can be computed from the load-compression curve based on an average area determined from considerations of constant volume i.e.,

$$A_0L_0 = A_FL_F \ , A_F = \frac{A_0L_0}{L_F}$$

$$\sigma_c\text{-average} = \frac{P_F}{A_F} \ \text{N/mm}^2$$

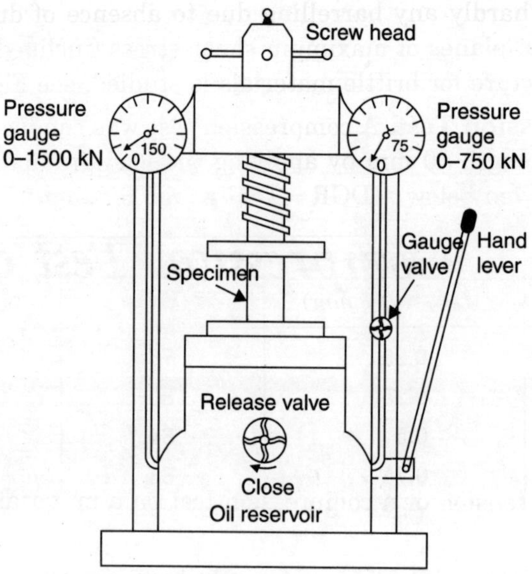

Figure 3.1. Compression testing machine

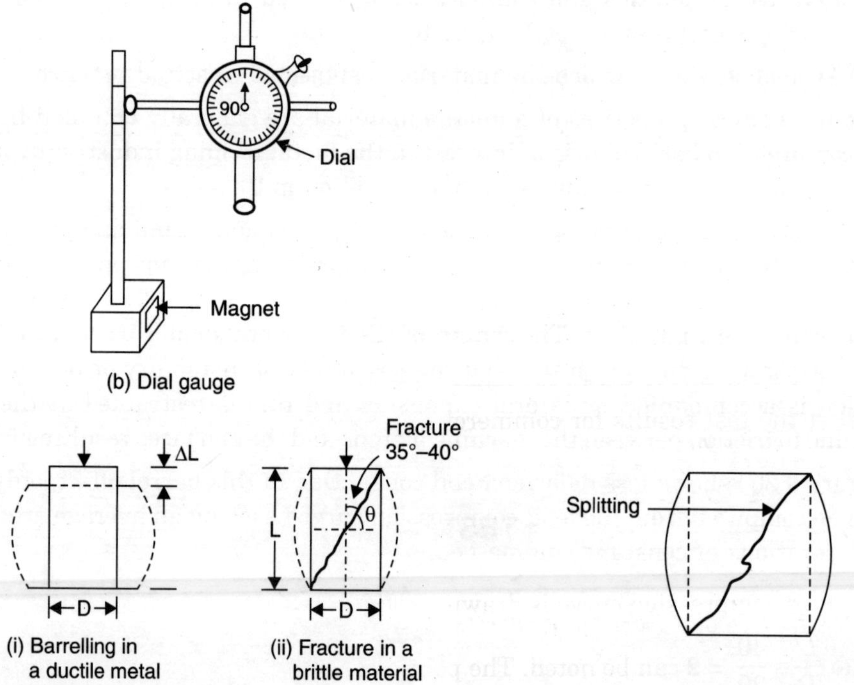

Figure 3.2. Deformation during compression test

Figure 3.3. Splitting in a ductile material

Failure of a ductile metal in compression only occurs owing to excessive barrelling causing axial splitting around the periphery, see Figure 3.3. For brittle materials such as flake cast iron, concrete, etc. (which would not normally be used in tension) the compression is used since there is hardly any barrelling due to absence of ductility in these materials. Fracture takes place on planes of maximum shear stress (inclined at 45° to the longitudinal axis). The shape of fracture for brittle materials is studied, see Figure 3.2 (*ii*).

Static Compression Test. A compression test was conducted on a brass specimen of 20 mm diameter and length 40 mm by applying the load gradually till failure occurred and the observations are given below : (DGR = Dial gauge reading).

Load (*kN*)	*DGR* (*L.C. = 0.1 mm*)	*Load* (*kN*)	*DGR* (*L.C. = 0.1 mm*)
4	0.25	76	9.0
8	0.25	80	10.5
12	0.5	84	12.5
16	0.75	88	15
20	1.0	92	17
24	1.5	96	24.5
28	1.75	100	30
32	2.0	110	36.5
36	2.25	120	42.5
40	2.5	130	49
44	2.75	140	55
48	3.0	150	62
52	3.5	160	68
56	4.0	170	74
60	4.5	180	80
64	5.25	190	87
68	6.25	206	98 Failure
72	7.5		

Report the test results for commercial acceptance.

TEST RESULTS

1. The load-compression curve is drawn (see Figure 3.4) and the shape of the curve for the ratio $\frac{L}{D} = \frac{40}{20} = 2$ can be noted. The points PL and YP cannot be identified on the curve.

2. An enlarged diagram of the load-compression curve for the initial portion is drawn on the same graph sheet.

$$A_0 = \frac{\pi}{4} D^2 = \frac{\pi}{4} (20^2) = 314.16 \text{ mm}^2$$

At the end of the initial straight portion of the curve (*A*),

$$PL = -\frac{48,000}{314.16} = \mathbf{152.6\ N/mm^2}$$

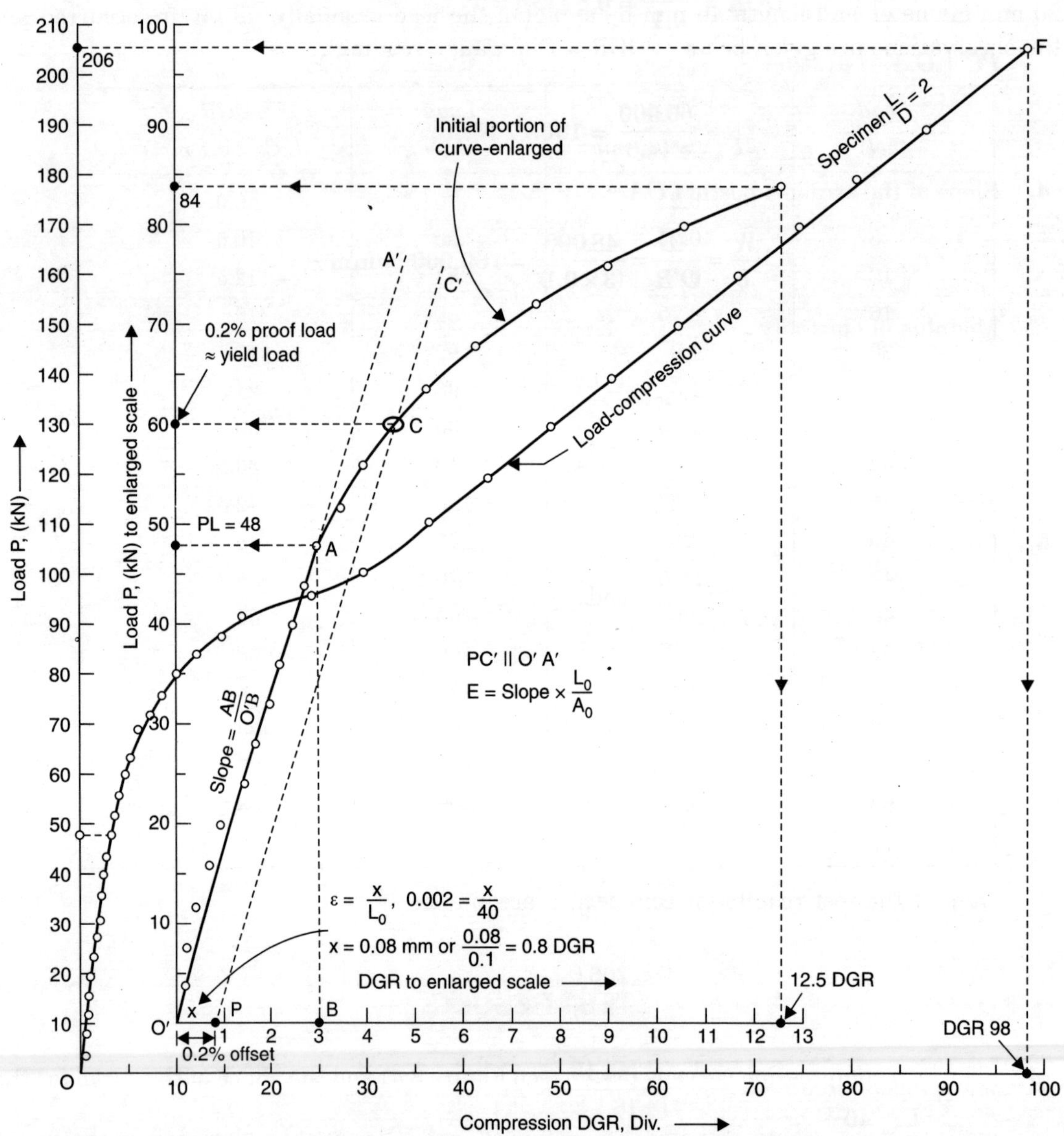

Figure 3.4. Static compression test for brass

3. 0.2% proof stress—since the yield point cannot be identified,

$$\text{Offset } OP = 0.002 \times L_0 = 0.002 \times 40 = 0.08 \text{ mm}$$

$$\text{DGR} = \frac{0.08}{0.1} = 0.8\ DGR = x$$

$PC' \parallel O'A'$ is drawn

$$\text{YP} \approx \frac{60{,}000}{314.6} = \mathbf{190.17\ N/mm^2}$$

4. Slope of the straight portion OA

$$\frac{W}{\delta} = \frac{AB}{O'B} = \frac{48{,}000}{(3 \times 0.1)} = \mathbf{160{,}000\ N/mm}$$

Modulus of elasticity

$$E = \text{Slope} \times \frac{L_0}{A_0} = 160{,}000 \times \frac{40}{314.6} = \mathbf{20{,}343\ N/mm^2}$$

which is a very low value for brass (common values of E-brass are in the range of 80,000-100,000 N/mm^2)

5. Compression strength,

$$\sigma_c = \frac{\text{Max. load at failure}}{A_0} = \frac{206{,}000}{314.16} = \mathbf{656\ N/mm^2}$$

6. Average compressive strength $= \dfrac{P_F}{A_F}$

$$L_F = 40 - (98 \times 0.1) = \mathbf{30.2\ mm}$$

$$A_0 L_0 = A_F L_F$$

$$A_F = \frac{A_0 L_0}{L_F} = \frac{314.16 \times 40}{30.2} = \mathbf{416.1\ mm^2}$$

$$\sigma_c\text{-average} = \frac{P_F}{A_F} = \frac{206{,}000}{416.1} = \mathbf{495\ N/mm^2}$$

7. % Expansion in area $= \dfrac{416.1 - 314.16}{314.16} \times 100 = \mathbf{32.45\%}$

% Reduction in length $= \dfrac{9.8}{40} \times 100 = \mathbf{24.5\%}$

Due to its low value of elastic modulus and high compression, the material is rejected for commercial use.

EXERCISE 3

1. A static compression test was conducted on a C.I. specimen of 20 mm dia. and length 42 mm till fracture and the observations are given below:

Load (*kN*)	*DGR* (*L.C.* = *0.01 mm*)	*Load* (*kN*)	*DGR* (*L.C.* = *0.01 mm*)
0	0	130	95
10	10	140	105
20	15	150	120
30	20	160	130
40	26	170	150
50	36	180	175
60	42	190	200
70	50	200	220
80	58	210	260
90	65	220	290
100	73	230	340
110	80	270	700 Failure
120	86		

Report the results for commercial acceptance.

(**Ans.** PL = 420, E = 18489, YP = 0.2%, PS = 485, σ_c = 873 N/mm^2, Expn. = 26.5%, Conpn. = 16.6%)

4 Compression Test on Wood

Timber has been used for engineering purposes from the dawn of civilization. It is relatively light in weight and can be cut to any desired shape. Although, it does not have a high ultimate strength in tension and compression, it can withstand a considerable amount of distortion before being seriously damaged and is often used for structural parts subjected to shocks. One weakness of wood is its lack of isotropy i.e., it has different mechanical properties in different directions. The strength of wood is effected by a large number of factors, say, grains. The tensile and compressive strengths are much greater in a direction parallel to the grains than in a direction perpendicular to the grains. The shearing strength along the grains is, however, much lower than that across the grains. Knots and shakes affect the strength of wood.

The important mechanical tests to find the suitability of wood for engineering purposes are the **static bending test** and **compression test**.

1. Compression Test on Wood-Along or Parallel to the Grains

The dimensions of the test piece are usually 50 × 50 × 75 mm height, see Figure 4.1 (*a*). The load is gradually applied at the rate of 0.6 mm/min till the specimen fails and the deformation is measured by a dial gauge above the specimen.

2. Compression Test on Wood-Perpendicular to the Grains

The test piece is in the form of a cube with 50 mm sides, see Figure 4.1 (*b*). The load is gradually increased at the rate of 0.6 mm/min till the compression reaches 2.5 mm. If the maximum load is reached at some lesser value of strain, the maximum load and deformation are recorded.

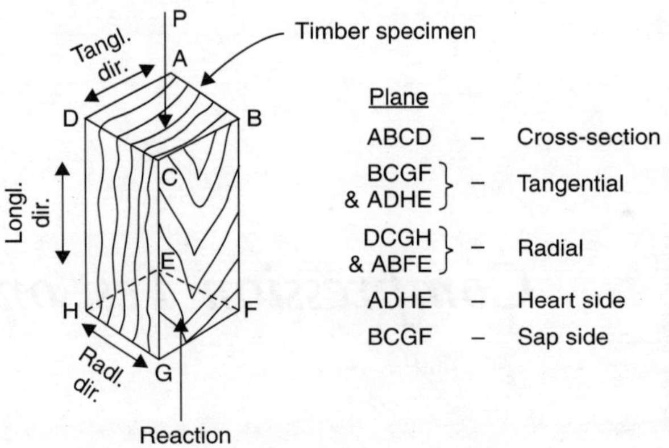

(a) Load along or parallel to grains

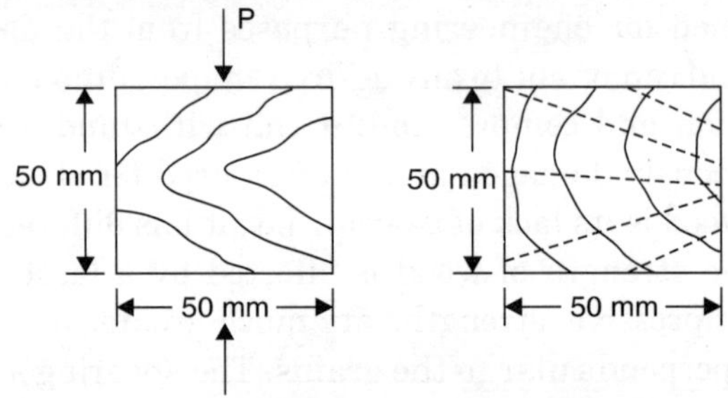

(b) Load perpendicular to grains

Figure 4.1. Compression test on wood

Static Compression Tests on Cubical specimens of wood of side 50 mm.

(*a*) Compression test parallel to grains till failure (DGR = Dial gauge reading)

Load (kN)	0	10	20	30	40	50	60	70	80	90	100	110
DGR (L.C. = 0.01 mm)	0	15	23	29	35	42	49	58	68	81	94	150

(*b*) Compression test perpendicular to the grains:

Load at failure = 95 kN

Report the results for certification.

TEST RESULTS

(*a*) Compression test parallel to grains

1. The load-compression curve is drawn, see Figure 4.2. It does not have a straight line relation from the beginning. A specific point *B*, on the curve at which 0.5% permanent strain, is produced during the test is selected by inspection,

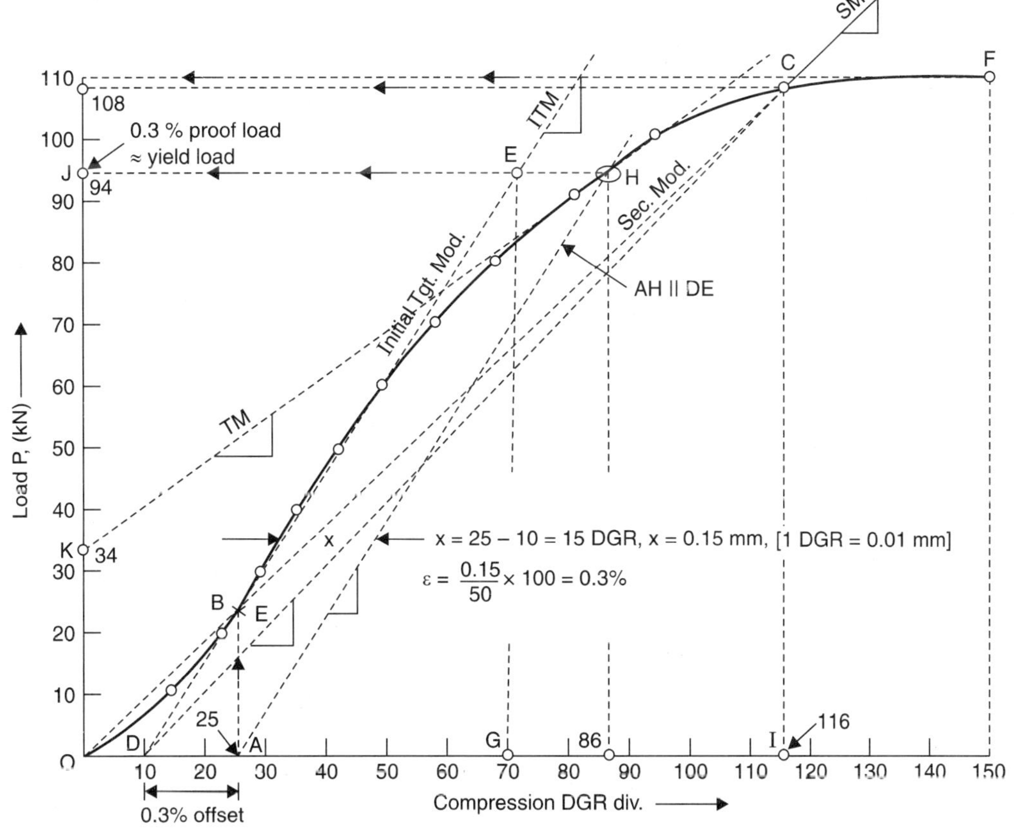

Figure 4.2. Load compression curve for wood (// to grains)

Offset $OA = 0.005 \times L_0 = 0.005 \times 50 =$ **0.25 mm**

$$\text{DGR} = \frac{0.25}{0.01} = 25 = x$$

At B, a tangent to the curve DBE is drawn. The slope of the tangent

$$= \frac{EG}{DG} = \frac{94 \times 1000 \text{ N}}{(70 - 10) \times 0.01 \text{ mm}} = \mathbf{156{,}667\ N/mm}$$

Initial tangent modulus (ITM) = Slope $\times \dfrac{L_0}{A_0}$

$$= 156{,}667 \times \frac{50}{(50 \times 50)} = \mathbf{3133\ N/mm^2}$$

2. The line OBC is the secant and its slope

$$= \frac{CI}{OI} = \frac{108 \times 1000 \text{ N}}{116 \times 0.01 \text{ mm}} = \mathbf{93{,}103\ N/mm}$$

Secant modulus (SM) = Slope × $\frac{L_0}{A_0}$.

$$= 93{,}103 \times \frac{50}{(50 \times 50)} = \mathbf{1862\ N/mm^2}$$

3. **Yield Stress.** Offset DA = 0.3% = 0.003 × 50 = **0.15 mm**

= 15 DGR, taken as x_1.

A line $AH \parallel DE$ is drawn to intersect the curve at H at which the proof load is 94 kN. The

yield stress at H $= \frac{94 \times 1000}{(50 \times 50)} = \mathbf{37.6\ N/mm^2}$

4. **Compressive Strength** = $\frac{\text{Max. load at failure}}{A_0}$

$$\sigma_{c_1} = \frac{110 \times 1000}{50 \times 50} \times 100 = \mathbf{44\ N/mm^2} \quad ...(i)$$

5. **% Decrease in Length** = $\frac{150 \times 0.01}{50} \times 100 = \mathbf{3\%}$

which is the compression strain at failure.

(*b*) Compressive strength perpendicular to the grains

$$\sigma_{c_2} = \frac{95 \times 1000}{(50 \times 50)} = \mathbf{38\ N/mm^2} \quad ...(ii)$$

From (*i*) and (*ii*), $\boldsymbol{\sigma_{c_1} > \sigma_{c_2}}$

6. **Tangent Modulus (TM).** Draw a tangent to the curve at H. The slope of the tangent

$$= \frac{KJ}{JH} = \frac{60 \times 1000}{86 \times 0.01} = \mathbf{69{,}767\ N/mm}$$

Tangent modulus (TM) = Slope × $\frac{L_0}{A_0}$

$$= 69{,}767 \times \frac{50}{50 \times 50} = \mathbf{1395\ N/mm^2}$$

Note. ITM > SM > TM. These are the approximations of the stress-strain constant E. Usually, the secant modulus is taken as the arbitrary value of E. In Figure 4.2, a fair estimate of E

$$= \text{Slope of } DC \times \frac{L_0}{A_0} = \frac{108 \times 10^3}{106 \times 0.01} \times \frac{50}{50 \times 50} = \mathbf{2038\ N/mm^2}$$

5 Direct Shear Test

Shear stresses are caused by forces which act parallel to the cross-sectional area and tend to produce sliding of one portion against another, see Figure 5.1.

If there is only one cross-section which resists the failure, the material is said to be in single shear see Figure 5.1 (*a*), and the average ultimate strength in shear τ, will be equal to the failure load P, divided by the area of cross-section A, i.e.,

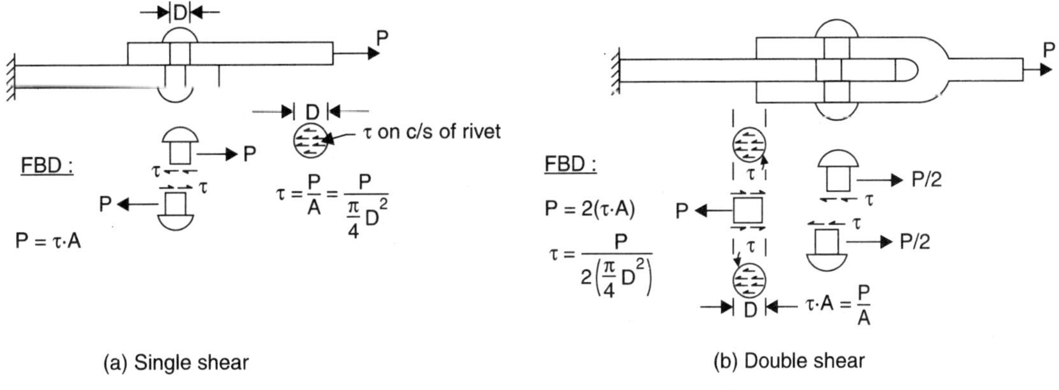

(a) Single shear

(b) Double shear

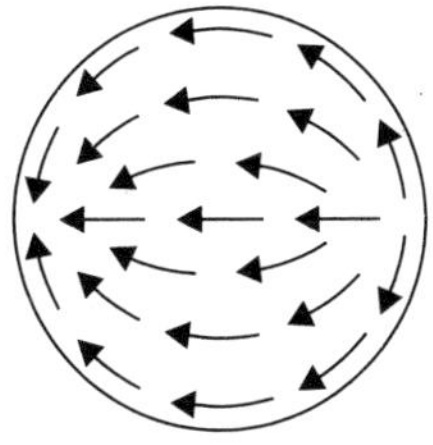

(c) Shear stress distribution over a section of a rivet

Figure 5.1. Direct shear

In single shear: $\tau = \dfrac{P}{A} = \dfrac{P}{\dfrac{\pi}{4} D^2}$ ...(5.1)

If two areas of cross-section resist the failure, then the material is said to be in double shear, see Figure 5.1 (*b*), and the average ultimate strength in shear will be equal to the failure load *P*, divided by twice the area of cross-section (2*A*), i.e.,

In double shear: $\tau = \dfrac{P}{2A} = \dfrac{P}{2\left(\dfrac{\pi}{4} D^2\right)}$ ...(5.2)

Determination of the Ultimate Shear Strength in Single and Double Shear. The dimensions (or diameter) of the specimen, are noted. The middle bush is placed in the shearing tool (see Figure 5.2), and it is aligned with the holes in the *U*-shaped portion of the shear tool attachment. The specimen (or rod) is then inserted and the side bushes are placed. It may be noted that the bush is placed on one side only if it is to be tested for single shear and on both the sides for double shear test.

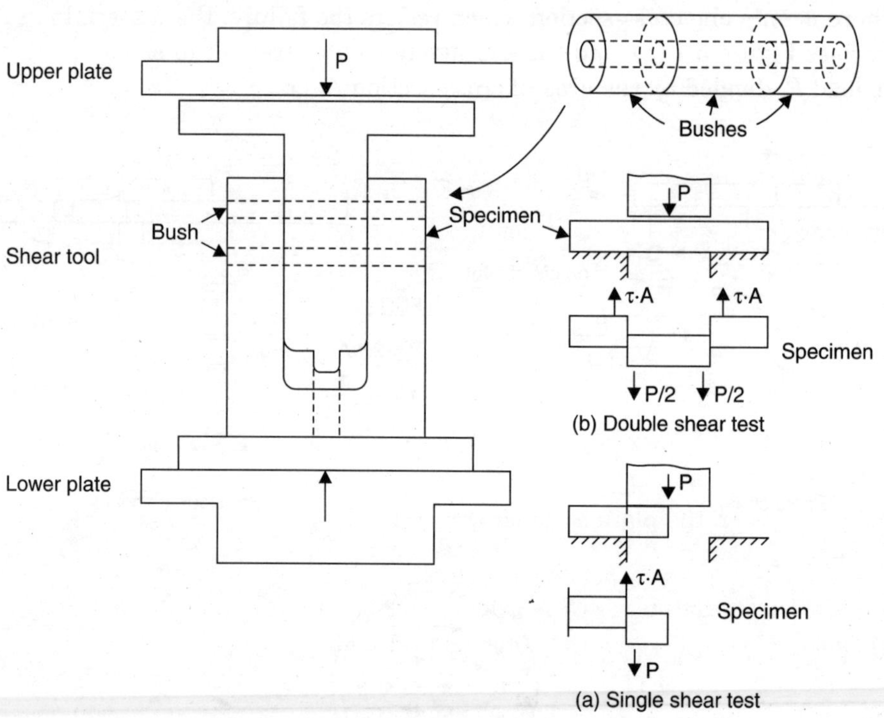

Figure 5.2. Shear tool assembly

The shearing tool assembly is then placed in the compression testing machine (or UTM). The load is increased gradually and the ultimate load at failure is noted. The ultimate shear strength is then calculated.

Observations of direct shear test on M.S. rod, M.S. plate, and timber specimens are given below:

Sl. No.	*Material*	*Type of test*	*Dimensions of specimen*	*Area of c/s A (mm^2)*	*Load at failure P (kN)*	*Ultimate shear strength $\tau = \frac{P}{A}$ (N/mm^2)*	*Remarks*
1.	M.S. rod	Single shear	15 mm dia	176.7	55	311.3	
		Double shear	15 mm dia	176.7 × 2 = 353.4	140	396.1	$\frac{396.1}{311.3} = 1.272$
2.	M.S. plate	Single shear	$t \times b$, mm 1.26 × 24.1	30.366	4.5	148.2	
		Double shear	1.26 × 24.1	30.366×2= 60.732	19	312.8	$\frac{312.8}{148.2} = 2.11$
3.	Timber	Single shear	$b \times d$, mm 45 × 48	2160	22	10.2	

Note. Double shear strength is more than single shear strength.

Example 5.1. *Calculate the minimum size of a hole that can be punched in a steel plate of 10 mm thickness given.*

Ultimate shear strength of steel = 300 N/mm²

Compressive strength of the material of the punch = 500 N/mm²

Solution. When the plate is under direct shear stress, the punch is under compressive stress.

$$\begin{pmatrix}\text{Compressive force on} \\ \text{the punch}\end{pmatrix} \geq \begin{pmatrix}\text{Shear force required to punch} \\ \text{the hole (dia } d\text{)}\end{pmatrix}$$

$$\sigma_c \times \frac{\pi d^2}{4} \geq \tau \times \pi d . t \qquad ...(i)$$

$$\therefore \quad d \geq \frac{\tau}{\sigma_c} 4t$$

$$\geq \frac{300}{500} (4 \times 10)$$

$$d \geq 24 \text{ mm}$$

i.e., the min. size of the hole is **24 mm.**

$$\text{Punching force} = 500\left(\frac{\pi}{4}\times 24^2\right) = 226195 \text{ N} \qquad ...(ii)$$

Note: If a 25 mm hole is to be punched,

$$\text{from } (i),\quad \sigma_c = \frac{4}{d}\,\tau \times t = \frac{4}{25} \times 300 \times 10 = 480 \text{ N/mm}^2.$$

$$\text{Punching force} = 480\left(\frac{\pi}{4}\times 25^2\right) = 2355619 \text{ N} \qquad ...(iii)$$

From (*ii*), and (*iii*), it can be seen that a smaller stress can produce more force if applied on a larger area.

EXERCISE 5

1. Calculate the maximum size of a hole that can be punched in a steel plate of 10 mm thickness with a punching force of 200 kN. Ultimate shear strength of steel = 300 N/mm². (**Ans.** 21.22 mm)
2. Estimate the force required to punch circular holes of 20 mm dia in a 5 mm plate. Ultimate shear strength = 300 N/mm². (**Ans.** 95 kN)
3. A M.S. rod failed under a load of 65 kN in single shear and under 145 kN in double shear. Find the ratio of double shear strength to single shear. (**Ans.** 1.12)

6 Static Bending Test on M.S. Beam

A gradually increasing load (W-total) is applied symmetrically at $\frac{1}{3}$-span on both ends of the beam, see Figure 6.1. With this arrangement, the central 1/3 length of the beam is subjected to a uniform bending moment $\left(M = \frac{WL}{6}\right)$ and the shear force (S.F.) is zero in the central portion. This facilitates the modulus of elasticity being calculated without the necessity of allowing for shear deflection. It may be noted that the shear deflection occurs over the whole length of a beam loaded at the centre, see Figure 6.1.

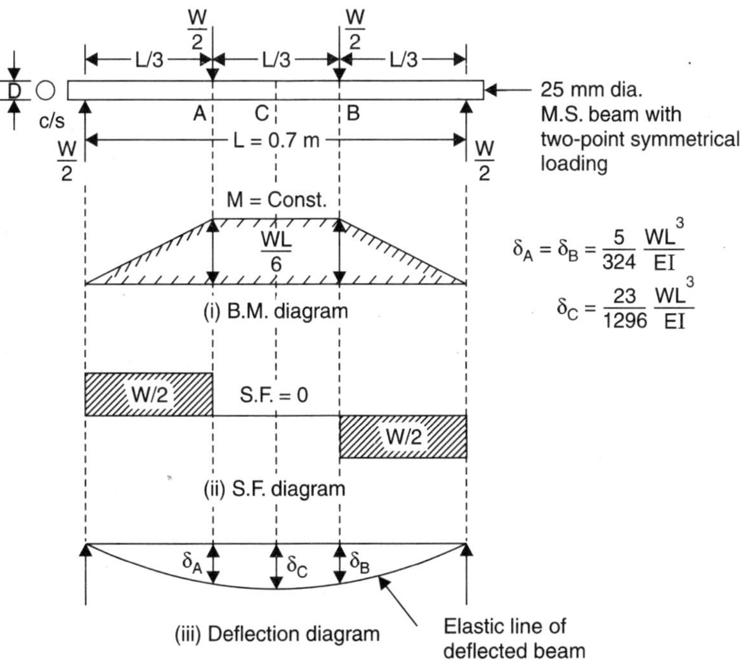

Figure 6.1. M.S. beam with 2-point loading

The dial gauge readings are noted at regular load increments and at failure. The load-deflection curves at $\frac{1}{3}$-span and mid-span and calculations are made. The values of theoretical deflections δ_T can be calculated by assuming a value of $E = 2.1 \times 10^5$ N/mm^2, and compared with the values obtained experimentally by plotting 'δ_{exp} vs δ_T curves' at $\frac{1}{3}$-span and mid-span, see Figure 6.3. The slopes of the straight portion of the plots i.e., the ratios of $\frac{\delta_{exp}}{\delta_T}$ are determined.

Static Bending Test. A 25 mm dia M.S. beam of span 0.7 m was tested for flexure by 2-point symmetrical loading at $\frac{1}{3}$-span and the following observations were made:

Total load W (kN)	*DGR (L.C. = 0.1 mm)*		
	at $\frac{1}{3}$-span		*at mid-span*
	A	*B*	*C*
0.4	6	6	8
0.8	11	12	14
1.2	16.5	17.5	20
1.6	22	22	26
2.0	27	28	32
2.4	32	33	38
2.8	38	38	44
3.2	44	43	50
3.6	48.5	49.5	56
4.0	55.5	55	62.5
4.4	60	60	68.5
4.8	66	68	76
5.2	78	74	88
5.6	96	92	113
6.0	108	108	195* – YP

*When excessive deflections were observed in the middle portion of the beam.

Report the results qualitatively for acceptance.

TEST RESULTS

1. Theoretical deflections (δ_T)

$$I = \frac{\pi D^4}{64} = \frac{\pi}{64}(25^4) = 19{,}175 \text{ mm}^4$$

$$E = 2.1 \times 10^5 \text{ N/mm}^2 \text{ (assumed)}$$

At $\frac{1}{3}$-span,
$$\delta_T = \frac{5}{324} \times \frac{WL^3}{EI} \quad \text{...(6.1)}$$

$$\delta_T = \frac{5}{324} \times \frac{W \times (700)^3}{(2.1 \times 10^5) \times 19{,}175} = 1.31452 \times 10^{-3} \text{ W} = \textbf{aW}$$

At mid-span,
$$\delta_T = \frac{23}{1296} \times \frac{WL^3}{EI} \quad \text{...(6.2)}$$

$$\delta_T = \frac{23}{1296} \times \frac{W \times (700)^3}{(2.1 \times 10^5) \times 19{,}175} = 1.5117 \times 10^{-3} \text{ W} = \textbf{bW}$$

2. Tabulation for 'Load-deflection' and 'δ_{exp} vs δ_T' curves

Total load W (N)	*At $\frac{1}{3}$-span*			*At mid-span C*	
	DGR (ave. of A and B)	δ_{exp} *= DGR × 0.1 (mm)*	δ_T *= aW (mm)*	δ_{exp} *= DGR × 0.1 (mm)*	δ_T *= bW (mm)*
400	6	0.6	0.53	0.8	0.6
800	11.5	1.15	1.05	1.4	1.21
1200	17	1.7	1.58	2.0	1.81
1600	22	2.2	2.10	2.6	2.42
2000	27.5	2.75	2.63	3.2	3.02
2400	32.5	3.25	3.15	3.8	3.62
2800	38	3.8	3.68	4.4	4.23
3200	43.5	4.35	4.21	5.0	4.80
3600	49	4.9	4.73	5.6	5.44
4000	55.25	5.53	5.26	6.25	6.05
4400	60	6.0	5.78	6.85	6.65
4800	67	6.7	6.31	7.6	7.25
5200	76	7.6	6.84	8.6	7.86
5600	94	9.4	7.36	11.3	8.46
6000	108	10.8	7.89	19.5	9.07

3. The 'load-deflection curve' for mid-span with DGR as abscissa (to get an enlarged diagram) is drawn in Figure 6.2.

Slope of the initial straight portion

$$= \frac{W}{\delta} = \frac{2000}{(30 \times 0.1)} = 666.7 \text{ N/mm}$$

$$\text{Modulus of elasticity } (E) = \frac{23}{1296} \times \frac{L^3}{I} \left(\frac{W}{\delta}\right)$$

$$E = \frac{23}{1296} \times \frac{(700)^3}{19{,}175} \times 666.7 = \mathbf{211{,}636\ N/mm^2}$$

which almost agrees with theoretical value of E assumed as 2.1×10^5 N/mm².

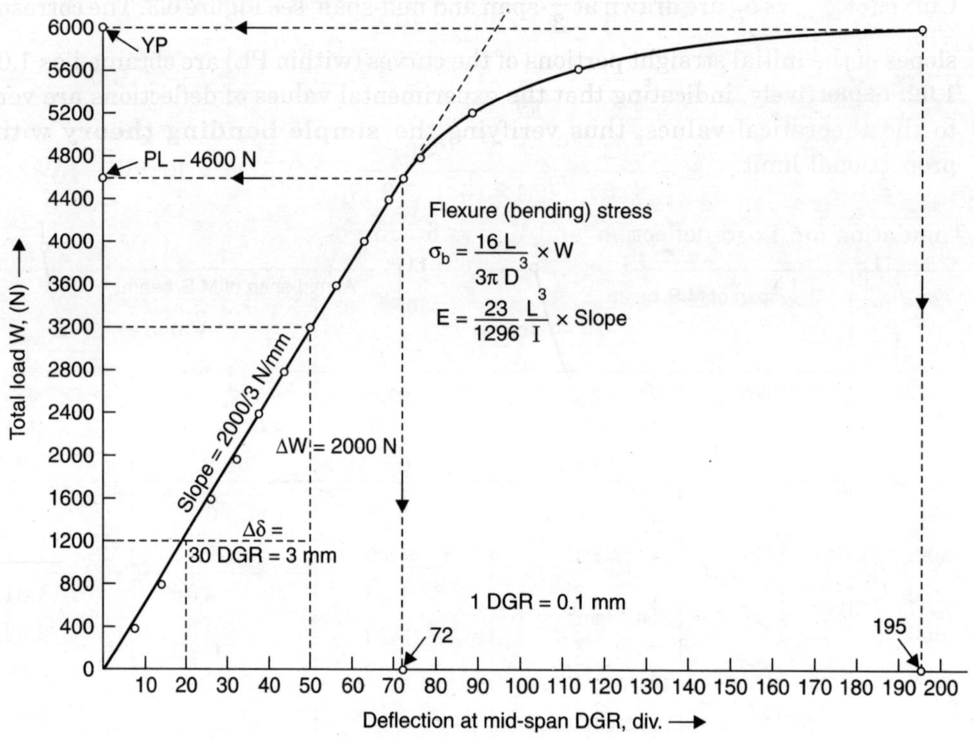

Figure 6.2. Load-deflection curve for M.S. beam

4. Flexure or bending stress $\boldsymbol{\sigma}_b$, due to bending moment M is given by

$$\frac{M}{I} = \frac{\sigma_b}{Y}, \text{ where } M = \frac{WL}{6}, Y = \frac{D}{2}$$

$$\therefore \qquad \sigma_b = M\frac{Y}{I} = \frac{WL}{6} \times \frac{D}{2I} = \frac{LD}{12\,I} \times W \qquad \ldots(i)$$

At PL, $W = 4600$ N, $\sigma_b = \frac{700 \times 25}{12 \times 19{,}175} \times 4600 = \mathbf{349.85\ N/mm^2}$

At YP, $W = 6000$ N, $\sigma_b = 349.85 \times \frac{6000}{4600} = \mathbf{456.32\ N/mm^2}$

Note. Putting $I = \frac{\pi D^4}{64}$ in (*i*), $\sigma_b = \frac{\mathbf{16\,L}}{\mathbf{3\,\pi D^3}}$ **× W.**

Also, $M = \sigma_b \times \frac{I}{Y}$, where $\frac{I}{Y} = z$, called the **section modulus**.

$\therefore$ $M = \sigma_b z$

5. Curves of 'δ_{exp} vs δ_T' are drawn at $\frac{1}{3}$-span and mid-span, see Figure 6.3. The corresponding slopes of the initial straight portions of the curves (within PL) are obtained as 1.025 and 1.02, respectively, indicating that the experimental values of deflections are very close to the theoretical values, thus verifying the **simple bending theory** within the proportional limit.

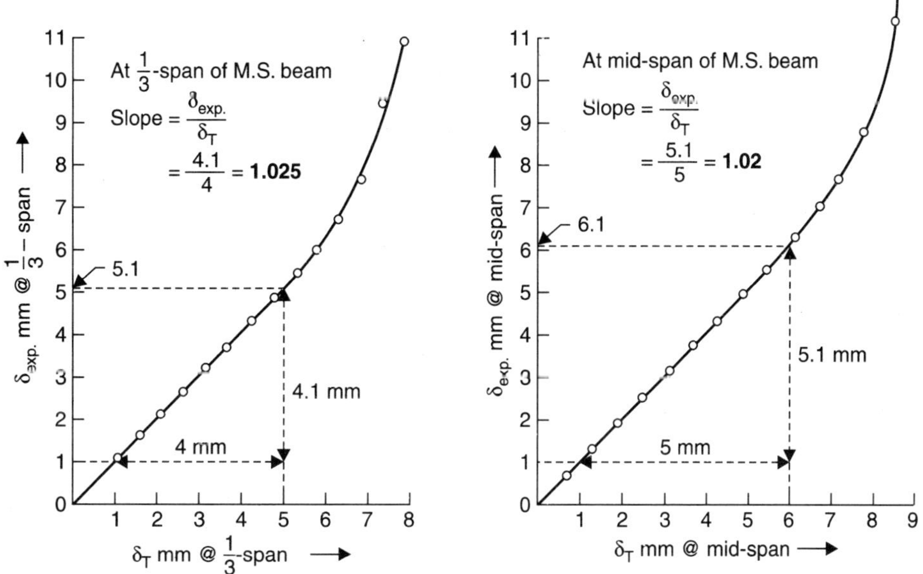

Figure 6.3. Curves of experimental vs. theoretical deflections

7 Static Bending Test on Timber Beam

A gradually increasing central concentrated load W, is applied till fracture and the corresponding deflections are noted on the dial gauge at the centre of the beam, see Figure 7.1. The fracture is studied in its appearance and development and the failure of the specimen is classified as cross-grain section, horizontal shear, etc. (Refer I.S.: 1708–1969).

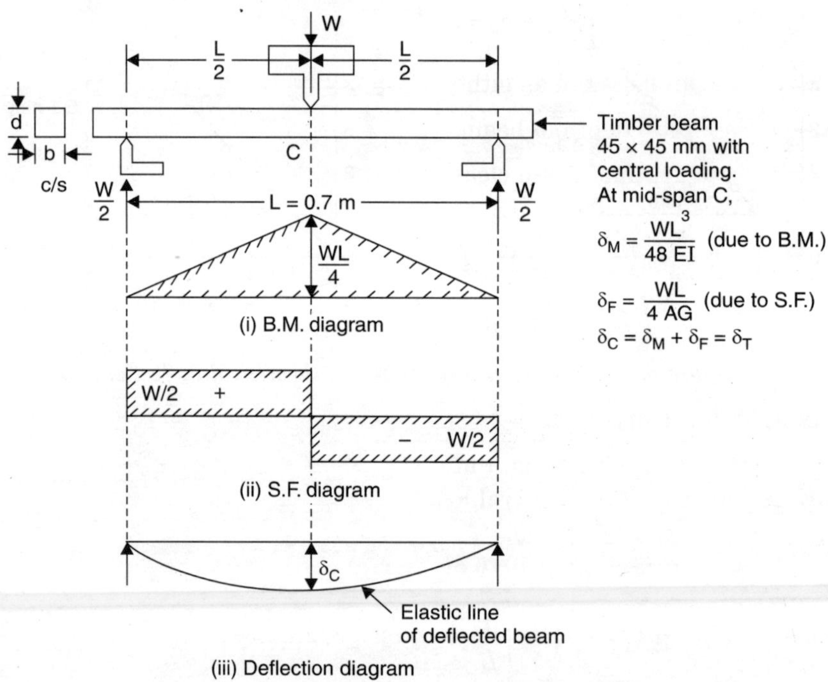

Figure 7.1. Timber beam with central loading

The B.M. and S.F. diagrams due to central load W, are shown in Figure 7.1. The maximum B.M. and maximum deflection δ_M, occur at the central section and their values are

$$M = \frac{WL}{4} \qquad \text{...(7.1)}$$

$$\delta_M = \frac{WL^3}{48\,EI} \text{ (due to } M) \qquad ...(7.2)$$

Apart from this, there is deflection δ_F, due to shear force

$$\text{S.F.} = \frac{W}{2} \qquad ...(7.3)$$

Average shear stress

$$\tau = \frac{\text{S.F.}}{\text{Area of cross-section}} = \frac{W/2}{A}$$

$$\text{Shear strain } (\gamma) = \frac{\tau}{G} = \frac{W}{2\,AG} \qquad ...(i)$$

$$\text{From Figure 7.1 } (iii), \text{ shear strain } (\gamma) = \frac{\delta_F}{L/2} \qquad ...(ii)$$

$$\text{From } (i) \text{ and } (ii), \ \delta_F = \frac{WL}{4\,AG}, \quad \text{due to S.F.} \qquad ...(7.4)$$

Hence, the total central deflection theoretically

$$\delta_T = \delta_M + \delta_F = \frac{WL^3}{48\,EI} + \frac{WL}{4\,AG} \qquad ...(7.5)$$

The value of A can be taken as either of

web thickness × total depth of beam

or, Web thickness × depth of beam clear of fillets.

Further,

$$\frac{\delta_T}{WL} = \frac{L^2}{48\,EI} + \frac{1}{4\,AG}$$

or

$$\frac{\delta_T}{WL} = AL^2 + B \text{ (straight line law)} \qquad ...(7.6)$$

where, $A = \dfrac{1}{48\,EI}$, $B = \dfrac{1}{4\,AG}$

To verify this theory, assuming a maximum bending stress σ_b, for the material of the beam, the maximum value of the central load W, is calculated from

$$\frac{M}{I} = \frac{\sigma_b}{Y}, \text{ where } M = \frac{WL}{4},$$

$$W = \left(\sigma_b \times \frac{I}{Y}\right)\frac{4}{L} \qquad ...(7.7)$$

The load deflections are observed for gradually increasing load upto the maximum value of W obtained from Eqn. (7.7) and the dial gauge readings are checked during unloading. The experiment is repeated for 2-3 shorter spans. A graph is plotted $\dfrac{\delta_T}{WL}$ vs L^2 (as abscissa) which yields a straight line plot, thus verifying the theory.

The theoretical deflection δ_T, can be calculated by assuming the elastic constants E and G suitably. From the plot of experimental values against theoretical values, the ratios $\frac{\delta_{exp}}{\delta_T}, \frac{\delta_M}{\delta_F}$ and $\frac{\delta_F}{\delta_T}$ can be determined and also their variation with span. The importance of flexure as the span is increased and that of shear as the beam is shortened, can be noticed.

Static Bending Test. A timber beam of cross-section 45 × 45 mm and span 0.7 m was subjected to central loading and the following observations were made:

Load (kN)	*DGR (L.C. = 0.1 mm)*	*Load (kN)*	*DGR (L.C. = 0.1 mm)*
0.4	10	3.6	104
0.8	20	4.0	118
1.2	31.5	4.4	134
1.6	40.5	4.8	153
2.0	51	5.2	172
2.4	61.5	5.6	194
2.8	78	6.0	233 Fracture
3.2	89		

Report the results qualitatively for certification.

TEST RESULTS

1. $I = \frac{bd^3}{12} = \frac{45 \times 45^3}{12} = 341{,}719 \text{ mm}^4$.

   Assuming $E = 8000$ N/mm² and $G = 600$ N/mm² for the timber beam, the theoretical deflection at mid-span

$$\delta_T = \delta_M + \delta_F = \frac{WL^3}{48\,EI} + \frac{WL}{4\,AG}$$

$$= \frac{W \times 700^3}{48 \times 8000 \times 341{,}719} + \frac{W \times 700}{4 \times (45 \times 45) \times 600}$$

$$= 2.614 \times 10^{-3}\, W + 1.44032 \times 10^{-4}\, W$$

$$\delta_T = a\,W + bW$$

2. Tabulation for 'load-deflection' and 'δ_{exp} vs δ_T' curves

Load W (*N*)	δ_{exp} = *DGR* × *0.1* (*mm*)	δ_M = aW (*mm*)	δ_F* = bW (*mm*)	δ_T = $\delta_M + \delta_F$ (*mm*)
400	1	1.05	0.06	1.11
800	2	2.09	0.12	2.21
1200	3.15	3.14	0.17	3.31
1600	4.05	4.18	0.23	4.41
2000	5.1	5.23	0.29	5.52
2400	6.15	6.27	0.35	6.62
2800	7.8	7.32	0.40	7.72
3200	8.9	8.36	0.46	8.82
3600	10.4	9.41	0.52	9.93
4000	11.8	10.46	0.58	11.04
4400	13.4	11.50	0.63	12.13
4800	15.3	12.55	0.69	13.24
5200	17.2	13.59	0.75	14.34
5600	19.4	14.64	0.81	15.45
6000	23.3	15.68	0.86	16.54

*It may be noted that the shear deflections δ_F, are very small.

3. The 'load-deflection curve' is drawn in Figure 7.2.

Slope of the initial straight portion

$$= \frac{W}{\delta} = \frac{2400}{(62 \times 0.1)} = 387.1 \text{ N/mm}$$

Deflection at mid-span due to bending moment (neglecting shear deflection)

$$\delta = \frac{WL^3}{48\,EI}$$

$$E = \frac{L^3}{48\,I}\left(\frac{W}{\delta}\right) = \frac{700^3}{48 \times 341{,}719} \times 387.1 = \mathbf{8094.75\ N/mm^2}$$

which is very near the assumed value of 8000 N/mm^2.

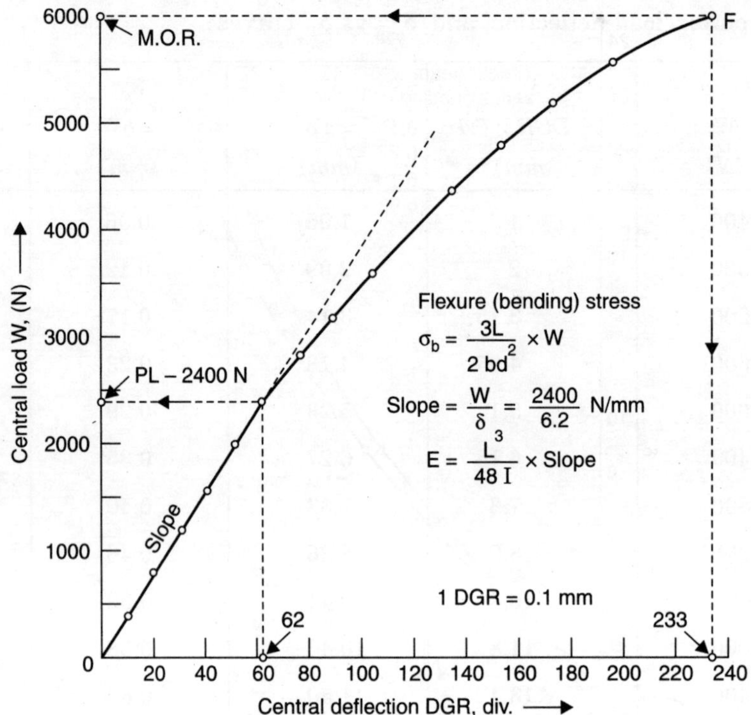

Figure 7.2. Load-deflection curve for timber beam

4. Flexure or bending stress σ_b, due to bending moment M, is given by

$$\frac{M}{I} = \frac{\sigma_b}{y}, \text{ where } M = \frac{WL}{4}, y = \frac{d}{2}$$

$$\therefore \qquad \sigma_b = M \times \frac{y}{I} = \frac{WL}{4} \times \frac{d/2}{I} = \frac{WLd}{8I} \qquad \ldots(i)$$

At PL, $W = 2400$ N

$$\sigma_b = \frac{2400 \times 700 \times 45}{8 \times 341{,}719} = 27.65 \text{ N/mm}^2$$

Load at fracture, W = 6000 N. Assuming the flexure equation to hold beyond PL, the breaking stress or modulus of rupture (MOR) from Eq. (i)

$$\text{MOR} = \sigma_b = \frac{6000 \times 700 \times 45}{8 \times 341{,}719} = \mathbf{69.13\ N/mm^2}$$

5. Curves of 'δ_{exp} vs δ_M' and 'δ_{exp} vs δ_T' are drawn in Figure 7.3. Slopes of the initial straight portion within PL are 0.98 and 0.93 which are very nearly equal to 1, thus verifying the 'simple bending theory' (shear deflections being small).

Note: Putting $I = \frac{bd^3}{12}$ in (i), $\boldsymbol{\sigma_b = \frac{3}{2} \times \frac{L}{bd^2} \times W}$

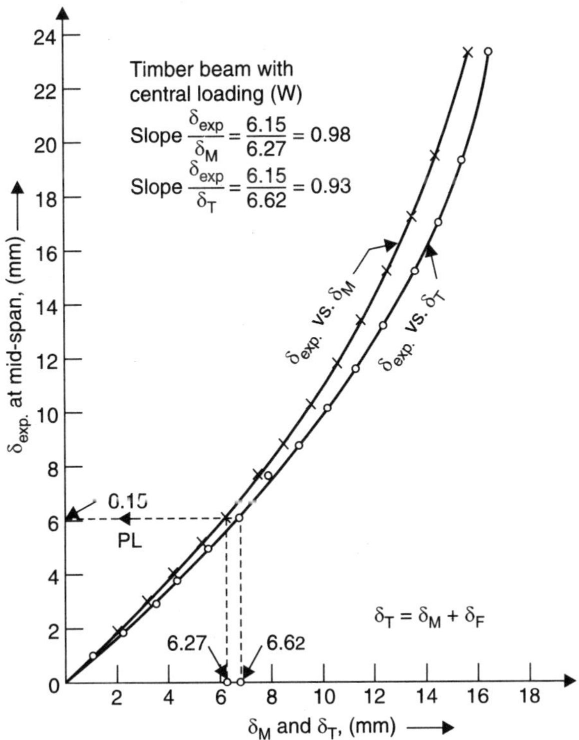

Figure 7.3. Curves of δ_{exp} vs. δ_M and δ_T

EXERCISE 7

1. A timber beam of cross-section 50 × 75 mm and length 0.75 m, span 0.6 m was subjected to loading at mid-span till fracture and the following observations were made:

Load (kN)	*DGR (L.C. = 0.1 mm)*	*Load (kN)*	*DGR (L.C. = 0.1 mm)*
0	0	14	60
3	10	17	80
5	20	19	100
7.5	30	21	120
10	40	22.5	140
12	50	24	160

Report the results qualitatively for certification.

[**Ans.** PL = 40, MOR = 77, E = 6200, N/mm^2, cross-grain structure.

$$\frac{\delta_{exp}}{\delta_M} = 1, \frac{\delta_{exp}}{\delta_T} = 0.84, \frac{\delta_M}{\delta_F} = 5.16, \frac{\delta_F}{\delta_T} = 0.167$$]

8 Torsion Test

Torsion test is conducted to study the behaviour of metal components like axles, shafts, twist drills, etc. and to determine their rigidity modulus in shear G. Torsion tests at elevated temperatures are supposed to give the forgeable nature of the metal. Laboratory investigations to study the effect of case hardening on steels also require torsion tests. In the case of tubes and pipes for torsion test, diameter to wall thickness ratio $\left(\frac{D}{t}\right)$ should not be greater than 10. The length of the specimen is recommended to be ten times the diameter. The specimen is fixed rigidly at one end and the other end is twisted. The torque is measured on a load indicator dial, see Figure 8.1. The angle of twist is measured by two pointers (on circular protractors) rigidly fixed to the specimen at the two ends of the gauge length, see Figure 8.2.

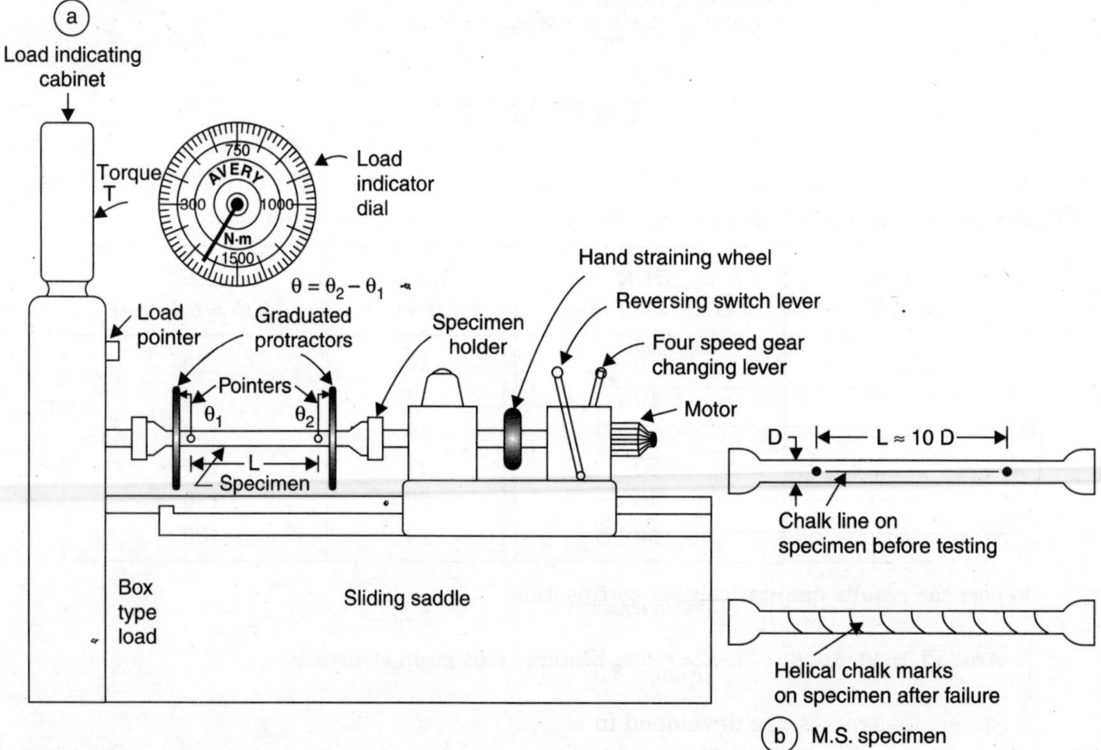

Figure 8.1. Avery's reverse torsion testing machine

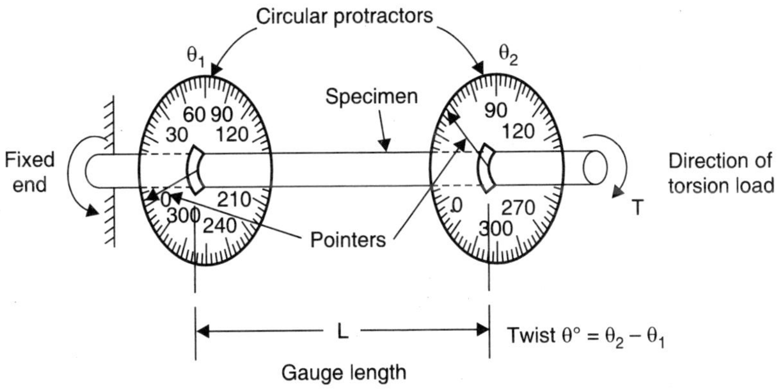

Figure 8.2. Torsion load on shaft

A solid shaft of dia. D, is subjected to torque T, and the angular twist θ, in a length L, is measured, see Figure 8.3. Then the torsion equation

$$\frac{T}{J} = \frac{\tau}{r} = \frac{G\,\theta}{L} \qquad \text{...(8.1)}$$

where J = polar moment of intertia of the c/s of the shaft

$$= \frac{\pi\, D^4}{32}$$

τ = torsional shear stress at radius r

L = length of the shaft (gauge length)

$$\text{Modulus of rigidity } (G) = \frac{T}{\theta} \times \frac{L}{J} \qquad \text{...(8.2)}$$

$\dfrac{T}{\theta}$ = slope of the plot **torque vs twist**; it is called the torsional stiffness k, defined as torque per radian of twist i.e.,

$$k = \frac{T}{\theta} = \frac{GJ}{L} \qquad \text{...(8.3)}$$

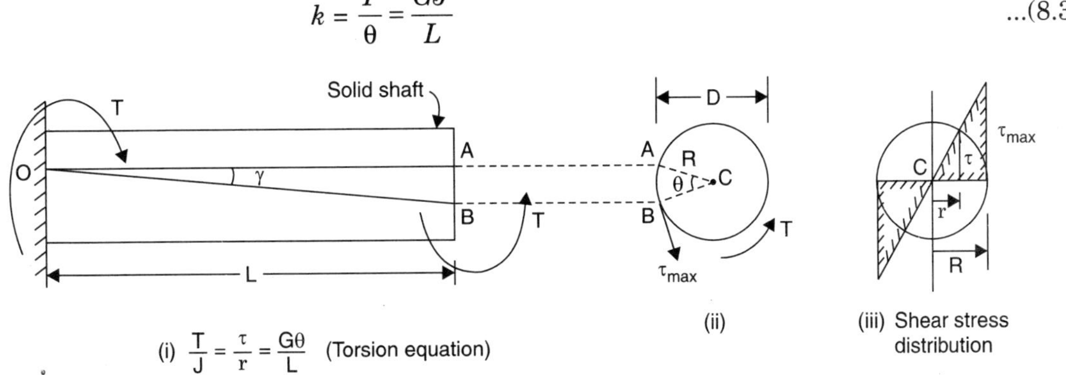

Figure 8.3. Torsion of solid shaft

Shearing stresses are developed in any cross-section of the shaft whose value increases from zero at the centre to a maximum at the outer periphery (see Figure 8.3 (*iii*)), unlike axial tension or compression test where the direct stress is uniform across the cross-section. In torsion test, at the end of the elastic range yielding commences in the outer fibres first while the core is still elastic (while in tension or compression test yielding occurs relatively evenly throughout the material).

The 'torque-twist plot' is similar to a 'load-extension plot' (see Figure 8.4) and work-hardening will occur at a gradually decreasing rate as straining proceeds, but the curve does not droop as in tension test, since 'necking' cannot take place.

For ductile metals fracture generally occurs in a plane of maximum shear stress perpendicular to the axis of the shaft; for brittle materials fracture occurs along a 45° helix to the axis of the shaft due to tensile stress across that plane, see Figure 8.1 (*b*).

Test Procedure. The average diameter and length of the shaft are measured. A straight line is drawn parallel to the axis of the bar with a piece of chalk. The specimen is fixed in position in the torsion testing machine, see Figure 8.1. For the elastic range torque is applied manually and the angle of twist recorded at regular intervals. After the yield point is reached, the torque is applied by an electric motor. The twist of the shaft is noted at regular increments of the torque applied. The test is continued till the shaft fractures and the torque at failure is noted. The type of fracture is studied.

Torsion Test. The following observations were made on a steel specimen 12.8 mm dia and gauge length 193 mm, in a torsion test conducted till fracture:

Torque T, (N·m)	*Angle of twist, θ°*	*Torque T, (N·m)*	*Angle of twist, θ°*
2	1	170	17
10	2	173	18
20	3	178	19
31	4	180	20
45	5	185	22
56	6	190	26
70	7	194	30
83	8	198	36
99	9	200	40
109	10	202	44
122	11	204.5	48
134	12	205.5	50 Fracture
143	13	205	51
152	14	202	52
160.5	15	196	53
165	16		

Report the test results qualitatively for certification.

TEST RESULTS

1. A ' torque-twist' curve is drawn in Figure 8.4.

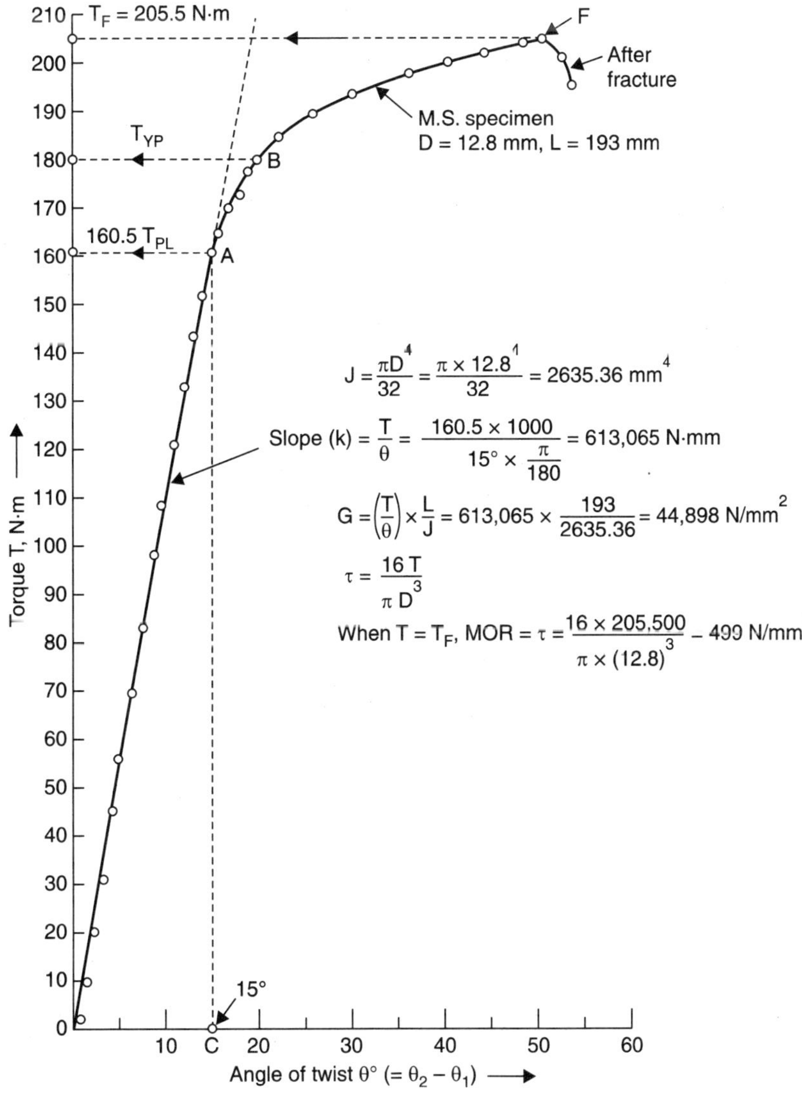

Figure 8.4. Torque-twist curve

$$J = \frac{\pi D^4}{32} = \frac{\pi (12.8^4)}{32} = \mathbf{2635.36\ mm^4}$$

2. Slope of the initial straight portion

$$k = \frac{T}{\theta} = \frac{AC}{OC} = \frac{160.5 \times 1000 \text{ N. mm}}{15^\circ \times \dfrac{\pi}{180^\circ} \text{ rad.}} = 613{,}065 \text{ N} \cdot \text{mm}$$

$$G = \left(\frac{T}{\theta}\right)\frac{L}{J} = 613{,}065 \times \frac{193}{2635.36} = \mathbf{44{,}898\ N/mm^2}$$

3. Shear stress τ at a torque $T = T \times \frac{r}{J}$, τ where $r = \frac{D}{2}$

(*i*) At PL = 160.5 N·m,

$$\tau_{PL} = \frac{(160.5 \times 1000)\dfrac{12.8}{2}}{2635.36} = \mathbf{390\ N/mm^2}$$

(*ii*) At YP = 180 N·m (approximately taken at *B*),

$$\tau_{YP} = \frac{(180 \times 1000)\dfrac{12.8}{2}}{2635.36} = \mathbf{437\ N/mm^2}$$

(*iii*) At fracture *F*, maximum T = 205.5 N · m,

$$\text{MOR} = \tau_F = \frac{(205.5 \times 1000)\dfrac{12.8}{2}}{2635.36} = \mathbf{499\ N/mm^2}$$

4. Safe or working shear stress in torsion

$$\tau = \frac{\text{YP}}{\text{F.S.}} = \frac{437}{2} = \mathbf{218\ N/mm^2}$$

Note. (*i*) Torsional shear stress for cylindrical bars or solid shafts

$$\tau = \frac{Tr}{J} = \frac{T.\dfrac{D}{2}}{\pi D^4 / 32} = \frac{16T}{\pi D^3}$$

(*ii*) For tubes or hollow shaft (dia—outer *D*, inner *d*)

$$= \frac{16\,TD}{\pi\,(D^4 - d^4)}$$

(*iii*) $$\frac{T_{\text{hollow}}}{T_{\text{solid}}} = \frac{D^4 - d^4}{DD_1^3} = \frac{d^3(n^4 - 1)}{n\,D^3{}_1} = \left(\frac{d}{D_1}\right)^3 \frac{n^4 - 1}{n}$$

where D_1 = dia of solid shaft, $\frac{D}{d} = n$ (for hollow shaft).

EXERCISE 8

1. A M.S. specimen 21.5 mm dia was tested for torsion in Avery's reverse torsion testing machine, see Figure 8.1. Graduated discs (protractors) and loose pointers are fitted for measuring the twist. The gauge length between the pointers = 94.5 mm. The specimen failed at a torque of 830 N·m. The 'torque-twist' observations are given below:

Torque (N·m)		0	40	80	120	160	225	265	300	345
Angle of twist (θ°)	*LP*	0	0	0	0	0.25	0.25	0.25	0.25	0.25
	RP	0	0.5	1.0	1.25	1.5	1.75	2.00	2.25	2.50

Report the results qualitatively for certification.

Hint : θ° = RP – LP, where RP and LP are Right and Left protractors respectively.

[**Ans.** G = 45430 N/mm^2, τ_{PL} = 177 N/mm^2 ≈ τ_{YP}, τ_{safe} = 90 N/mm^2, τ_F = 425 N/mm^2]

9 Brinell Hardness Test

Hardness may be defined as the resistance of a material to indentation (penetration), impact (rebound), scratch, wear or abrasion. In industrial testing, indentation hardness and rebound hardness are generally measured.

Indentation Hardness Tests. In these tests, the plastic deformation caused when a loaded steel ball or diamond is impressed to the surface of the material, is taken as the **hardness index**. The common indentation hardness tests are Brinell, Vickers, and Rockwell methods. The factors which affect the depth of penetration (indentation) are

(*i*) the load applied,

(*ii*) the shape and dimensions of the indentor,

(*iii*) the time for which the load is applied.

Brinell Hardness Test (1901). In this method, a hardened steel ball (dia. D) is pressed into the surface of the material (specimen) under a specified load P, which is held for a fixed period t, and then released. A permanent impression is left on the surface of the material and the Brinell hardness number (BHN or HB)

$$\text{BHN} = \frac{\text{Load applied } (P) \text{ in kg}_f}{\text{Spherical area of impression, } (A) \text{ in mm}^2}$$

Surface area of impression (spherical surface)

$$A = \pi Dh = \frac{\pi D}{2}[D - \sqrt{D^2 - d^2}]$$

(see Figure 9.1)

where, h = depth of impression, mm = $\dfrac{D}{2} - x$

d = projected diameter of impression on the surface of the material, mm

D = ball diameter, mm

The diameter of impression d, is read from a low power Brinell microscope, see Figure 9.2.

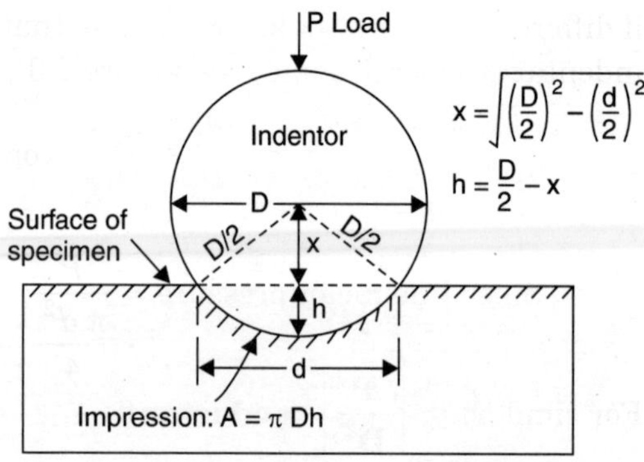

Figure 9.1. Brinell indentation, $BHN = \dfrac{P}{A}$

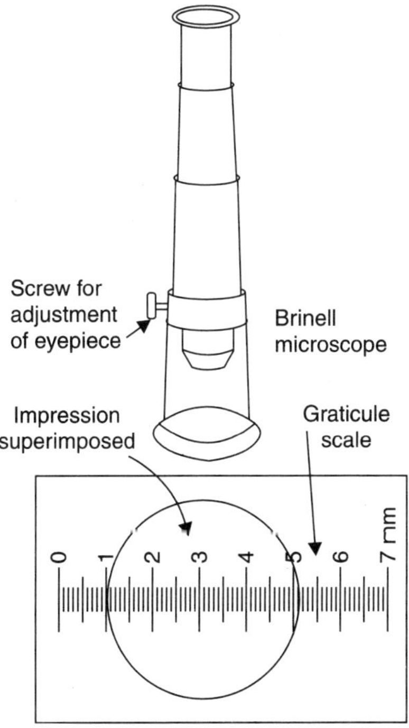

Figure 9.2.

$$\text{BHN} = \frac{P}{\frac{\pi D}{2}\,[D - \sqrt{D^2 - d^2}\,]},\ \text{kg}_f/\text{mm}^2 \qquad \text{...(9.1)}$$

BHN can also be directly read from the standard charts knowing the load P, applied and d measured.

Brinell's Indentation Geometry. Similarity can only be obtained under different loads if different diameter balls are used so that the angle subtended at the centre of the ball for the indentations is the same (see Figure 9.3), i.e.

$$\frac{d_1}{D_1} = \frac{d_2}{D_2} = \text{constant}$$

Since, the mean pressure = $\dfrac{P}{\dfrac{\pi d^2}{4}}$, where d = a constant × D

For similarity, $\left|\dfrac{\mathbf{P}}{\mathbf{D^2}}\right|$ **= constant.** ...(9.2)

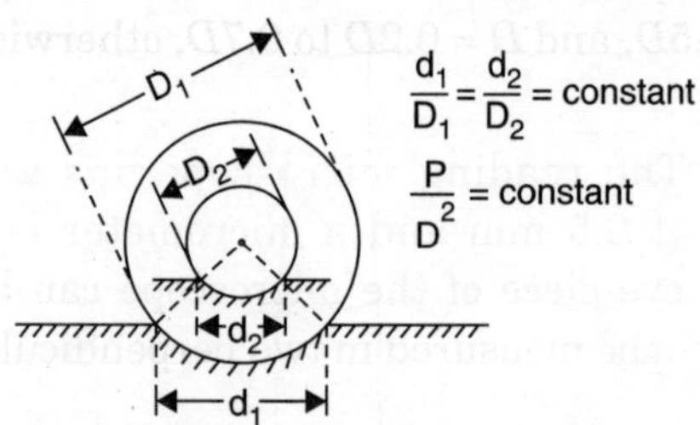

Figure 9.3. Indentation geometry in Brinell test

Thus

(*i*) for M.S. and C.I. : $\frac{P}{D^2} = 30$ ...(9.3)

i.e., for a 10 mm ball, $P = 3000 \text{ kg}_f$

for a 5 mm ball, $P = 750 \text{ kg}_f$

(*ii*) for medium-hard materials like brass, bronze, and other non-ferrous alloys of copper and aluminium : $\frac{P}{D^2} = 10$ ...(9.4)

(*iii*) for soft materials like pure aluminium, copper, magnesium, zinc and cast brass:

$$\frac{P}{D^2} = 5 \qquad ...(9.5)$$

i.e., for 10 mm ball, $P = 500 \text{ kg}_f$

(*iv*) for lead, tin and their alloys : $\frac{P}{D^2} = 2$ ...(9.6)

The diameter of the impression

$$d = 0.2D \text{ to } 0.7D \qquad ...(9.7)$$

Otherwise the load is suitably altered. The size of the Brinell indentation is such that the test is generally employed for checking raw stock or unmachined components rather than finished products.

The load is allowed to act for a duration of 10-15 sec. for ferrous material, and 30 sec. for non-ferrous materials.

The material of the ball (10 mm or 5 mm) is selected depending on the hardness range, as follows:

BHN upto 525 — high carbon steel ball

upto 600 — heat treated high carbon steel ball (Hultgren ball)

upto 725 — tungsten carbide ball.

For extremely hard materials (BHN 600), Brinell method is not suitable because of the distortion of the steel ball.

Test Specimen. The thickness of the specimen should be at least 10*d*, distance between any two successive indentations at least 4*d*, the distance of the centre of impression from the edge of the specimen at least 2.5D, and D = 0.2D to 0.7D, otherwise the load should be suitably altered.

Brinell Microscope. The reading microscope has a 25-fold magnification. The graduations of the scale are at 0.5 mm and a micrometer is arranged sideways to read 0.01 mm, see Figure 9.2. The eye-piece of the microscope can be turned by 90° so that the diameter of the impression d, can be measured in two perpendicular directions and the average is taken for calculation.

Correlation between BHN and UTS. It has been established that there is some relation between UTS and BHN for various steels as

$$\text{UTS} = k \times \text{BHN} \quad \text{...(9.8)}$$

Due to this relation, an estimate of the UTS of the specimen can be made by a hardness test without expense and time of preparing a 'tensile test specimen'. The ratio $k \left(= \dfrac{\text{UTS}}{\text{BHN}}\right)$ for various steels is given below to serve as a guide :

Class of steel	$k = \dfrac{UTS\ in\ N/mm^2}{BHN\ in\ kg_f/mm^2}$ *ratio*
(*i*) Heat treated alloy steels BHN = 250-400	3.3
(*ii*) Heat treated carbon steels and alloy steels BHN < 250	3.38
(*iii*) Plain carbon steels (upto 1% carbon)	3.45
(*iv*) Medium carbon steels as rolled, normalised or annealed	3.46
(*v*) Mild steels as rolled, normalised or annealed	3.62

Test Procedure. In the B.H. testing machine (see Figure 9.4), the load is applied by a lever mounted on knife edges and carrying a hanger for putting the required load. The surface of the material to be tested is first cleaned of dirt, oil, scale, etc. The surface is rubbed with sand paper to facilitate reading of the impression. The selected load P, is put on the hanger. The ball indentor of suitable diameter is put into its position. The specimen is kept on the supporting table and the large hand wheel is turned in the clockwise direction till the gap between the surface of the specimen and the clamping bush is about 5 mm. The hand lever is brought into position I, from the initial position 0. This operation rises the supporting table, and the specimen is clamped against the bush. Now the hand lever is put into position II, to press the indentor onto the surface of the specimen under the load P. The load is allowed to act for 10-15 sec. and then the hand lever is put back into its original position. The specimen is taken out after turning the hand wheel. The diameter of the impression is read from the microscope in two perpendicular directions.

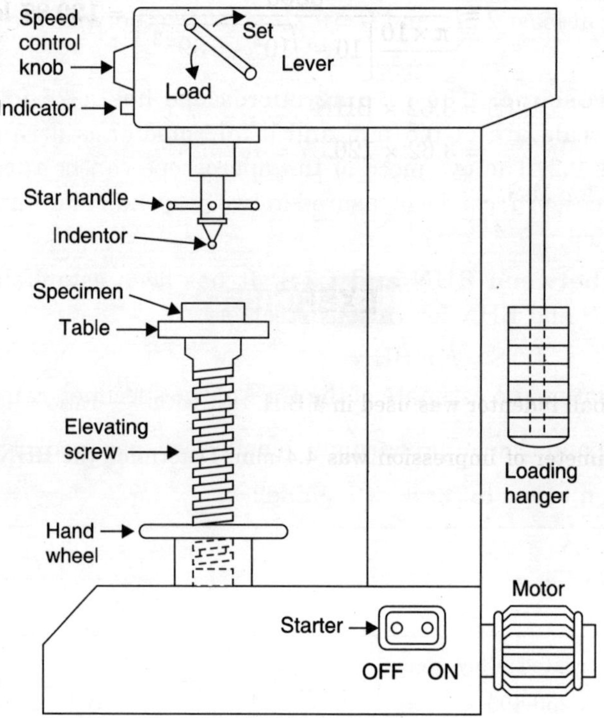

Figure 9.4. Brinell hardness testing machine

B.H. Test Observations. Dia of ball indentor (D) = 10 mm, D^2 = 100 mm^2

Sl. No.	*Material*	*Trial*	$\frac{P}{D^2}$	*Load P (kg_f)*	*Dia of impression, (mm)*			*BHN (kg_f/mm^2)*
					d_x	d_y	$d_{(ave)}$	
1.	M.S.		30	3000	5.41	5.38	5.395	120.87*
2.	C.I.		30	3000	4.49	4.51	4.50	178.54
3.	Brass		10	1000	3.49	3.51	3.50	100.70
4.	Copper		5	500	3.19	3.21	3.20	60.50
5.	Aluminium		5	500	4.62	4.58	4.60	28.40

*Specimen calculations : Sl. No. 1—M.S. specimen

$$D = 10 \text{ mm}, \frac{P}{D^2} = 30, P = 30 \times 10^2 = 3000 \text{ kg}_f$$

$$\text{BHN} = \frac{P}{\frac{\pi D}{2}\left[D - \sqrt{D^2 - d^2}\right]}$$

$$= \frac{3000}{\frac{\pi \times 10}{2}\left[10 - \sqrt{10^2 - 5.395^2}\right]} = \mathbf{120.87\ kg_f/mm^2}$$

$$\text{UTS} = 3.62 \times \text{BHN}$$

$$= 3.62 \times 120.87 = 437.55 \text{ N/mm}^2$$

which is a fair value for M.S.

EXERCISE 9

1. A 10 mm steel ball indentor was used in a B.H. test with $\frac{P}{D^2}$ ratio = 30 on a M.S. specimen and the average diameter of impression was 4.4 mm. Determine the BHN and estimate the UTS.

(**Ans.** 187.23 kg_f/mm^2, 677 N/mm^2)

10 Vickers Hardness Test

The Vickers hardness test (1920) is similar to Brinell except that the indentor is a polished square-based diamond pyramid, see Figure 10.1(a). The angle between the opposite faces of the pyramid is 136° and this was chosen to obtain close correlation between Vickers pyramidal hardness number (VPN) and BHN. The angle of 136° corresponds to the geometry of an impression given by a $\left(\frac{d}{D}\right)$ ratio of 0.375 (Brinell). The impression is diamond-shaped, (see Figure 10.1(b)) and both diagonals are measured to allow for any asymmetry. If 'd' is the average length of the diagonal of the impression in the plane of the surface of the specimen, then the area of the surface impression, see Figure 10.1(b),

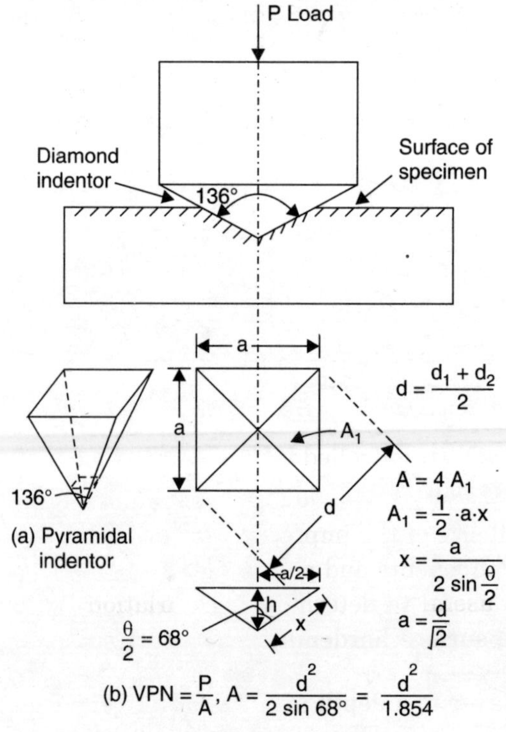

Figure 10.1. Vickers indentation

$$A = \frac{d^2}{2\sin\left(\frac{136^\circ}{2}\right)} = \frac{d^2}{1.854} \qquad ...(10.1)$$

$$\text{VPN (or HV)} = \frac{\text{Load applied } (P) \text{ in kg}_f}{\text{Area of impression } (A) \text{ in mm}^2}$$

$$\text{VPN} = \frac{1.854 \times P}{d^2} \qquad ...(10.2)$$

The Vickers machine is more accurate and versatile than the B.H. Tester. Instead of choosing the indentor as well as the loads depending upon the nature of the material tested, only the load is changed in the Vickers hardness tester. The indentor is capable of giving geometrically similar impressions under different loads; therefore the VPN for a given material is vertually independent of the applied load, see Figure 10.2. Both the Brinell and Vickers tests give more or less than same numbers upto 300 BHN and after that the Brinell test tends to give softer results, see Figure 10.3.

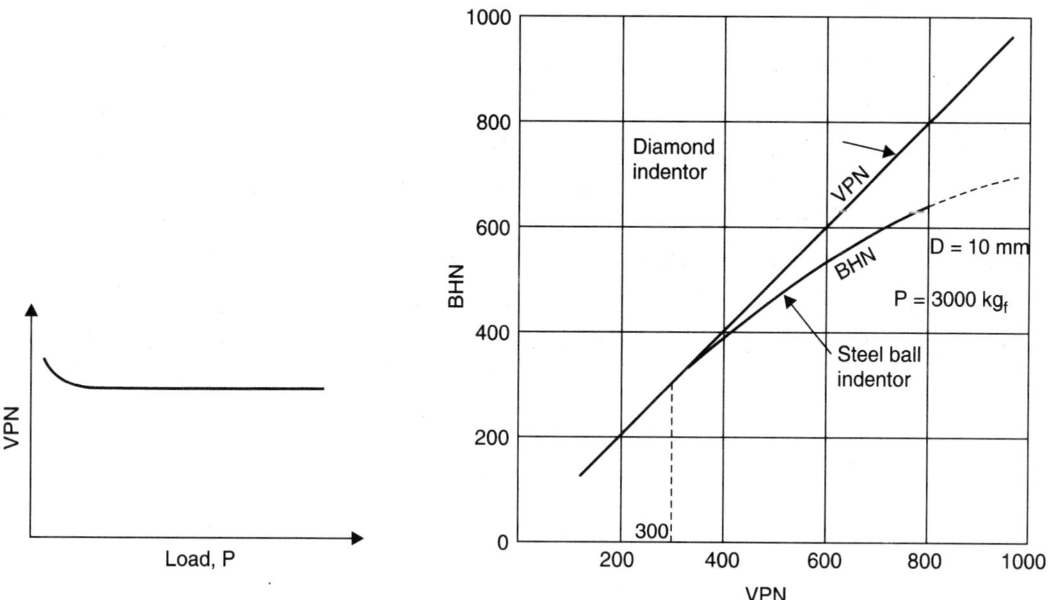

Figure 10.2. VPN vs. load applied

Figure 10.3. Values of BHN vs. VPN

The upper limit of VPN is controlled by the diamond and can be upto 1500 which is not possible by the Brinell steel ball.

The extremely small size of the impression in the Vickers is an advantage in checking the hardness of finished components and polished surfaces at various points (without leaving a severe mark). It is also useful to determine the variation in hardness through the cross-section particularly in the surface hardening treatments such as nitriding, case hardening, etc. in which the surface layer to a depth of $1\frac{1}{4}$ mm is much harder than the core.

The Vickers test can be used for thin sheets (limiting thickness = $1\frac{1}{2}$ times the diagonal of the impression) and for very hard metals. The load is selected in the range of 1-120 kg_f depending upon the thickness of the test specimen. Normally either 10 kg_f or 30 kg_f is selected. The load is allowed to act for 15-20 sec. The indentor is inserted in the thrust piece of the machine and screwed on, see Figure 10.4(a).

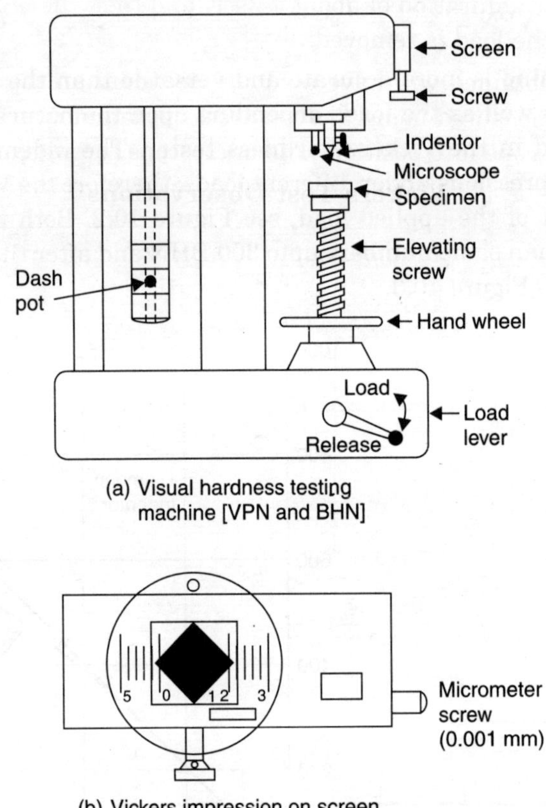

(a) Visual hardness testing machine [VPN and BHN]

(b) Vickers impression on screen

Figure 10.4. Vickers or diamond pyramid hardness test

A Brinell-type test can also be conducted using a different holder with 1-2 mm hardened steel balls supplied with the machine (Avery Denison type 6406). Standard charts are supplied with the Vickers machine which give the VPN for the measured mean length of the diagonal 'd', of the impression and load applied 'P'.

Test Procedure. The specimen is cleaned from dirt, oil, scale, etc. and the test area is made even and polished. The distance from the centre of any impression, to the edges of the test piece or to the edges of any other impression should be at least 2.5 times the diagonal of the impression.

In the Visual Hardness Testing Machine manufactured by W and T Avery Ltd. both the Vickers and Brinell tests (using 2.5 mm steel ball) can be performed.

The prepared specimen is placed on the supporting table on the top of the elevating screw and the screw is raised by turning the hand wheel to the right. The reflector throws light on the surface of the specimen and the surface qualities can be sharply viewed on the ground glass screen. The clamping sleeve is turned to the left to clamp the specimen.

The dashpot is set at the selected rate of loading and the load lever is raised nearly 30° from which position it automatically moves further till it stops when the full load will act. The load is allowed to act for a duration of about 15 sec. and then the lever is pushed down to its original position when the load is removed.

The impression obtained will have a square sectional area, see Figure 10.4 (*b*). Each diagonal is accurately measured using a crosswire focussing device in the optical equipment by swivelling the microscope into position. The microscope screw gives readings in 0.001 mm.

Vickers Test Observations

Sl. No.	*Material*	*Trial*	*Load P* (kg_f)	*Diagonal of impression (mm)*			*VPN* $= \frac{1.854\,P}{d^2}$ (kg_f/mm^2)	*VPN from chart*
				d_1	d_2	$d_{(ave)}$		
1.	Stainless steel		30	0.386	0.386	0.386	373.3*	373
2.	M.S.		20	0.42	0.425	0.4225	207.72	205
3.	Brass		10	0.36	0.36	0.36	143.06	143
4.	Copper		5	0.28	0.28	0.28	118.24	118.3
5.	Carbon steel		100	0.48	0.468	0.474	825.19	825
6.	M.S.	Steel ball $D = 1$ mm	$P/D^2 = 30$	Dia of impression $d_{(ave)} = 0.46$ mm			BHN = 170.36	

Specimen calculation : Sl. No. 1—Stainless steel

$$\text{VPN} = \frac{1.854 \times P}{d^2} = \frac{1.854 \times 30}{(0.386)^2} = \mathbf{373.3\ kg_f/mm^2.}$$

EXERCISE 10

1. The diagonals of the diamond-shaped impression in a Vickers test under a load of 30 kg_f for 15 sec. on a stainless steel specimen were measured by adjusting the micrometer screws as 0.36 and 0.34 mm. Estimate the VPN. (**Ans.** 454 kg_f/mm^2)

11 Rockwell Hardness Test

This test was introduced in the U.S.A. at about the same time as the Vickers test in England. Rockwell uses the 'depth of impression' to measure hardness and the hardness number can be read directly on the dial indicator of the machine, (see Figure 11.1) having two scales *B* and *C*. The inner scale *B* is marked in red and the outer scale *C* in black, see Figure 11.2. Each scale is graduated with 100 divisions and 30 of the *B*-scale is against '0' of the *C*-scale and thus there is a hardness number difference of 30 at any point.

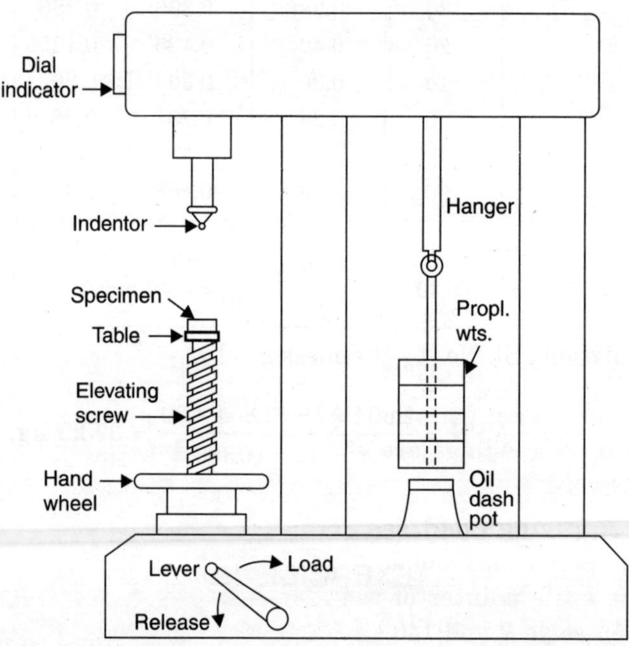

Figure 11.1. Rockwell hardness testing machine (Direct reading)

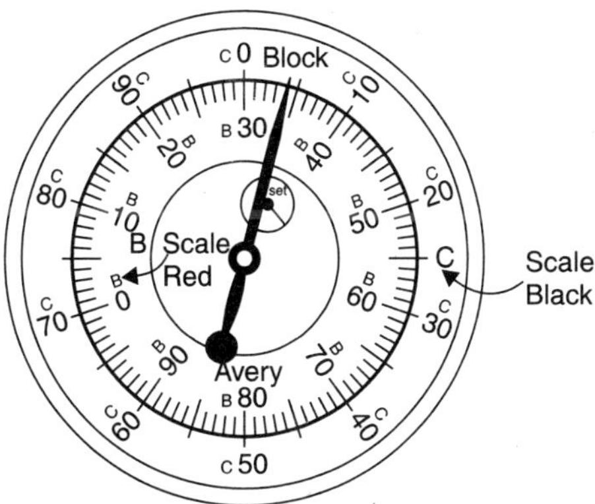

Figure 11.2. Dial indicator

Two types of penetrators are used, see Figure 11.3,

(*i*) Steel ball indentor 1.6 mm for soft and moderately hard materials,

(*ii*) Diamond cone of 120° apex angle and the apex rounded at 0.2 mm radius, for hard materials ('Brale' indentor).

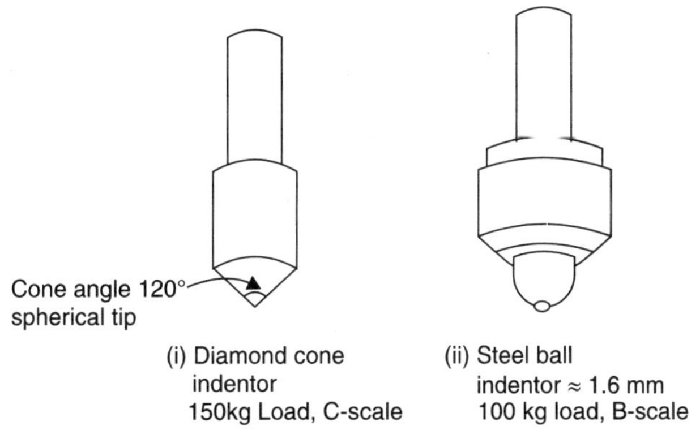

Figure 11.3. Indentors

B-scale readings are taken with the ball indentor and *C*-scale readings with the cone indentor and the hardness readings are prefixed with the scale used as HRB and HRC (or R_B and R_C), respectively.

Procedure for Applying Load, see Figure 11.4. First a minor load (10 kg_f) is applied to provide for surface imperfections and a datum for hardness measurement. Let the depth of impression be h_1. The scale pointer of the dial is then set to zero. An additional load (major load) of 90 kg_f for scale-*B* and 140 kg_f for scale-*C* is applied so that the total load now is 100 or 150 kg_f respectively, and let the depth of impression now be h_2. The major load is then removed (the minor load still acting) and there is a partial elastic recovery in the depth of impression. If h_3 is the total depth of impression under this condition ($h_3 < h_2$), then the hardness number directly read on the dial (on the corresponding scale) due to the application of major load corresponds to the depth $h = h_3 - h_1$. The Rockwell hardness is defined as (see Figure 11.4),

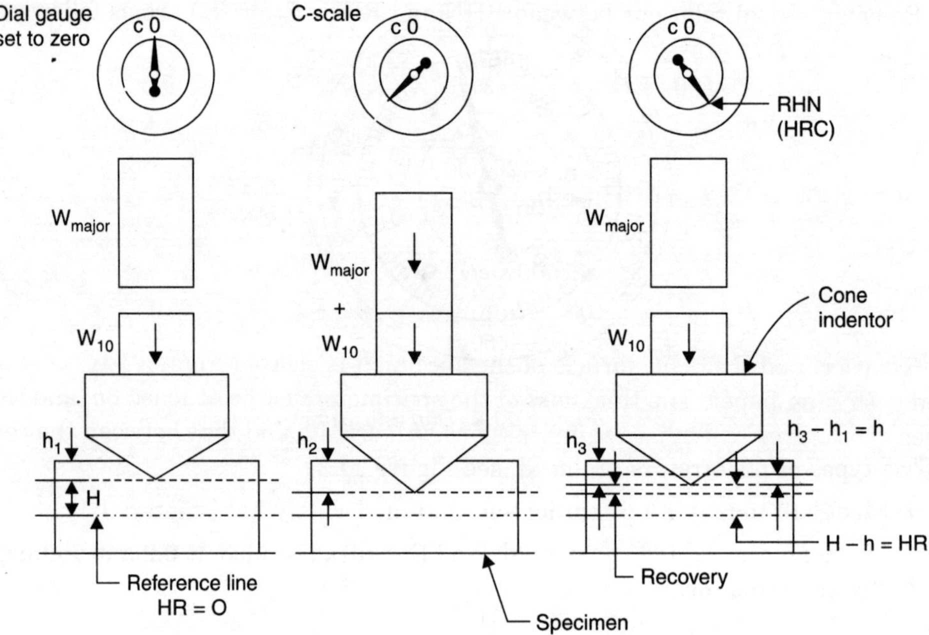

Figure 11.4. Rockwell hardness reading RHN or HR = H – h

$$\text{RHN} \quad \text{or} \quad \text{HR} = H - h \qquad \text{...(11.1)}$$

where H = an arbitrary constant depending on the type of penetrator. The scales are divided into 100 divisions, and each division is equivalent to 0.002 mm indentation.

For the cone penetrator

$$\text{HRC} = 100 - \frac{h}{0.002} \qquad \text{...(11.2)}$$

where h is in mm.

For the ball penetrator,

$$\text{HRB} = 100 - \frac{h}{0.002} \qquad \text{...(11.3)}$$

where h is in mm.

The Rockwell hardness is expressed as a dimensionless number. The dial is calibrated inversely so that the hardness number increases with increasing hardness, and can be read directly on the corresponding scale.

The time taken to carryout the operations is about 10 sec. The direct reading eliminates time consuming operation (of measurement with a microscope and calculation). The Rockwell hardness test is quite popular as it has wide range of versatility, rapid and useful for finished parts. It is frequently used for shop floor for quality control checks. There is a greater latitude for soft to hard materials. The use of the initial minor load avoids the errors arising out of the uneven surface of the metal, machine and table errors. In this test, even small differences in hardness of hardened steels can be observed.

Some empirical relations between BHN and RHN (R_B or R_C) are as follows:

(*i*) For R_B = 35–100 : $\text{BHN} = \dfrac{7300}{130 - R_B}$

(*ii*) For R_C = 20–40 : $\text{BHN} = \dfrac{1,420,000}{(100 - R_C)^2}$

(*iii*) For $R_C > 40$: $\text{BHN} = \dfrac{25,000}{100 - R_C}$.

Test Procedure. The surface of the specimen is cleared from oil, dirt and scales, and rubbed with sand paper. The thickness of the specimen must be at least '8*h*' and the distance between the centres and adjacent impressions at least '4*d*' and that between the centre of the impression and edge of the specimen at least '2.5*d*' (*d* = dia. of indentation).

Depending on the material to be tested either the ball indentor or cone indentor is inserted and screwed to the thurst member. The lamps for the dial gauge and signal lamp are switched on.

The test piece is placed on the supporting table and the hand wheel is turned to the right until the specimen contacts the clamping sleeve and the minor load of 10 kg_f is applied. This is indicated by the signal lamp being extinguished.

To apply the major load, the button is pulled out. Now the pointer of the dial gauge will start moving in the anticlockwise direction. Two seconds after the pointer comes to rest, the major load is removed pushing down the hand lever. Now the pointer will move in the clockwise direction to the extent of partial recovery in the depth of indentation.

Depending on the indentor used the RHN is directly read on the dial on the relevent scale of graduation.

R.H. Test Observations

Sl. No.	*Material*	*Trial No.*	*Indentor*	*Minor load (kg_f)*	*Major load (kg_f)*	*Total load (kg_f)*	*Scale used*	*RHN*
1.	M.S.		Diamond cone	10	140	150	*C*	R_C-19
2.	Spring steel		Diamond cone	10	140	150	*C*	R_C-46
3.	C.I.		Diamond cone	10	140	150	*C*	R_C-81
4.	Brass		Steel ball	10	90	100	*B*	R_B-67
5.	Copper		Steel ball	10	90	100	*B*	R_B-10
6.	Aluminium		Steel ball	10	90	100	*B*	R_B-4

BHN for M.S. Specimen

For R_C = 20–40 : $\text{BHN} = \dfrac{1,420,000}{(100 - R_C)^2} = \dfrac{1,420,000}{(100 - 19)^2} = \mathbf{216.43\ kg_f/mm^2}$

which is a comparable value.

BHN for Brass Specimen

$$\text{For } R_B = 35\text{–}100 : \text{BHN} = \frac{7300}{130 - R_B} = \frac{7300}{130 - 67} = \mathbf{115.87\ kg_f/mm^2}$$

which is a comparable value.

Rockwell Superficial Hardness Test. This test employs a lighter load and is used for thin sections where only a very shallow impression is permissible and surface hardness is of primary importance. The test is particularly adopted for testing thin sheets such as razor blades, case hardened surfaces, electroplatings, etc. The standard minor load is 3 kg_f and the major loads are 15, 30 and 45 kg_f. The indentors are 1.6 mm hardened steel ball (*T*-scale) and diamond cone for *N*-scale (*N*-Brale) meant for hard materials such as nitrided components.

12 Utility of Hardness Test and Rebound Hardness

Utility of Hardness Tests. A hardness test is fast, simple and is useful for the following purposes:

(*i*) To grade steel according to their composition.

(*ii*) To find the effectiveness of heat treatment like case hardening, annealing, tempering, coldworking, etc.

(*iii*) For quality control of products like armour plates, steel rails and tyres, etc. where hardness is an important property.

(*iv*) To find out the mechanical properties of finished products (where it is not possible to make usual tensile and other tests).

(*v*) To determine the UTS in certain cases as there exists a relation between the UTS and BHN.

Rebound Hardness Test (Shore Scleroscope Method). Here the resistance offered to strike and rebound is taken as the index of hardness. It consists of a diamond tipped hammer weighing about 2.6 gm_f moving in a graduated glass tube, see Figure 12.1. The whole length of the glass tube is graduated into 100 or 140 equal parts. The hammer is allowed to fall from a height of 25 cm and strike the surface of the specimen. The height of rebound observed on the glass tube is the index of hardness. If the metal is hard, the indentation made will be small (since it yields less) and the height to which the hammer rebounds will be more, and vice versa.

The indentation is so minute that it does not affect the surface finish. The scleroscope is portable and is used as a standard comparison or quality control tool. It is commonly employed for testing rolls of rolling mills, crankshafts and other components with difficult profiles. Very thin sections can also be tested.

The glass tube is aligned to be perfectly vertical. Two models of the shores scleroscope are available — 'C_2' and 'D'.

In the 'C_2' type, the height of rebound may be noted as follows:

(*i*) By watching the top of the hammer itself.

(*ii*) By means of a magnifying glass and pointer attached to the pipe which are moved to the expected height of rebound.

This is done by preliminary tests.

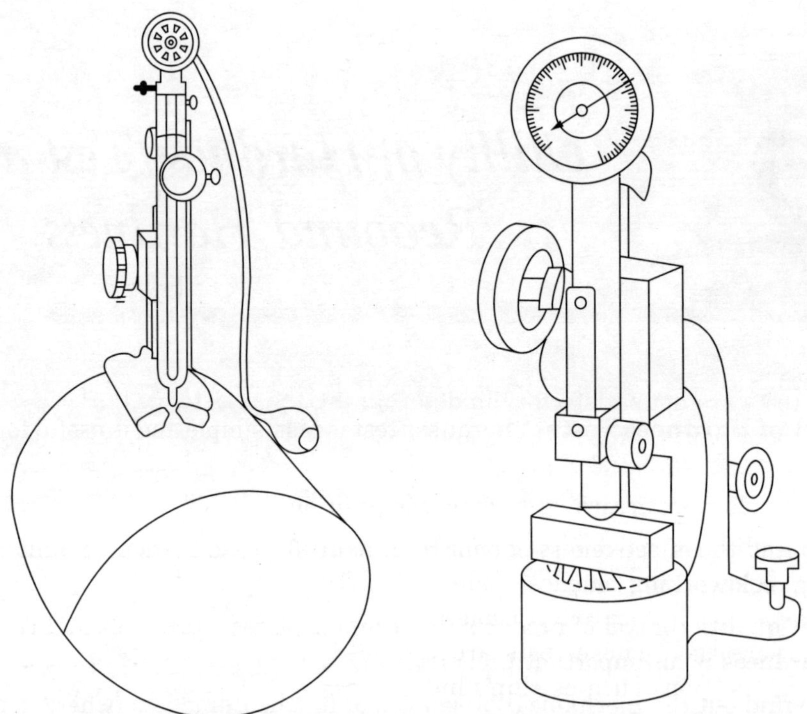

Figure 12.1. Shore scleroscope, model–C_2 *Figure 12.2. Shore scleroscope, model–D*

In the model '*D*', the scleroscope number is indicated on the dial at the top of the instrument in addition to giving conversion into other types of hardness scales, see Figure 12.2.

13 Impact Tests

Static tests are not satisfactory in determining the resistance to shock or impact loads such as automobile parts, ship's hull, etc. are subjected to. In the impact test, a notched specimen of the material is fractured by a single blow from a heavy hammer. The energy absorbed for fracture gives the 'impact value' or toughness of the material.

Notches of specific geometry are made in the specimen to create a tendency for brittle fracture. In the presence of notches, many ductile materials behave in a brittle manner; this property is called 'notch sensitivity'. Generally most of the steels (except austenitic steels) are highly notch sensitive. These tests are useful in specifying alloys particularly for low temperature applications such as ship's hull plates, pressure vessels, steels used for welded structures and cryogenic (low temperature components).

The notched-bar impact bend test is most commonly adopted for checking toughness of a material (Toughness is the property to resist impact or shock loads). The machine used for the test is of pendulum type; it consists of a hammer which swings about a pivot situated above an anvil holding the test piece, see Figure 13.1. The hammer is raised to a preset position, thus acquiring potential energy, see Figure 13.2. When released by a trigger, it swings down, hits the specimen to fracture, and continues to swing up. The height it reaches, is a function of the energy remaining. The energy absorbed during fracture, see Figure 13.2,

$$E_L = Wh_1 - Wh_2$$

$$E_L = WL\,(\cos\beta - \cos\alpha) \qquad \text{...(13.1)}$$

where W = weight of hammer,

L = pendulum arm,

α, β = initial and final angles of swing of the pendulum.

If the material is brittle, E_L would be small and the hammer would swing to a high position.

A correction is necessary to allow for friction of the machine. A test is carried out without a specimen and the energy expended in overcoming frictional resistance of the bearing is calculated from the two positions of the hammer as before. This value has to be subtracted to get the actual energy absorbed by the specimen.

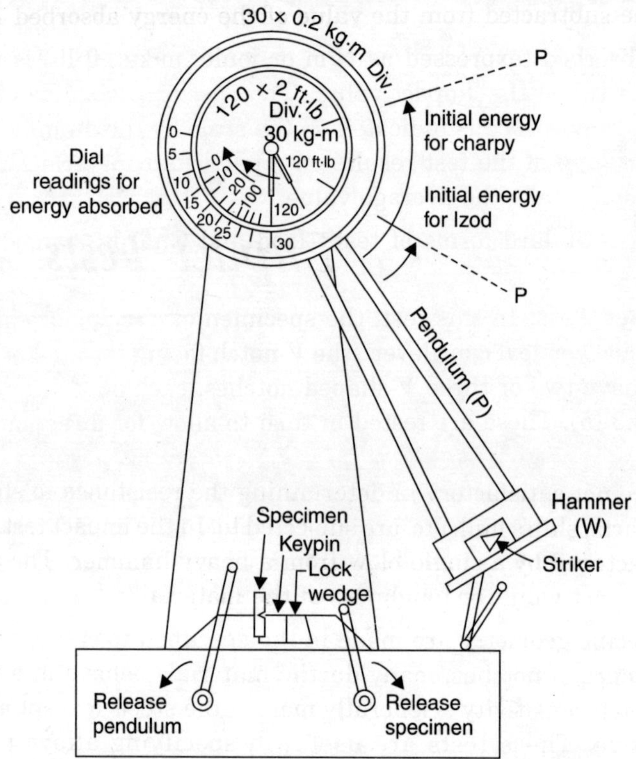

Figure 13.1. Charpy and Izod impact testing machine

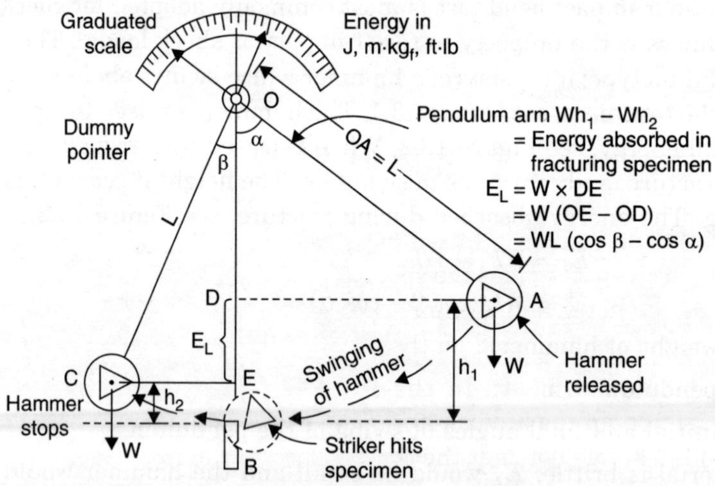

Figure 13.2. Calibration of scale

The hammer (striker) not only breaks the specimen but also throws out the fragment of weight 'w', by a distance 's', from the rotational axis of the pendulum. The energy expended in the throw,

$$e_L = ws\,(1 - \cos \beta) \qquad ...(13.2)$$

This should be subtracted from the value of the energy absorbed E_L, during fracture.

The energy absorbed (expressed as N·m or joule, m·kg$_f$, ft·lb) is read directly on the calibrated dial and is called the 'impact value'. The specimen bends and fractures at a high strain rate. The energy absorbed is indicated on the scale by the dummy pointer. Since, there is a considerable variation of the test results on a particular material, it is necessary to test two or three samples and take the average value.

There are two standard forms of test—Izod and Charpy, named after the men who invented them.

(*i*) **Izod-Impact Test.** In this test, the specimen of circular or square cross-section is fixed in the anvil like a 'vertical cantilever', the *V*-notch facing the striker, see Figure 13.3 (*a*). The test piece may have two or three *V*-shaped notches, each on a different side and 28 mm apart, see Figure 13.3 (*b*). These are tested in turn to allow for directionality in properties.

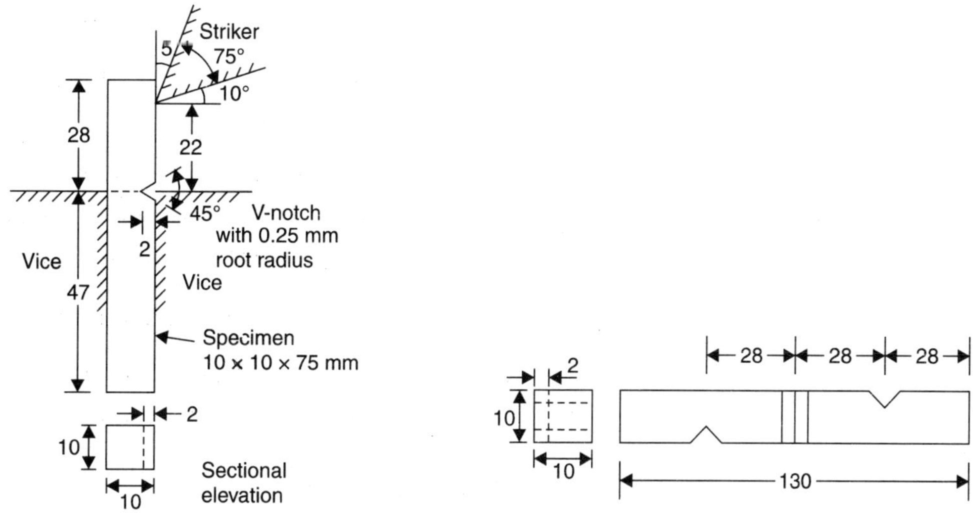

Figure 13.3. (a) Cantilever – Izod impact test specimen V-notch

Figure 13.3. (b) Izod test specimen

The pendulum and striker is raised to its position of initial energy or 120 ft·lb (164 J) and is released by a trigger. It strikes the specimen fixed with a velocity of 12.5 ft/sec (3.75 m/s) on the same side as the notch and 22 mm above it, and swings to the other side. The energy absorbed during fracture is indicated on the scale by the dummy pointer.

(*ii*) **Charpy-Impact Test.** In the Charpy test, the test piece is a square bar 10 × 10 × 55 mm placed on the anvils (40 mm apart) as a horizontal beam simply supported at each end, and has a single notch at the centre facing away from the striker. The notch may be of *V*-or *U*-shape or a key hole, but the *V*-shape is most common, see Figure 13.4.

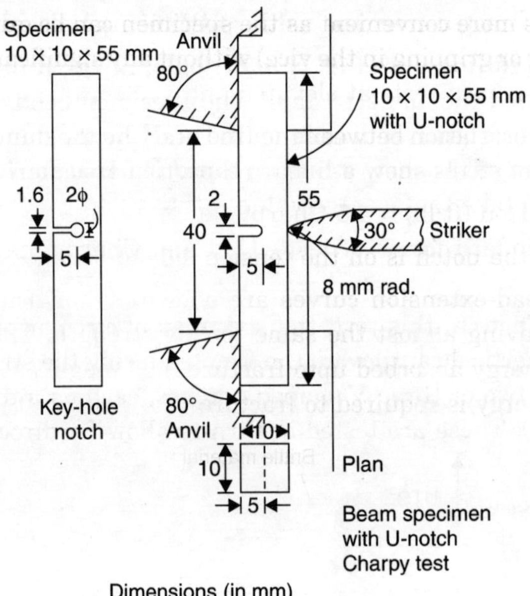

Figure 13.4.

The pendulum is raised to its position of initial energy of 30 $kg_f \cdot m$ (300 J). The hammer (16 kg_f) on releasing swings down and strikes the specimen behind the notch with a velocity of 5 m/s and swings to the other side. The energy absorbed is indicated on the scale by the dummy pointer.

The I.S. method of the Charpy impact test uses a central *U*-notch 5 mm deep. The impact value is preceded by KCU and is the ratio of the energy absorbed in $kg_f \cdot m$ to the cross-sectional area of the test piece behind the 5 mm U-notch in cm^2, with the normal striking energy of 30 $kg_f \cdot m$ (300 J). Testing machines with different striking energies are permitted in which case the value of KCU is supplemented by an appropriate index as:

KCU 100/3 — for striking energy of 100 J and 3 mm deep notch

KCU 300/3 — for normal striking energy of 300 J and 3 mm deep notch.

$$\text{Charpy-Impact Value KCU} = \frac{\text{Energy absorbed}}{c/s \text{ area below notch}} \text{ J/mm}^2$$

$$\text{Standard Test: KCU} = \frac{\text{Energy absorbed}}{(10 \times 5)} \text{ J/mm}^2$$

For striking energy of 30 $kg_f \cdot m$ (= 300 J) with 3 mm deep notch

$$\text{KCU 300/3} = \frac{\text{Energy absorbed}}{(10 \times 3)} \text{ J/mm}^2$$

For striking energy of 10 $kg_f \cdot m$ (= 100 J) with 3 mm deep notch

$$\text{KCU 100/3} = \frac{\text{Energy absorbed}}{(10 \times 3)} \text{ J/mm}^2$$

If during the test, the specimen is not completely broken, the impact value obtained is indefinite.

The Charpy test is more convenient as the specimen can be quickly positioned (without the necessity of clamping or gripping in the vice) without any significant change of temperature and is ideally suited for impact tests at elevated and sub-zero temperatures.

There is no direct correlation between the Izod and Charpy values, however experimental results on a wide range of steels show a linear relation in Izod range of 15 to 70 ft·lb as

$$\text{Izod (ft·lb)} = 4.5 \text{ Charpy (kg·m)} - 10 \qquad \text{...(13.3)}$$

In both the tests, the notch is on the tension side of bending, thus initiating fracture.

In Figure 13.5, load-extension curves are shown (1) for brittle material and (2) for ductile material, both having almost the same tensile strength. The area under the stress-strain curve gives the energy absorbed upto fracture (i.e., toughness). It can be seen that for brittle materials, less energy is required to fracture the specimen than for ductile materials.

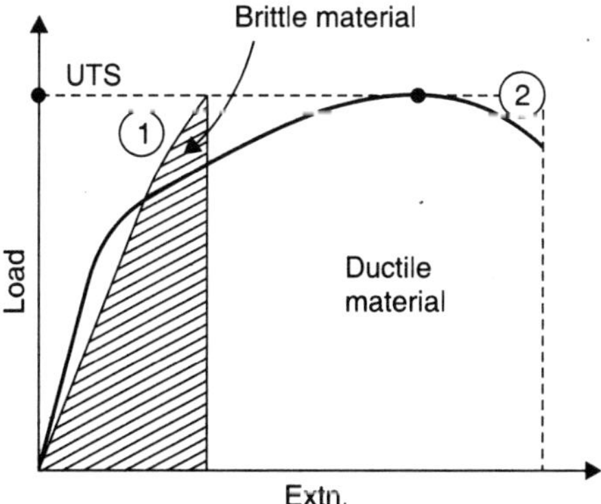

Figure 13.5. Energy to fracture (= area under curve)

The impact strength of notched bars is greatly affected by the temperature. The change of temperature from brittle to ductile is a gradual process and occurs over a wide range of temperature. This is called the transition temperature range; the metal breaks very easily below the transition temperature range (brittle failure), see Figure 13.6. Steels are found to loose toughness remarkably when held between 350°C and the transition temperature. This is referred to as the **temper embrittlement.** Tempering or slow cooling in this temperature range should be avoided to get rid of this phenomenon.

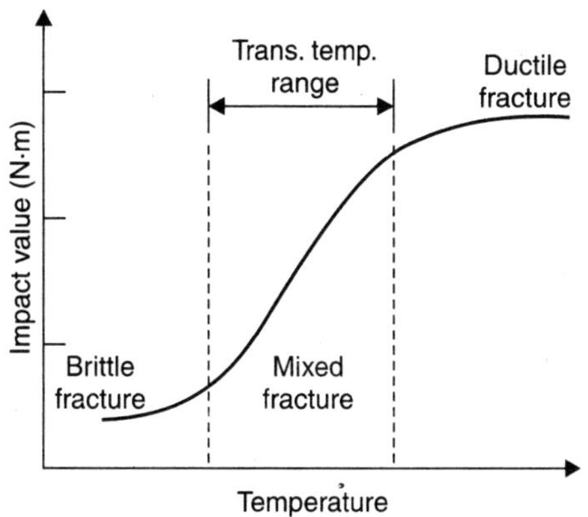

Figure 13.6. Impact strength vs. temperature

Impact tests are useful in quality control checks such as heat treatment operations like tempering of steels; a low impact value indicates that the tempering has not been carried out correctly. The impact test reveals the 'temper brittleness' in heat-treated nickel chrome steels, and also the resistance to fracture due to stress concentrations in a component. The notch sets up conditions of stress concentration.

Impact Test Procedure

1. The hammer is brought to its initial energy level and clamped.
2. The specimen is fixed as a cantilever or beam in the anvil, according to the type of test.
3. The scale is adjusted to read zero.
4. The hammer is released from the clamp.
5. The energy absorbed by the specimen (during fracture) is read on the scale.

Impact Test Observations

Sl. No.	*Material*	*Type of test*	*Trial No.*	*Initial energy*	*Notch type*	*Impact value*	*Average*
1.	M.S.	Izod	1	120 ft·lb (164 J)	45°-V	100 ft·lb	101 ft·lb
			2			102 ft·lb	
2.	C.I.	Izod	1		45°-V	25 ft·lb	26 ft·lb
			2			27 ft·lb	
3.	M.S.	Charpy	1	30 $kg_f \cdot m$ (300 J)	45°-V	5.8 $kg_f \cdot m$	6.3 $kg_f m$
			2			6.8 $kg_f \cdot m$	
4.	C.I.	Charpy	1		45°-V	4.6 $kg_f \cdot m$	5.1 $kg_f m$
			2			5.6 $kg_f \cdot m$	
5.	C.I.	Charpy KCU		30 $kg_f \cdot m$ (300 J)	5 mm-U	$\frac{1.6\ kg_f \cdot m}{(1 \times 0.5)\ cm^2}$	= 3.2 KCU
6.	M.S.	Charpy KCU 300/3		30 $kg_f \cdot m$ (300 J)	3 mm-U	$\frac{7.2\ kg_f \cdot m}{(1 \times 0.7)\ cm^2}$	= 10.3 KCU 300/3

Note. Mild steels and low carbon steels have impact values ranging from 5 to 12 $kg_f \cdot m$. Plain carbon case hardened steels possess impact values ranging from 6 $kg_f \cdot m$ onwards.

14 Fatigue Test

Materials fail under repeated loading and unloading, or under reversal of stress, at stresses smaller than the ultimate strength of the material under static loads. The magnitude decreases as the number of cycles of stress increases. The behaviour of materials under rapidly fluctuating loads (causing repeated stress or stress reversal) is known as **fatigue.** Some examples of components subjected to fatigue are springs, gears, rotating shafts, railway axles, aircraft wing structures and girder bridges.

Stress Cycle and Range. In practice, the shape of the cycle stress reversals may vary; however a sinusoidal cycle with constant upper and lower stress limits throughout the life is assumed when considering fatigue behaviour. The range of stress R, over which fluctuations occur is the algebraic difference between the maximum and minimum stresses, treating compression as negative.

Range of stress $R = \sigma_{max} - \sigma_{min}$

Mean stress $\sigma_{mean} = \dfrac{\sigma_{max} + \sigma_{min}}{2}$

Stress ratio $r = \dfrac{\sigma_{min}}{\sigma_{max}}$

Three types of stress cycles are shown in Figure 14.1.

(*a*) In a 'fluctuating cycle', an alternating stress is imposed on a mean stress with upper and lower limits positive or negative, see Figure 14.1 (*a*).

(*b*) In a 'reversed stress cycle' the fluctuations are of equal tension or compression, see Figure 14.1 (*b*).

$$| \sigma_{max} | = | \sigma_{min} | \qquad ...(14.1)$$

$$\sigma_{mean} = \frac{\sigma_{max} + (-\sigma_{min})}{2} = 0 \qquad ...(14.2)$$

$$R = \sigma_{max} - (-\sigma_{min}) = 2\sigma_{max} \qquad ...(14.3)$$

(*c*) In a 'repeated stress cycle', see Figure 14.1 (*c*).

$$\sigma_{min} = 0, \quad R = \sigma_{max} \qquad ...(14.4)$$

$$\sigma_{mean} = \frac{\sigma_{max}}{2} = \frac{R}{2} \qquad ...(14.5)$$

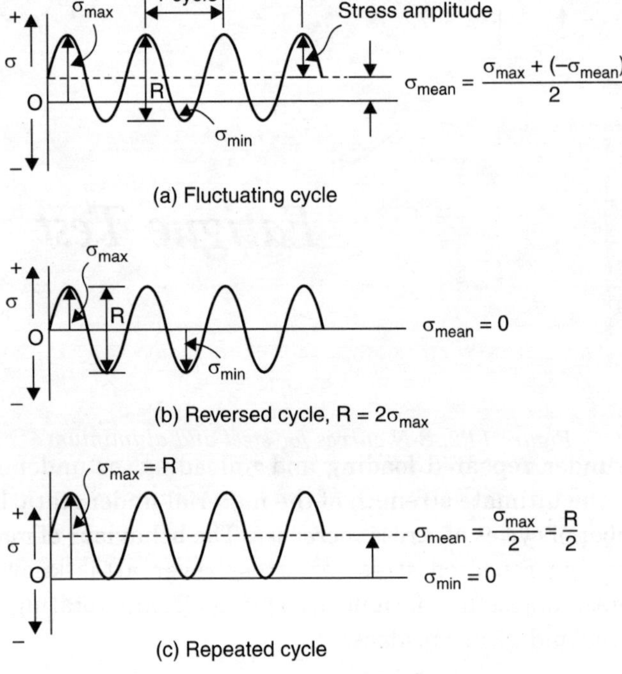

Figure 14.1. Types of stress cycles

The cycle is completely defined if the range of stress and maximum stress are given.

Fluctuating stresses occur in practice due to

(*i*) direct stresses under axial loading (tension or compression)

(*ii*) flexure stresses due to bending moment (positive or negative)

(*iii*) shear stresses due to cyclic torsion, or any combination of these.

Testing machines have been developed to reproduce all these types of stress cycles and the design stresses should conform to the conditions under which the component has to operate. The stress level at which a metal fails by fatigue is called **fatigue strength.**

Endurance or Fatigue Limit. 'Fatigue life' is the number of cycles *N*, of stress reversal that the material can withstand before failure takes place and is a function of the cyclic stress level ($S = \sigma_{max}$) and the relationship is shown as a *S-N* curve. As there may be many millions of cycle for failure, a log scale is used for the life (*N*).

For most steels and ferrous alloys as the cyclic stress level (*S*) decreases, the life *N*, increases, and there is a stress level (S_1) below which fatigue cracks are not initiated and the life is indefinite, see Figure 14.2. This limiting stress below which a load may be repeatedly applied an indefinitely large number of times without causing failure is called **endurance limit** or **fatigue limit.** The ratio of endurance limit to the static strength (UTS) is called the **endurance ratio.** Its value is around 0.5 for steels with tensile strength less than 1035 N/mm^2. On the other hand, non-ferrous metals, aluminium alloys and thermosetting plastics do not show a definite endurance limit and can fracture after several hundred million cycles of stress. For these metals, the fatigue limit or fatigue strength is usually quoted as the stress value to give a specific large number of cycles *N*, usually 50×10^6.

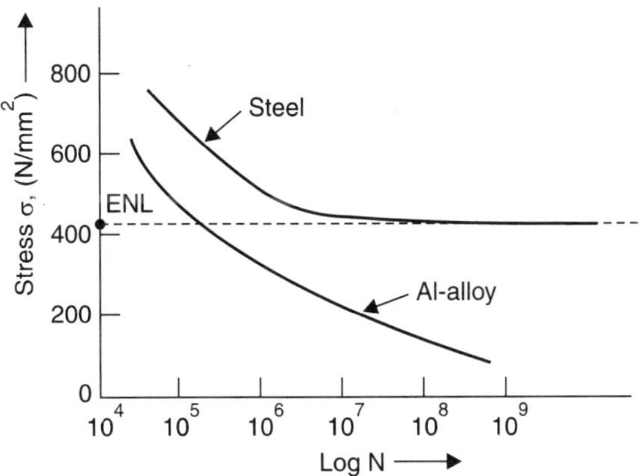

Figure 14.2. S-N curves for steel and aluminium

Presence of a notch causes stress concentration making it easier to initiate the fatigue crack, see Figure 14.3. The failure can be interpreted as a short life (small N, say 10^5) to start the crack, or the maximum stress for a given life is lower.

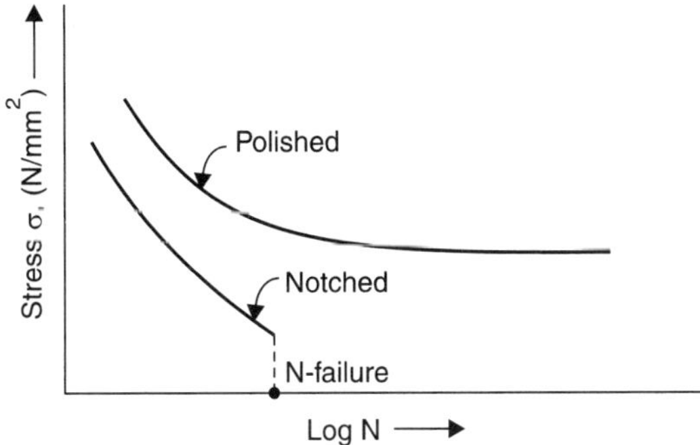

Figure 14.3. S-N curves for polished and notched steel bars

Fatigue limit is invariably lowered by surface imperfections. Since sharp corners, deep scratches, punch marks, oil holes, fillets and keyways act as 'stress raisers' precautions should be taken in the design and manufacture of components such as screw threads subjected to dynamic loading to minimise fatigue cracks developing.

Factors Affecting the Fatigue Strength. Fatigue cracks usually originate at the surface. Fatigue strength can be increased by fine surface finish (smooth ground surfaces) and surface finish treatment like surface hardening (carburising and nitriding), work hardening, cold rolling and shot peening, which have been found to increase the endurance limit upto 20%. In general, steels of quenched and tempered structures have a higher fatigue strength than annealed or normalised steels.

The frequency of stress reversals also influences the fatigue limit, which is higher for increased frequency. Stress concentrations caused by sudden changes in cross-sections decreases fatigue strength.

In corrosive atmosphere, fatigue strength is greatly lowered and there is no definite endurance limit (corrosion fatigue).

Fatigue Testing Machine, see Figure 14.4. The specimen holder with specimen gripped in it, is driven by the motor at a particular speed, say 6000 rpm. The number of rotations is counted by the cycle counter. The specimen will be acted upon by a bending moment M, by adding weights to the pan. The flexure stress on the top and bottom fibres of the specimen will vary from a tensile stress to a compressive stress as the specimen undergoes half a revolution and the cycle repeates for each revolution.

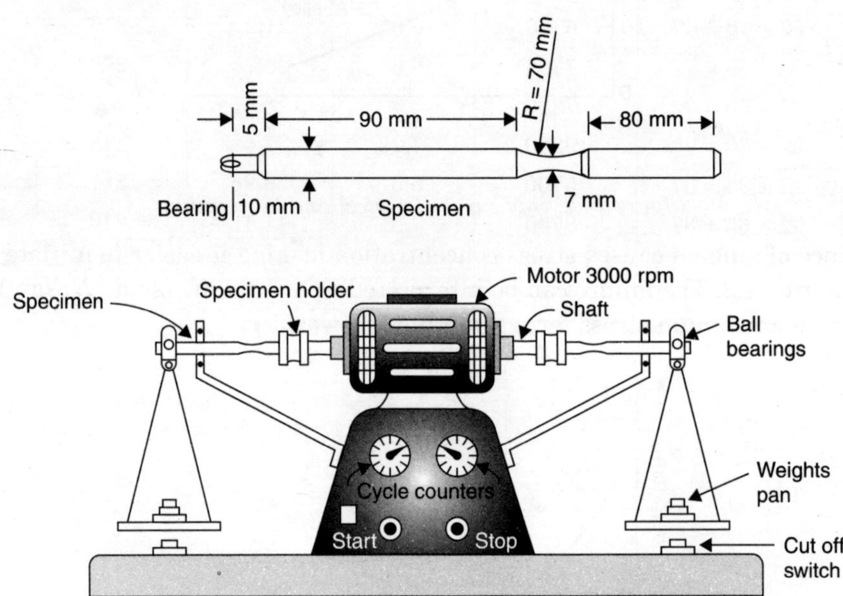

Figure 14.4. Fatigue testing machine (for rotating cantilever bending tests)

After a certain number of revolutions, the specimen will break when the pan will fall on the automatic cut off switch and the motor is switched off. The number of cycles of stress reversal is read on the counter.

Tests are conducted on specimens subjected to a gradually decreasing range of stress when the number of cycles of stress required to fracture each specimen increases, and as the fatigue limit is approached, it will take some hundreds of millions of reversal before fracture, as indicated by the cycle counter. A curve of 'σ_{max} vs log N' is plotted and the **endurance limit** (ENL) is indicated by a definite discontinuity in the curve.

Fatigue Test Observations. L and D of the cylindrical specimen are measured to the critical section to initiate failure by fatigue.

Load applied at the bearings = W

$$W = \text{Self weight of pan (22 N)} + \text{Weights } (P) \text{ put on the pan}$$

Bending moment at the critical section

$$M = WL,\ W = (22 + P) \text{ newton},\ L = 100 \text{ mm}$$

$$\frac{M}{I} = \frac{\sigma}{Y}, \quad \sigma = \frac{M}{I/Y} = \frac{M}{z}$$

where the section modulus $z = \dfrac{I}{Y} = \dfrac{\pi D^4/64}{D/2} = \dfrac{\pi D^3}{32}$.

No. of cycles of stress reversals before fracture recorded by the counter = N

Sl. No.	*Load W = 22 + P (N)*	*M = 100 W (N·mm)*	*Average dia D (mm)*	$z = \frac{\pi D^3}{32}$ *(mm³)*	$\sigma_{max} = \frac{M}{z}$ *(N/mm²)*	*N (cycles)*
1.	22 + 45 = 67	6700	6.0	21.2	316	208,500
2.	22 + 55 = 77	7700	6.6	28.2	273	325,000
3.	22 + 50 = 72	7200	6.5	26.96	267	400,000
4.	22 + 70 = 92	9200	7.25	37.4	246	650,000
5.	22 + 35 = 57	5700	6.5	26.96	211	2,200,000
6.	22 + 65 = 87	8700	7.5	41.4	210	5,250,000
7.	22 + 40 = 62	6200	7.0	33.7	184	Unbroken
8.	22 + 30 = 52	5200	7.0	33.7	154	Unbroken

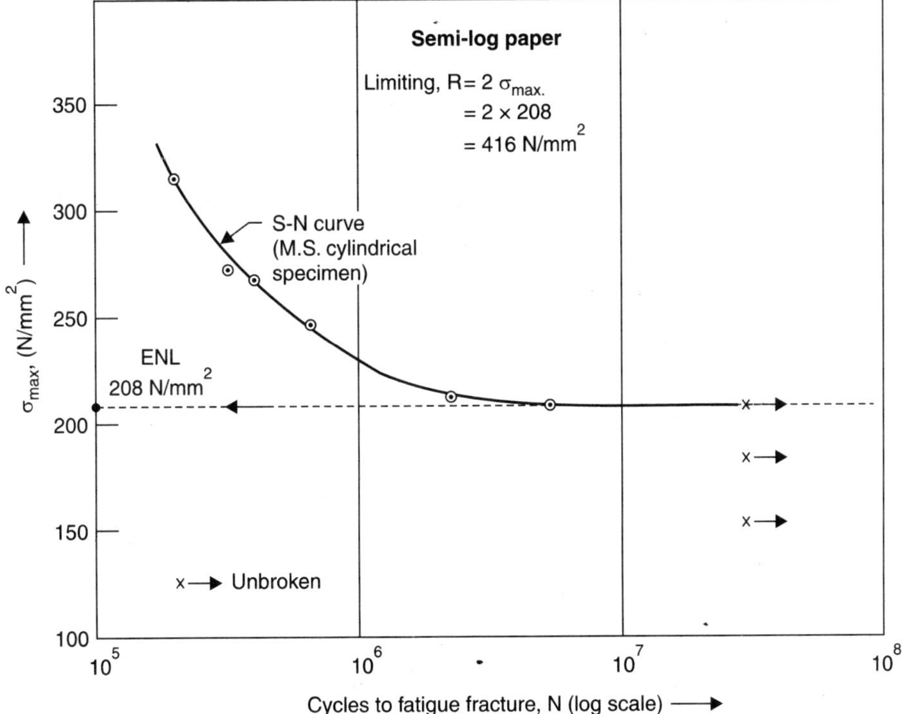

Figure 14.5. Fatigue test on M.S. specimen

The *S-N* curve 'σ_{max} vs Log N', is drawn in Figure 14.5, and the endurance limit.

ENL = **208 N/mm²**

Limiting range of stress (R) = 2 σ_{max} = 2 × 208 = **416 N/mm²**

Assuming UTS = 500 N/mm^2,

$$\text{Endurance ratio} \quad \text{ENR} = \frac{\text{ENL}}{\text{UTS}} = \frac{208}{500} = \mathbf{0.4}$$

For most structural materials, this ratio varies from 0.2 to 0.6.

Stress Concentration. When a sharp change in cross-section occurs in a component subjected to tension, Compression or bending or due to some discontinuity such as a hole, keyway, grooves, edge fillets, gear teeth, etc. The stress will no longer be uniform but concentrates towards a point as can be seen on a photoelastic stress pattern, see Figure 14.6. In such cases, the stress cannot be calculated by the normal procedure.

$$\text{The ratio } k = \frac{\text{Maximum boundary stress at the discontinuity}}{\text{Average stress at the minimum section.}}$$

is called the **stress concentration factor**, see Figure 14.7. The values of the factor k, range depending on the size and shape of the changes in cross-section or hole and usually varies from 1.25 to 3.0, see Figure 14.8. At such abrupt changes in section, the stresses are highly localised and can be determined by the mathematical theory of elasticity or by photoelastic method.

Figure 14.6. Stress concentration at contact of gear teeth

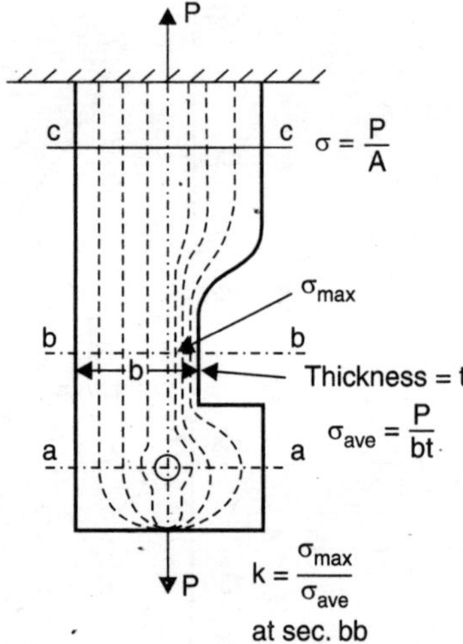

Figure 14.7. Stress concentration in lift bar

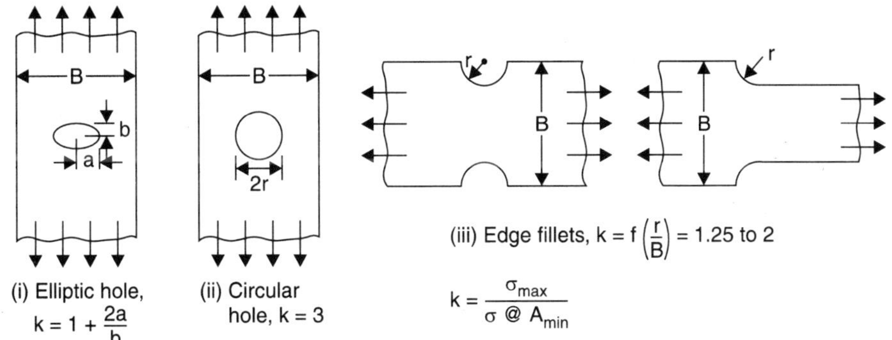

Figure 14.8. Stress concentration in tensile members

Stress concentrations are always a danger under 'fatigue' conditions and in brittle materials. The material reaction to stress concentration under fatigue loading is represented by a notch sensitivity factor q, defined as

$$q = \frac{k_f - 1}{k - 1} \qquad \text{...(14.6)}$$

where the fatigue strength reduction factor

$$k_f = \frac{\text{Plain fatigue strength at } N \text{ cycles or fatigue limit}}{\text{Notched fatigue strength at } N \text{ cycles or fatigue limit}}$$

and generally $k > k_f > 1$.

For a component which yields $k_f = k$, $q = 1$ which shows maximum sensitivity; when there is no strength reduction, $k_f = 1$ and $q = 0$, showing no sensitivity.

Also see page 167, Figs. 1.22, 1.23 and 1.24.

15 Creep Test

It can be observed from the stress-strain curve during a tension test at room temperature, at a point *A*, (see Figure 15.1), the material enters the plastic range. If the loading is held constant at *A*, the metal continues to stretch for a few sec. or min. till equilibrium is reached at *B*. This **time-dependent** increase in strain (*AB*) at constant load is known as **creep** and is measured as the **rate of strain per hour** under a certain stress at a given temperature.

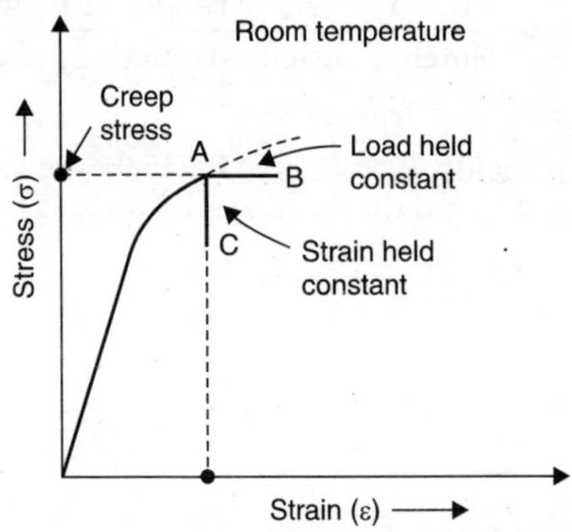

Figure 15.1. Creep stress and relaxation

If on the other hand, instead of load, the strain is held constant at *A*, it can be observed that there is a drop in load with time till equilibrium is reached at point *C*. This behaviour is termed **creep-stress relaxation.**

Creep Curve. Creep takes place over long periods of time and may be almost imperceptible to the casual observer. A typical 'creep strain vs log time' curve is shown in Figure 15.2, and it can be seen that the creep progresses in three stages. The first or primary stage consists of a short period during which strain increases rapidly i.e., strain hardening due to increase in temperature. This primary strain is denoted by ε_0. The secondary stage is over a long period during which the strain is almost constant with severe distortion of the intercrystalline structure. The final or tertiary stage during which the creep strain increases due to increase in stress level owing to the reduction in area (necking) and continues till the material fractures.

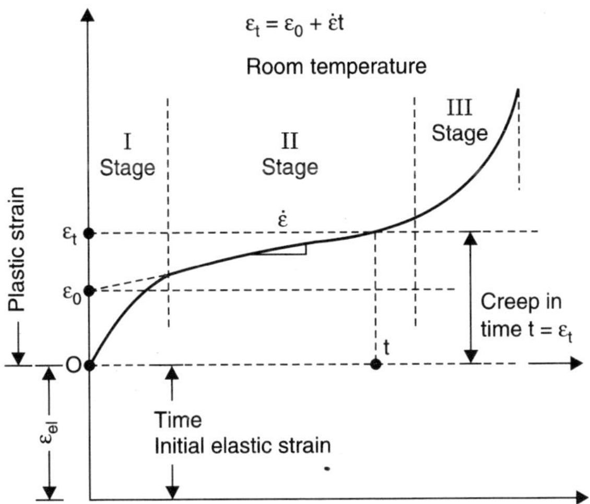

Figure 15.2. Typical creep curve

Whether or not this tertiary stage exists but depends on the stress level. By plotting 'creep strain vs log time' curves for a number of stress levels (see Figure 15.3), it can be seen that there is a critical stress level, below which the tertiary stage is absent $\left(\frac{\partial \varepsilon}{\partial t} = 0\right)$; although creep occurs, the material may not fracture even under an indefinite period. This critical stress has the same sort of significance as the fatigue limit under cyclic stress reversals. Above this critical stress value, fracture occurs within an increasingly shorter time as the stress level is raised nearer to the elastic limit. The 'limiting creep stress' is now a days based on a **permissible creep strain after a given time.**

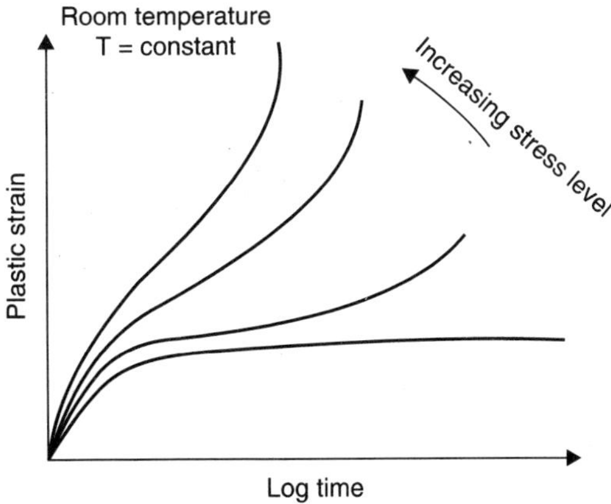

Figure 15.3. Creep curves at various stress levels

It has been found experimentally that the minimum creep rate $\left(\dot{\varepsilon} = \frac{\partial \varepsilon}{\partial t}\right)$ and the stress σ, at a constant temperature are related as

$$\dot{\varepsilon} = A\,\sigma^n \qquad \text{...(15.1)}$$

The constants A and n can be determined by a plot of '$\dot{\varepsilon}$ vs σ' on a log-log paper which yields a straight line.

As per Andrade's simplified creep curve (see Figure 15.4), the total creep strain after time (t)

$$\varepsilon_t = \varepsilon_0 + \dot{\varepsilon}t \qquad ...(15.2)$$

from which the time to reach a specified value of total creep strain can be obtained as

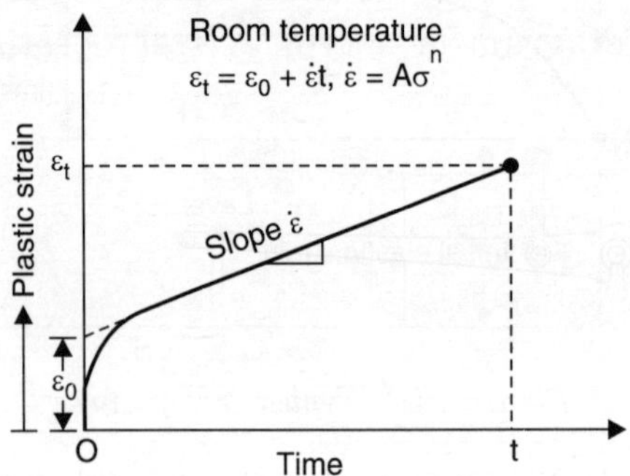

Figure 15.4. Simplified creep curve (Andrade's concept)

$$t = \frac{\varepsilon_t - \varepsilon_0}{A\sigma^n} \qquad ...(15.3)$$

The creep rate also increases with temperature and for a given stress

$$\dot{\varepsilon} = Be^{-\frac{Q}{RT}} \qquad ...(15.4)$$

where Q = activation energy for creep,

R = gas constant,

T = absolute temperature (K),

B = a constant for the material.

Creep strength is usually defined as the stress for 1% strain in 10,000 hr (≈ 1 yr) as used in jet turbine design, or 1% strain in 100,000 hr (11.4 yr) as used in steam turbine design. This limiting creep stress may frequently be less than half of the UTS at that temperature. The effect of creep is particularly observed in metals operating at elevated temperatures.

Creep resisting materials such as special alloy steels containing small percentages of molebdenum, vanadium, cobalt or tungsten have been used for gas turbine blades and high pressure steam fitting. Only the allowable stress will fix the life of the material.

Creep Testing Equipment. Any tensile testing machine could be used for this test and since a constant loading on the specimen is required over very long periods, a dead weight loading system by lever mechanism is usually employed, see Figure 15.5 (a).

The specimen of 6, 12 or 16 mm diameter with gauge length more than four times the diameter (or even larger for accurate measurement of strain) is surrounded by an electric furnace to keep the temperature of the specimen constant as far as possible, see Figure 15.5 (a). An extensometer arrangement is made to record extensions carefully and accurately. The temperature is very accurately recorded by a thermocouple and a galvanometer installed.

During the test, the specimen is fixed on the machine and is heated till the temperature of the test is attained. When the temperature of the specimen is steady, the gauge length is observed and the predetermined load is applied quickly without shock.

The resulting initial extension is noted which is essentially the elastic strain. Thereafter, strain measurements are made at fixed time intervals. At least fifty observations are made to plot the creep curve. At each observation of strain, the temperature is noted and the average of all the recorded temperature readings is reported as the actual temperature of the test.

An alternative line diagram of a typical creep testing equipment is shown in Figure 15.5 (*b*).

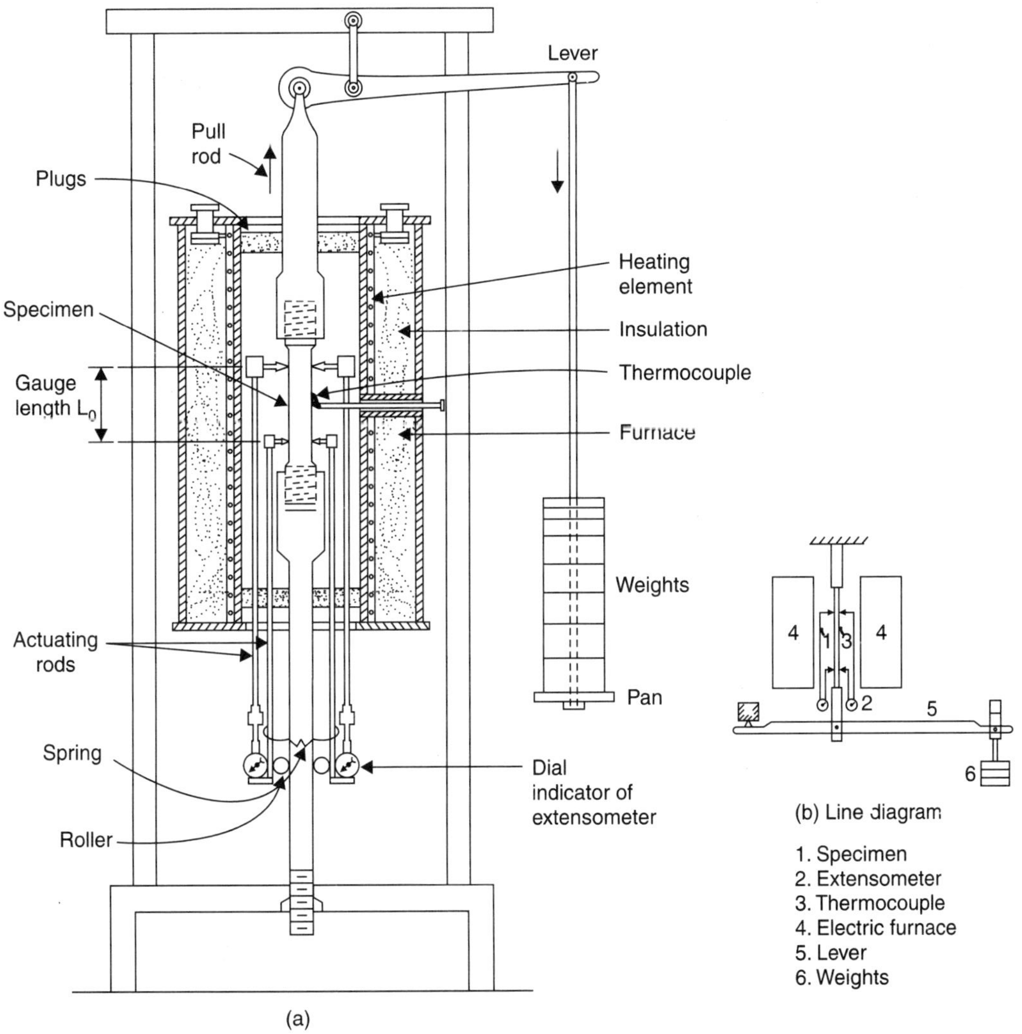

Figure 15.5. Tension creep testing equipment

Since creep testing takes a very long time, it is customary to conduct creep tests at various stresses and temperatures such that a part of the uniform creep rate of the secondary stage is obtained, see Figure 15.6 (*i*). The curves can then be extrapolated to the required time period and the total creep strain can be determined by Eqn. (15.2). Extrapolations can be made upto a limiting creep strain of 1% at various stress levels and operating temperatures for a period 100,000 hr. (≈ 10 yr), see Figure 15.6 (*ii*).

The creep rate is known to diminish with time and a typical stress value obtained by conducting a creep test is that which will cause 'a creep of 1 millionth per hour after 40 days', see Figure 15.6 (*iii*).

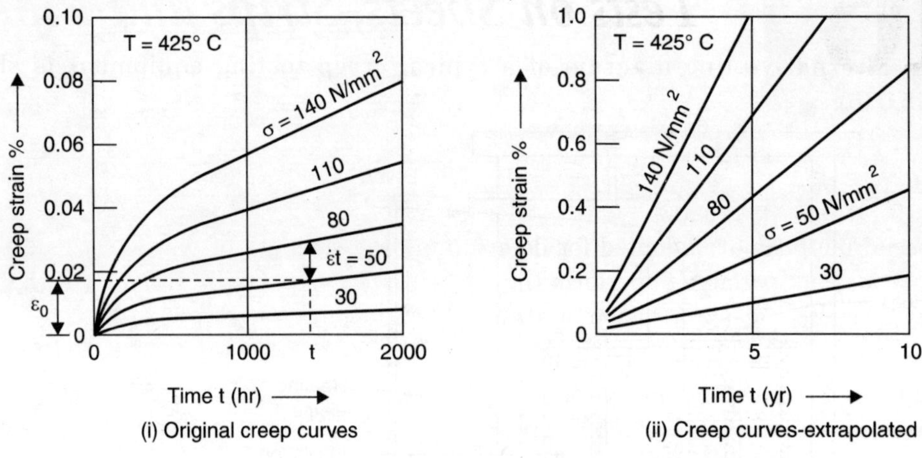

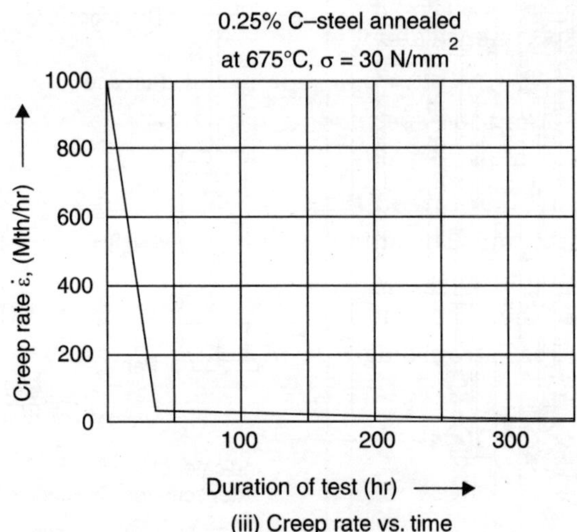

Figure 15.6.

16 Tests on Sheets, Strips and Pipes

Special methods are adopted for determining the strength of plates, strips, tubes and wires, since it is not possible to perform the standard tests because of their small thicknesses and heavy duty machines (extension or strain readings will lack sensitivity if they are used).

Sheet metal is mostly employed in fabrication processes where ductility is as important as strength. Hence 'bending' and 'cupping' tests are conducted on the sheet metal, since the conditions of testing are similar to the actual 'working' conditions in the sheet metal industry.

The practical difficulties in testing sheets, strips, plates and tubular products are:

(*i*) Standard test specimens cannot be used because of their small dimensions (say on a 5 mm sheet or 3 mm wire).

(*ii*) Problems of gripping the specimen in the testing machine. For example, a wire is cast into an artificial lead grip and then tested.

Tensile Strength of Sheets and Plates. A **Hounsfield Tensometer** is usually used. It is a small bench type machine. The standard sheet specimens for tensile testing are shown in Figure 16.1. The specimen (sheet or plate) is gripped by pin-type grips which slide on horizontal rods during loading. The load is applied by a hand lever upto breaking. There is provision for plotting the stress-strain diagram during the test. The tensile strength is calculated as usual.

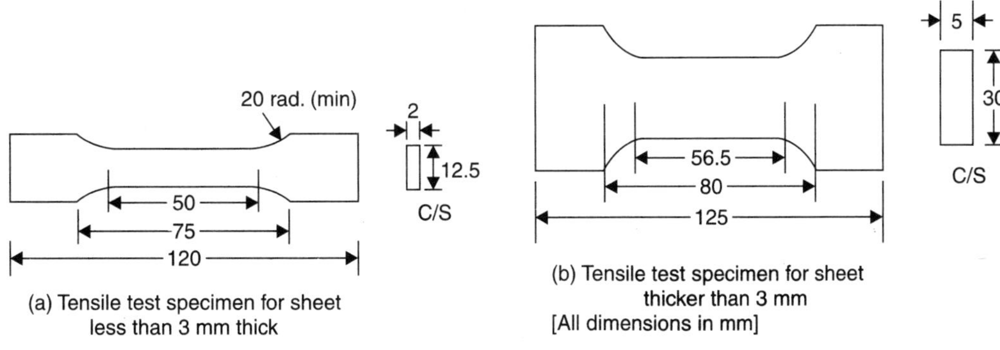

Figure 16.1. Tension test on sheets

$$\text{UTS} = \frac{\text{Maximum load}}{\text{Cross-sectional area of the specimen}}$$

Hardness values, if required, are obtained by the Rockwell Tester. It should be noted that the thickness of metal beneath the indentation should be at least five times the diameter of the indentor.

Bend Tests. The various types of bend tests commonly employed consist of supporting a sheet of material at its ends and applying load at mid-span by a former whose radius is equal to twice the thickness of the plate, see Figure 16.2. The material is considered acceptable if the bar or sheet can be bent through 180° without cracking. Bend tests are used as quality-control checks on the ductility of a material after heat treatment.

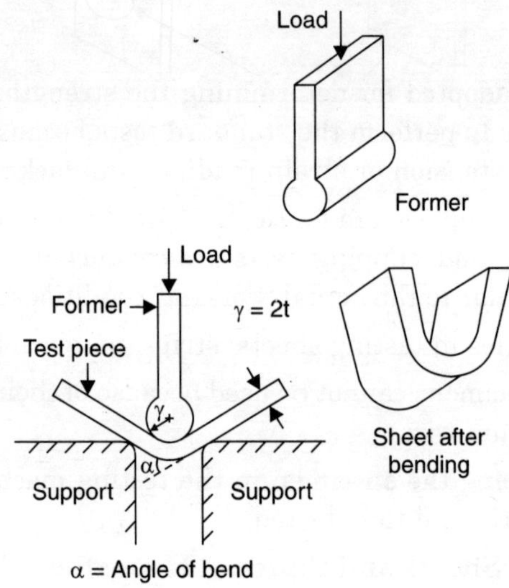

Figure 16.2. Bend test

Cupping Tests. Ductility in sheet metal can be interpreted as its ability to be stretched into the shape. A number of cupping tests have been developed which involve forcing a hemispherical plunger into one face of the sheet and measuring the size of the cup which can be produced before the metal tears. The failure of the metal can be observed in a mirror attached to the cup side of the machine.

The best known as this type is the **Erichsen test**. The sheet to be tested (usually 89 mm square) is held between two ring clamps. A round-nosed ram is pressed into the sheet, which deforms to give a cup-shaped depression, see Figure 16.3. The test is stopped as soon as a crack appears. The nature of cracks and the depth of the cup are used to interpret the metal's ability to be stretch-formed or pressed. Some metals develop an 'orange-peel' effect which are undesirable for pressings that are to be painted, say, car body panels.

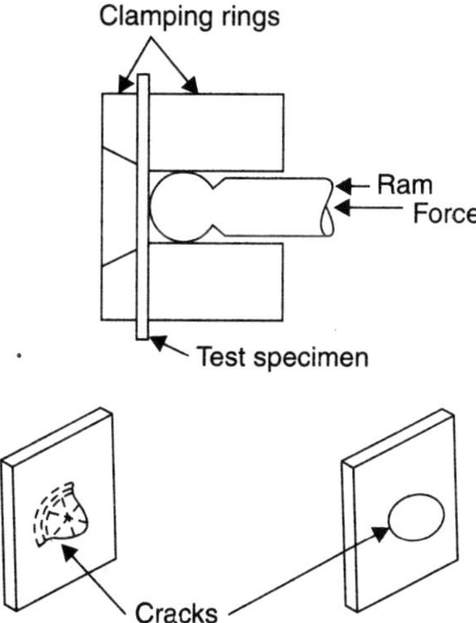

Figure 16.3. Erichsen cupping test

Tension Test on Tubes and Pipes. A tube is similar to a pipe except that the thickness of a tube is very small when compared to its diameter while a pipe may have an appreciable wall thickness.

The tension test is conducted as usual and the only problem which arises, is the gripping of the specimen in the tensile testing machine. One method is to flatten the two ends with a hand hammer and grip them as if it were a plate. Sometimes threads (internal or external) may be cut on the two ends and screwed to the fixtures containing threads specially made for the purpose. It should be noted that the gauge length portion should not be subjected to any deformation before the test.

17 Tests on Leaf Springs

Springs are elastic bodies or resilient members; they distort when loaded and recover to their original shape when load is removed. The principal function of a spring is to absorb energy, store it for a long period (as in a watch) or for a short period (as in an engine valve), and then transfer to the surrounding material. There are many types of springs for specific purposes and the most common forms are close and open coiled, conical, flat, spiral and leaf.

Leaf Springs. Also named as semi-elliptical, carriage springs, and built up springs. They are commonly used in carriages such as cars, lorries, railway wagons, etc. They consist of a number of thin curved plates or leaves of uniform cross-section but of varying lengths secured together at the centre with a bolt and clamped at distances for compactness, see Figure 17.1. The springs rest on the axle of a vehicle and are pin-jointed to the chassis. Any jerk, shock or vibration form the wheels is immediately taken by the spring without affecting the carriage.

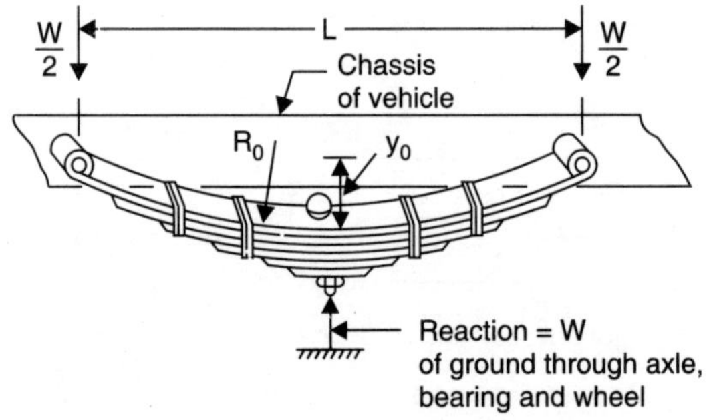

Figure 17.1. Carriage or leaf springs

Beam of Uniform Strength. If W is the load shared by a spring in a carriage, the load acting on each hinge is $\frac{W}{2}$. Span of the longest leaf of the spring is L. Number of plates (leaves) is n.

The loaded spring is similar to that of a simply supported beam carrying a load W at mid-span and the bending moment diagram BMD, is shown in Figure 17.2. The length of each strip may be obtained by dividing the BMD into the number of strips n, required at the centre.

Both M and I of each strip are constant over the central portion and decrease near the ends (assuming contact at ends only). For this reason, the widths are tapered to a point. Consequently for the whole leaf, $\frac{M}{I}$ or $\frac{M}{EI}$ is constant. If the leaves are initially curved to circular arcs of the same radius R_0 and in the strained state if the radius is R, from the 'simple theory of bending of curved bars'.

$$\frac{M}{EI} = \frac{1}{R} - \frac{1}{R_0} \qquad \text{...(17.1)}$$

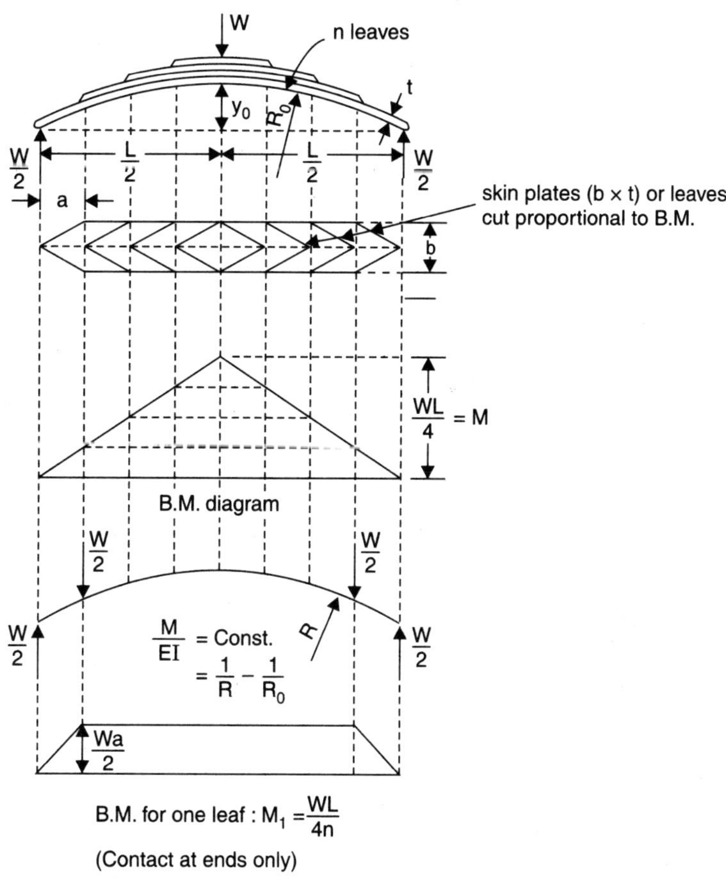

Figure 17.2. Beam of uniform strength

The spring consists of a number of plates n, each of constant width b, and thickness t, and is so designed that each plate has the **same bending stress at all cross-sections** i.e., the spring is equivalent to **a beam of uniform strength**.

Theory. Considering the central section

$$M = \frac{WL}{4},$$

tending to decrease the curvature i.e., changing the original 'rise' y_0 to y.

Resisting moment of each plate $= \frac{WL}{4n}$

If σ is the maximum bending stress developed in each leaf, from Eqn. (17.1).

$$\frac{\sigma}{y} = \frac{M}{I} = E\left(\frac{1}{R} - \frac{1}{R_0}\right) \qquad ...(17.2)$$

$$\Rightarrow \quad \sigma = M \times \frac{y}{I} = \frac{WL}{4n} \times \frac{t/2}{bt^3/12} = \frac{3}{2}\frac{WL}{nbt^2} \qquad ...(17.3)$$

The deflection may be obtained by considering the longest plate, since each is an arc of a circle, see Figure 17.3. Since, the rise y, is very small compared with R, neglecting y^2.

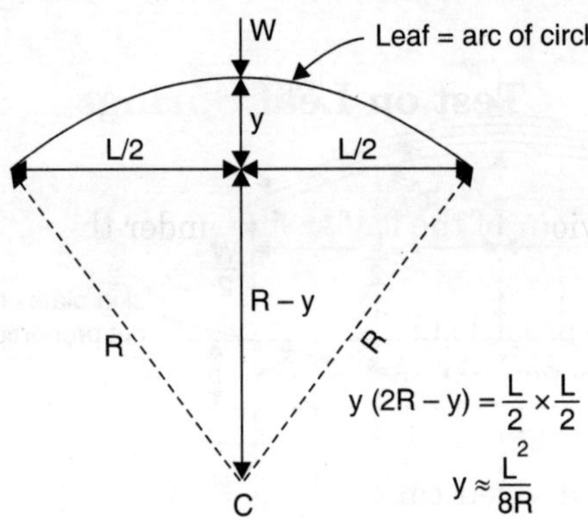

Figure 17.3. Deflection under load

$$y = \frac{L^2}{8R}, \qquad y_0 = \frac{L^2}{8R_0} \qquad ...(17.4)$$

$$\frac{M}{EI} = \frac{1}{R} - \frac{1}{R_0}, \qquad \delta = y_0 - y = \text{central deflection}$$

$$\frac{WL}{4nE\,(bt^3/12)} = \frac{8}{L^2}(y - y_0) = \frac{8}{L^2}\,\delta$$

Ignoring negative sign for δ which indicates the decrease of y_0.

$$\therefore \quad \text{Deflection } (\delta) = y_0 - y = \frac{3}{8} \times \frac{WL^3}{nEbt^3} \text{ or } \frac{WL^3}{32\,EIn} \qquad ...(17.5)$$

$$\text{Modulus of elasticity } (E) = \frac{3}{8} \times \frac{L^3}{nbt^3}\left(\frac{W}{\delta}\right) \qquad ...(17.6)$$

$\frac{W}{\delta}$ is the slope of the 'load-deflection' plot and is called the spring stiffness k. It is the load required to produce unit deflection. Sometimes, the reciprocal of k is called the spring constant

$$k = \frac{W}{\delta}, \qquad k' = \frac{1}{k} = \frac{\delta}{W} \qquad ...(17.7)$$

The load required to make the spring flat is called the **proof load** W_p, i.e., when $y = 0$ or $\delta = Y_0$. From Eqn. (17.5),

$$W_P = \frac{8}{3} \times \frac{nbt^3 \; Ey_0}{L^3} \quad \text{or} \quad \frac{32 \, EI \, ny_0}{L^3} \qquad \text{...(17.8)}$$

Strain energy (S.E.) stored in the spring upto a certain limiting stress σ,

$$\therefore \qquad \text{S.E.} = \frac{\sigma^2}{6E} \times \text{Volume of the spring} \qquad \text{...(17.9)}$$

Volume of the spring = $(b \times t)$ × Total length of all leaves

$$\text{Resilience } (U) = \frac{\sigma^2}{6E} \text{ per unit volume.} \qquad \text{...(17.10)}$$

Test on Leaf Springs

Object

(*i*) To study the behaviour of the leaf spring under the action of a gradually increasing load.

(*ii*) To determine the 'proof load', 'the skin stress at proof load', 'the stiffness of the spring' (or spring constant) and the 'proof resilience'.

(*iii*) To determine the 'modulus of elasticity, of the material of the spring'.

Test Procedure. For a given carriage spring, the width b, and the thickness of the plates t, are measured, and the number of plates (leaves) are counted. The spring is placed on the roller supports, the span L, and the central rise Y_0, are measured. The load is gradually applied over the centre and the central deflection δ, of the spring is read on the ivory scale (IVS) for regular load increments till the spring becomes flat. The value of the proof load is noted. The spring is gradually unloaded, the deflection of the spring being noted for the same load stages.

Test Observations. A carriage spring with 4 leaves, each of 8 mm thickness and 51 mm width, span 1.01 m and central rise 108 mm was tested in a UTM of 400 kN capacity and the following 'load-deflection' observations were made.

Load (kN)	*Deflection on IVS (mm)*		*Load (kN)*	*Deflection on IVS (mm)*	
	Loading	*Unloading*		*Loading*	*Unloading*
0.4	3	1	4.0	60	62
0.8	10	4	4.4	66	69
1.2	16	11	4.8	73	76
1.6	22	17	5.2	79	83
2.0	29	28	5.6	86	91
2.4	34	34	6.0	92	97
2.8	40	41	6.4	98	103
3.2	47	48	6.8	105	107
3.6	53.5	55	7.0	107.5	107.5

Report the results qualitatively for certification.

Test Results. The 'load-deflection curve' is drawn, see Figure 17.4. For any particular load, the deflection during the loading cycle will be less than the deflection during the unloading cycle and the plot will form a loop. A mean straight line is drawn and the stiffness of the spring k, is obtained.

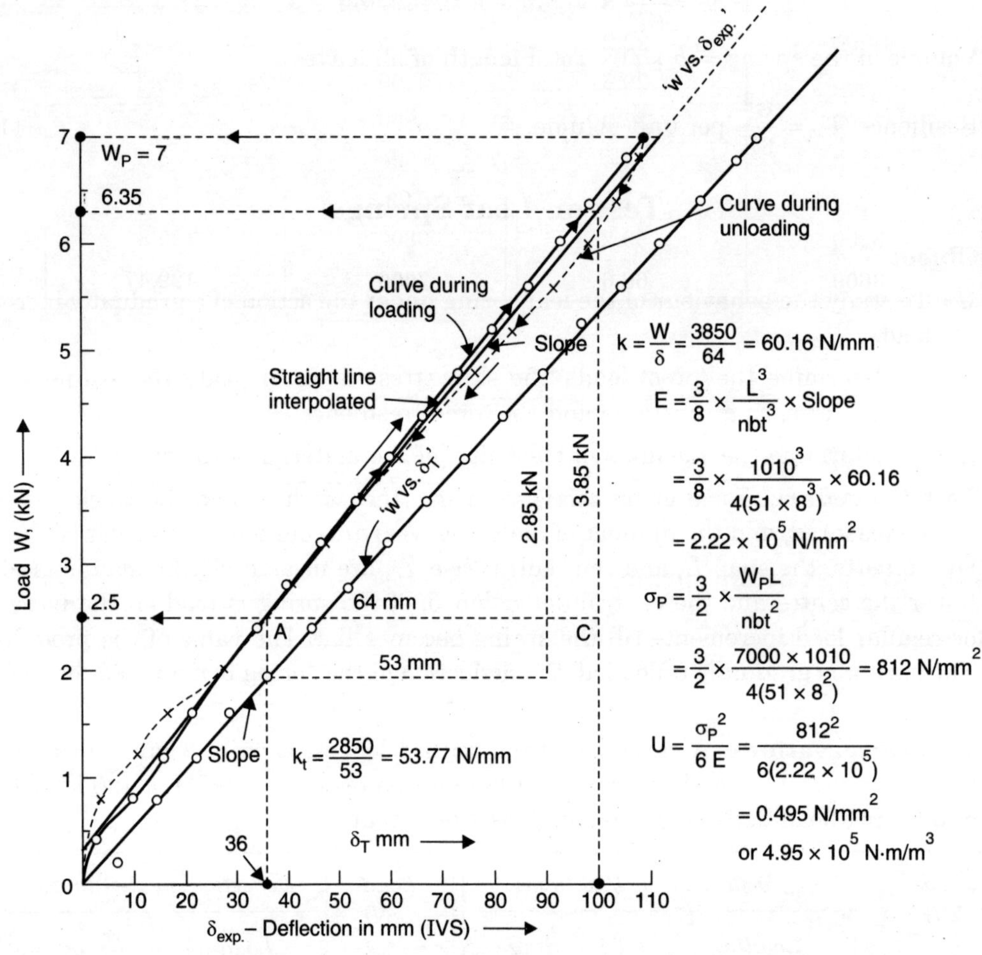

Figure 17.4. Load-deflection curve for leaf springs

A graph of load vs theoretical deflection of the spring δ_T, is drawn, taking, $E = 2 \times 10^5$ N/mm^2 in the Eqn. (17.5).

$$\delta_T = \left(\frac{3}{8}\frac{L^3}{nEbt^3}\right) W$$

$$= \frac{3}{8} \times \frac{1010^3 \times W}{4(2 \times 10^5)(51 \times 8^3)} = 0.0185\ W$$

which yields a straight line and its slope gives the theoretical stiffness of the spring k_t.

W (N)	T (mm)	W (N)	T (mm)
400	7.4	4000	74
800	14.8	4400	81.4
1200	22.2	4800	88.8
1600	29.6	5200	96.2
2000	37.0	5600	103.6
2400	44.4	6000	111.0
2800	51.8	6400	118.4
3200	59.2	6800	125.8
3600	66.6	7000	129.47

1. From the plot of W vs δ_T

$$\text{Slope } (k_t) = \frac{W}{\delta_T} = \frac{2850}{53} = \mathbf{53.77\ N/mm}$$

From the mean line of the plot of W vs δ_{exp},

$$\text{Slope } (k) = \frac{W}{\delta} = \frac{3850}{64} = \mathbf{60.16\ N/mm}$$

which is the **stiffness of the spring.**

$$\text{Spring constant } (k') = \frac{1}{k} = \frac{1}{60.16} = \mathbf{0.0166\ mm/N}$$

$$\text{Ratio } \left(\frac{k_t}{k}\right) = \frac{53.77}{60.16} = \mathbf{0.894}$$

The loop is the result of the frictional forces acting between the leaves of the spring. The area of the loop is a measure of the energy lost in overcoming these frictional resistances and of the self damping properties of the spring.

2. Modulus of elasticity $(E) = \frac{3}{8} \times \frac{L^3}{n\,bt^3}\left(\frac{W}{\delta}\right)$

$$\Rightarrow \qquad E = \frac{3}{8} \times \frac{1010^3}{4(51 \times 8^3)} \times 60.16 = \mathbf{2.22 \times 10^5\ N/mm^2}$$

3. Proof load $(W_p) = \frac{8}{3} \times \frac{nb\,t^3\,Ey_0}{L^3}$

$$W_P = \frac{8}{3} \times \frac{4(51 \times 8^3)\,(2.22 \times 10^5)\,108}{1010^3} = \mathbf{6482\ N}$$

while the actual proof load is 7000 N.

4. Skin stress at the actual proof load, Eq. (17.3),

$$\sigma_P = \frac{3}{2} \times \frac{W_P L}{nbt^2} = \frac{3}{2} \times \frac{7000 \times 1010}{4(51 \times 8^2)} = \mathbf{812\ N/mm^2}$$

5. Proof resilience $(U) = \dfrac{\sigma_P^2}{6E}$, per unit volume

$$U = \frac{812^2}{6(2.22 \times 10^5)} = \mathbf{0.495\ N/mm^2} \quad \text{or} \quad \mathbf{0.495\ N{\cdot}mm/mm^3}$$

The carriage spring is certified for commerical use.

Example 17.1. *A carriage spring (semi-elliptic) has to be designed for a certain vehicle given the following data:*

Span = 1 m

Bending stress not to exceed 300 N/mm².

Central deflection not to exceed 50 mm.

Load = 5 kN

Modulus of elasticity (E) = 2 × 10⁵ N/mm².

Assume width of plates = 10 times thickness.

(*a*) *Determine the number of plates, their thickness and width.*

(*b*) *To what radius the plates should be formed to carry a proof load of 10 kN ?*

(*c*) *What will be the radius of curvature at a load of 5 kN ?*

Solution:

Design (*a*) $M = \dfrac{WL}{4n}, \quad \dfrac{M}{I} = \dfrac{\sigma}{Y} \quad or, \quad M = \sigma \times \dfrac{I}{Y}$

$$\frac{5000 \times 1000}{4n} = 300 \times \frac{bt^3/12}{t/2} = 300 \times \frac{bt^2}{6}$$

$$bt^2 n = 25{,}000 \qquad \text{...}(i)$$

$$\delta = \frac{WL^3}{32\ EIn}$$

$$50 = \frac{5000 \times 1000^3}{32(2 \times 10^5) \times \left(\dfrac{bt^3}{12}\right) \times n}, \quad bt^3 n = 187{,}500 \qquad \text{...}(ii)$$

(*ii*) ÷ (*i*) gives, t = **7.5 mm**, $b = 10 \times t$ = **75 mm**

From (*i*), $n = \dfrac{25{,}000}{75(7.5^2)} = 5.926$, say **6 plates.**

(*b*) At proof load, the spring becomes flat

$$\delta = Y_0 = \frac{W_p L^3}{32\ EIn} = \frac{(10 \times 1000) \times 1000^3}{32(2 \times 10^5)(75 \times 7.5^3/12) \times 6} = 98.8 \text{ mm}$$

$$R_0 = \frac{L^2}{8Y_0} = \frac{1000^2}{8 \times 98.8} = 1265.6 \text{ mm or } \mathbf{1.2626\ m}$$

(*c*) Since, $W \sim \delta$, at W = 5 kN, $Y = \dfrac{Y_0}{2} = \dfrac{98.8}{2} = 49.4$ mm

$$R = \frac{L^2}{8Y} = \frac{1000^2}{8 \times 49.4} = 2530.4 \text{ mm or } \mathbf{2.53 \text{ m}}$$

Note. $\delta = Y_0 - Y = 98.8 - 49.4 = 49.4$ mm.

EXERCISE 17

1. A carriage spring with 9 leaves of width 44 mm and total thickness at centre 48 mm, span 910 mm, and central rise 56 mm was tested in a UTM and it became flat at a load of 2.9 kN. The following load-deflection observations were made:

Load (kN)	*Deflection (mm)*	*Load (kN)*	*Deflection (mm)*
0	0	1.6	28
0.25	4	1.85	32
0.6	8	2.05	36
0.8	12	2.3	40
1	16	2.5	44
1.2	20	2.7	48
1.45	24	2.9	56

Report the results qualitatively for certification

(**Ans.** $k = 50$ N/mm; $E = 2.37 \times 10^5$; $\sigma_P = 353$ N/mm^2 ; $U = 0.0876$ N·mm/mm^3)

2. A carriage spring has to be built from the available plates 6 mm thick and 72 mm width to a span of 0.75 m. Determine the number of plates required if the stress in the material is not to exceed 300 N/mm^2 under a design load of 6 kN. What would be the deflection if $E = 2 \times 10^5$ N/mm^2 ?
If the proof load is 10 kN, to what initial radius the plates have to be formed ?

(**Ans.** 9, 33.9 mm; 1.244 m)

18 Tests on Helical Springs

A length of wire, wound into a helix (on cylinder) is called a helical spring. There are two types of helical springs —

(*i*) Close-coiled in which the wire is wound quite closely such that the pitch distance between the two consecutive turns is small.

(*ii*) Open-coiled in which the pitch is large as compared to that in the closed-coiled springs.

Theory. Consider a close-coiled helical spring subjected to an axial load W, see Figure 18.1 (*i*).

Let n = number of turns of the coil,

D = mean coil diameter,

L = length of wire of the spring = πDn,

d = diameter of the spring wire,

δ = deflection i.e., axial elongation or compression due to load W.

Since, the angle of the helix is small, a torque $(T) = W \times \dfrac{D}{2}$ acts on the wire, (see Figure 18.1 (*ii*)), the bending and shear forces being neglegible. The wire is therefore twisted like a shaft and if θ is the total angle of twist (radians) of one end cross-section with respect to the other end cross-section, then

$$\delta = \frac{D}{2} \times \theta, \quad \text{or} \quad \theta = \frac{2\delta}{D} \qquad ...(18.1)$$

1. Therefore, the torsion equation

$$\frac{T}{J} = \frac{\tau}{r} = \frac{G\theta}{L} \text{ may be written as}$$

$$\frac{W \times \dfrac{D}{2}}{\pi d^4/32} = \frac{\tau}{d/2} = \frac{G \times \dfrac{2\delta}{D}}{\pi Dn} \qquad ...(18.2)$$

or

$$\frac{8WD}{\pi d^4} = \frac{\tau}{d} = \frac{G \times \delta}{\pi D^2 n} \qquad ...(18.3)$$

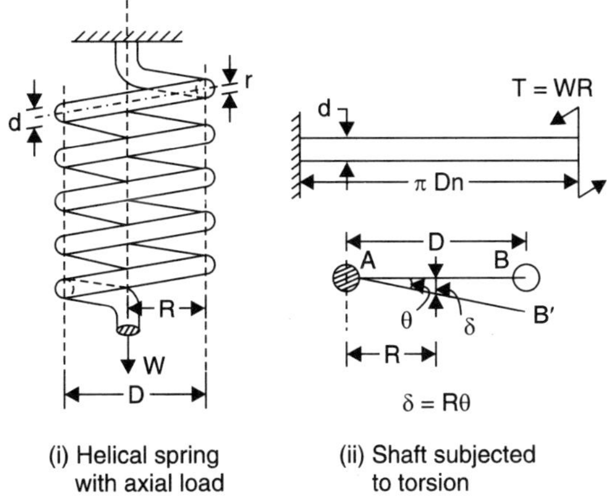

Figure 18.1.

from which,
$$G = \frac{8D^3n}{d^4}\left(\frac{W}{\delta}\right) \qquad ...(18.4)$$

where $\frac{W}{\delta}$ = slope of the 'load-deflection' curve and is called the stiffness of the spring k. The spring stiffness is the load required to produce unit axial deflection (compression), see Figure 18.2.

Sometimes a spring constant $k' = \frac{1}{k} = \frac{\delta}{W}$ is used.

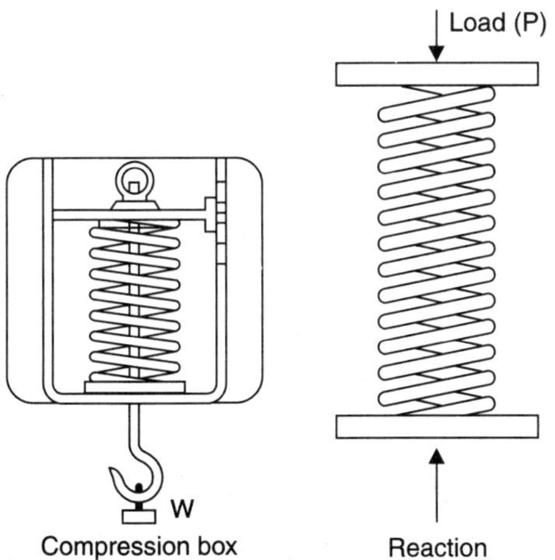

Figure 18.2.

2. Strain energy (S.E.) stored in the spring,

$$\text{S.E.} = \frac{1}{2}W\delta \qquad ...(18.5)$$

Substituting for W and δ from Eqn. (18.3) in terms of τ,

$$\text{S.E.} = \frac{\tau^2}{4G}\left(\frac{\pi d^2}{4} \times \pi Dn\right) \qquad ...(18.6)$$

or $$\text{S.E.} = \frac{\tau^2}{4G} \times \text{Volume of the spring} \quad ...(18.7)$$

Torsional resilience of the spring,

$$U = \frac{\tau^2}{4G}, \text{ per unit volume} \quad ...(18.8)$$

Note. When the wire is of square cross-section of side '*a*'

$$\text{S.E.} = 0.909 \times \frac{\tau^2}{G} \times \text{Volume of the spring}$$

$$U = 0.909\,\frac{\tau^2}{G}, \text{ per unit volume.}$$

3. Proof load and proof resilience. The proof load W_P, is the maximum load which a spring can carry without any permanent deformation and the corresponding stress is called the **proof stress** τ_P, the strain energy stored under the proof load is called the **proof resilience** of the spring.

When the helical spring is subjected to compression, the maximum load which reduces the gap between the successive coils to zero is called the proof load W_P, and the proof stress from Eqn. (18.3)

$$\tau_P = \frac{8 W_P D}{\pi d^3} \quad ...(18.9)$$

Test on Helical Spring (to determine k, G, W_P and U). The mean coil diameter *D*, and the wire diameter *d*, are measured. The number of coils *n*, is counted. The spring is placed in the UTM (50 kN capacity) and the load is gradually increased. Compression of the spring is observed on the ivory scale (IVS) at regular load increments. The load is increased till the proof load is reached. Then the load is gradually decreased and the compression is observed for the same load stages.

Test Observations. A closely coiled helical spring of mean coil diameter 93 mm consisting of 9 coils, wire dia. 16 mm, was subjected to loading and the 'load-compression' observations are given below:

Load W (kN)	*Load-compn. on IVS (mm)*		*Load W (kN)*	*Load-compn. on IVS (mm)*	
	Loading	*Unloading*		*Loading*	*Unloading*
0.5	8	11	4.0	48	51
1.0	13	17	4.5	54	56
1.5	20	23	5.0	60	62
2.0	25	29	5.5	66	67
2.5	31	35	6.0	71	72
3.0	37	40	6.5	76	77
3.5	43	46	7.0	80	80

Report the results qualitatively for certification.

TEST RESULTS

1. The 'load-compression' curve is drawn (see Figure 18.3), which is in the form of a loop. A mean line is drawn whose slope gives the stiffness of the spring k.

$$k = \frac{W}{\delta} = \frac{BC}{AC} = \frac{4.8 \times 1000}{55} = \mathbf{87.3\ N/mm}$$

$$k' = \frac{1}{k} = \frac{\delta}{W} = \frac{1}{87.3} = \mathbf{0.0114\ mm/N}$$

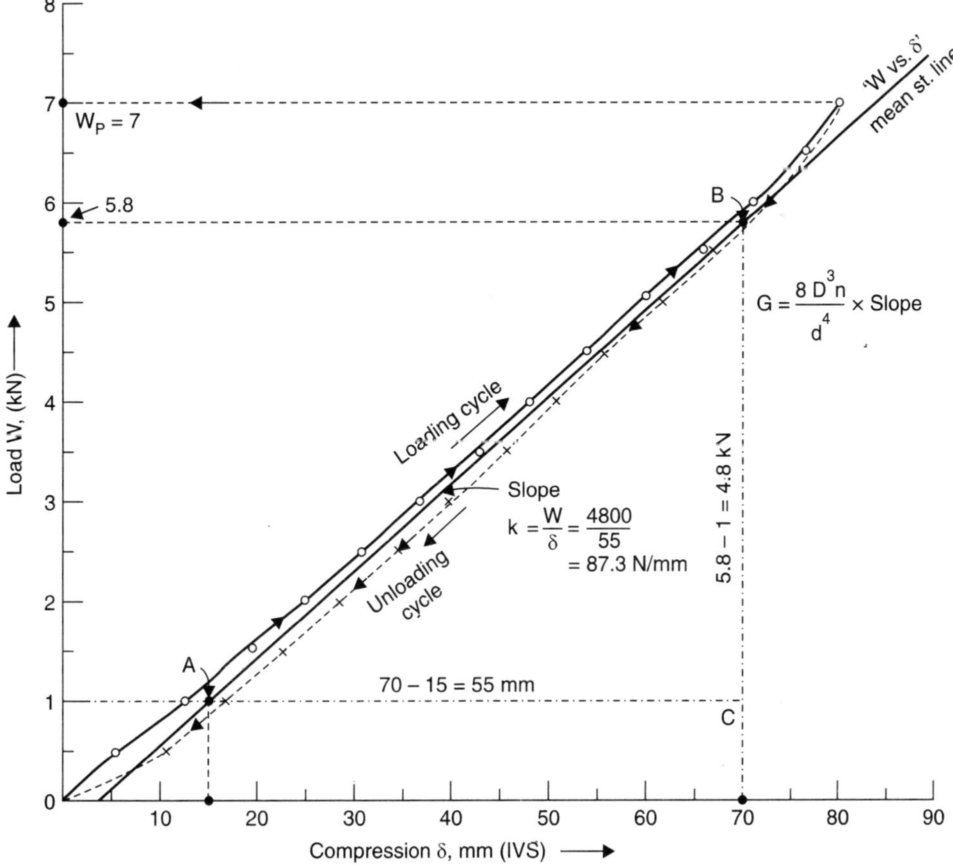

Figure 18.3. Load-compression curve for helical spring

2. Modulus of rigidity $(G) = \frac{8D^3n}{d^4}\left(\frac{W}{\delta}\right)$

$$G = \frac{8 \times 93^3 \times 9}{16^4} \times 87.3 = \mathbf{77{,}146.4\ N/mm^2}$$

3. Maximum shear stress at proof load (W_P = 7000 N)

$$\tau_p = \frac{8W_PD}{\pi d^3} = \frac{8 \times 7000 \times 93}{\pi \times 16^3} = \mathbf{405\ N/mm^2}$$

4. Proof resilience

$$U = \frac{\tau^2}{4G} = \frac{405^2}{4 \times 77{,}146.4} = \mathbf{0.5315\ N/mm^2} \text{ or } \mathbf{0.5315\ N{\cdot}mm/mm^3}$$

The helical spring is certified for commercial use.

Example 18.1. *A close-coiled helical spring of mean coil diameter 12 times the wire diameter is to be designed to absorb 300 N·m of energy when compressed by 150 mm. If the maximum shear stress is not to exceed 140 N/mm², determine the mean dia. of the coil, dia of the wire, and the number of turns. Assume G = 8 × 10⁴ N/mm².*

Also find the load which produces a compression of 50 mm in the spring.

Solution:

Design:
$$\text{S.E.} = \frac{1}{2} W\delta$$

$$300 = \frac{1}{2} W \times 0.150, \quad W = 4000 \text{ N}$$

$\therefore$
$$\tau = \frac{8\,WD}{\pi d^3}, \text{ where } D = 12\,d$$

$$\tau = \frac{96\,W}{\pi d^2}$$

$$140 = \frac{96 \times 4000}{\pi d^2}, \quad d = 29.55, \text{ say } \mathbf{30\ mm}$$

Since,
$$D = 12\,d = 12 \times 30 = \mathbf{360\ mm}$$

$$G = \frac{8D^3 n}{d^4}\left(\frac{W}{\delta}\right)$$

$$8 \times 10^4 = \frac{8 \times 360^3 \times n}{30^4} \times \left(\frac{4000}{150}\right), \quad n = \mathbf{6.51\ turns}$$

$\because$
$$W \sim \delta, \quad \frac{4000}{150} = \frac{W_1}{50}, \quad W_1 = \mathbf{1333.3\ N}$$

EXERCISE 18

1. A close-coiled helical spring of mean coil diameter 66 mm consisting of 20 coils wire dia. 9 mm, was subjected to gradual compression and the maximum load when the coils were touching was 940 N. The 'load-compression' observations are given below:

Load (*N*)	*Compn.* (*mm*)	*Load* (*N*)	*Compn.* (*mm*)
0	0	460	28
60	4	540	32
140	8	600	36
220	12	660	40
260	16	720	44
320	20	800	48
380	24	940	52

Report the results qualitatively for certification

(**Ans.** k = 16 N/mm, $G = 11 \times 10^4$ N/mm^2, τ_P = 210 N/mm^2, U = 0.1 N·mm/mm^3)

2. A close-coiled helical spring is to have a stiffness of 40 N/mm. Determine the number of turns of the 10 mm wire of the spring if the coil diameter is eigth times the wire diameter. [Take $G = 8 \times 10^4$ N/mm^2.] (**Ans.** 4.88 ≈ 5)

3. A close-coiled helical spring is to have a stiffness of 900 N/m in compression, which a maximum load of 45 N and a maximum shear stress of 120 N/mm^2. The solid length of the spring (i.e., when the coils are touching) is to be made 45 mm. Find the diameter of the wire, the coil diameter and the number of turns. [Take $G = 4 \times 10^4$ N/mm^2]

[**Hint.** Solid length = nd] (**Ans.** 3.22 mm, 35.2 mm, 14)

PART B
EXPERIMENTAL STRESS ANALYSIS

19 Nondestructive Testing

The term 'nondestructive testing' is a general name given to all test methods which permit testing or inspection of material without impairing its future usefulness. Nondestructive testing procedures are based upon some property of the material. The response of the material to the physical, electrical or any other property and how this response gets modified in the presence of a defect are studied. Depending upon the material property being made use of in the test, the various techniques of nondestructive testing can be classified into the following major types:

1. Visual examination
2. Pressure and leak testing
3. Penetrant methods
4. Thermal methods
5. Radiography X-ray and gamma-ray
6. Acoustic methods—Sonic and ultrasonic
7. Magnetic particle inspection
8. Electrical and electrostatic methods
9. Electromagnetic induction
10. Surface thickness measurements.

The method of application or technique of a nondestructive testing method will produce varied results. Although most of the tests are simple in principle, the techniques used and the interpretation of the results require skill and experience. Results are on the surface of the test specimen and are qualitative. It is essential that interpretation be made by an experienced person. During routine testing, periodic destructive tests must be made to ascertain that the functioning technique and the instrumentation are functioning properly and at the desired sensitivety and resolution.

The type of defects detected by nondestructive testing may be

(*i*) inherent or original defects in the raw material

(*ii*) defects during processing or during service.

The various steps of any nondestructive testing are:

(*i*) **Application** of a testing or inspection medium.

(*ii*) **Modification** of the medium by defects or variations in the structure or properties of the material.

(*iii*) **Detection** of this change by a suitable detector.

(*iv*) **Conversion** of this change into a form suitable for interpretation.

(*v*) **Interpretation** of the information obtained.

The benefits from nondestructive testing to industry are increased productivity, increased serviceability, safety and identification of materials. To cite an example, detection of a minute crack by 'Ultrasonic Flaw Detector' in a railway line serves as a preventive maintenance, avoids costly and laborious mechanical testing procedures, saves time, avoids accidents in major breakdowns causing derailment, damage, loss to life, property and vital equipment.

The major types of nondestructive tests are briefly discussed in the following:

1. Visual Examination. In this method, the test specimen is illuminated with light, usually in the visible region. The specimen is then examined with the eye, aided or unaided or with other light sensitive devices such as a photocell.

2. Pressure and Leak Tests. These tests are characterised by the flow of a liquid or gas into or through defects. Techniques using radioactive material are employed.

3. Penetrant Methods. Penetrant testing can be used to find defect only open to the surface of the specimen. A liquid having low viscosity and surface tension enters into defects by capillary action and is then partially removed from the defect by the blotting action of the developer.

4. Thermal Methods. In these tests, the test specimen is heated and the resulting temperature distribution revealed by heat sensitive chemicals, phosphores, thermocouples, thermometers of various types and infrared sensitive cells.

5. Radiography-X-ray and Gamma Radiation. Here the so-called 'penetrating radiation' is used. Variations in thickness and density modify the passage of radiation through the test specimen. This variation in the intensity of the transmitted radiation can be detected in a variety of ways, by use of film, semiconductors, photoconductors, and scintillation crystals.

6. Acoustic Method. In this testing, acoustic energy above the audiable range is named as **ultrasonics**, and with frequencies in the audiable range is entitled **dynamic testing.** The ultrasonic testing is described at the end.

7. Magnetic Method. Here the specimen is magnetised, and the distortion of the magnetic field is produced by defects. The magnetic field distortion can be measured in a number of ways, the most widely used employs 'magnetic particles' such as iron oxide. The Magnaflux machine manufactured by the Magnaflux Corporation, Chicago, U.S.A. is the one most commonly used, see Figure 19.1.

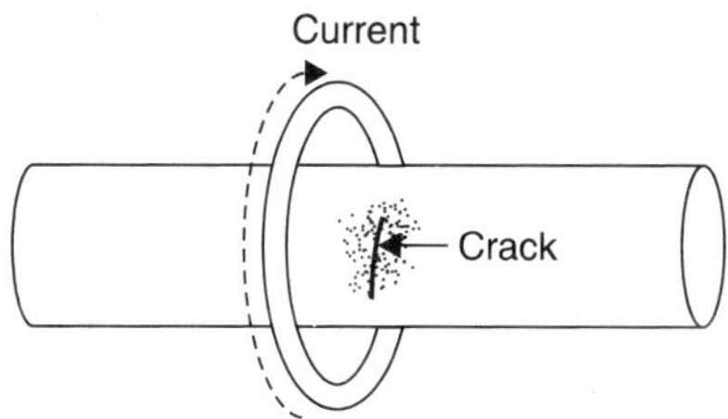

Figure 19.1. Magnaflux indication of a transverse crack in a cylindrical component

8. Electrical Method. These methods depend primarily on the differences in electrical resistivity ρ, and is related to the length L, and area of cross-section A, of the specimen as

$$\text{Resistance } (R) = \frac{\rho L}{A} \qquad \text{...(19.1)}$$

In the electro-static method, variations in the static electric field are detected by the use of charged particles.

9. Electromagnetic Induction. When a coil carrying a high frequency alternating current is brought into the neighbourhood of a conductor, eddy currents are induced in the conductor. There is a magnetic field associated with the induced eddy currents. Flaws as well as physical, chemical and other structural variations causes resistivity changes in the specimen and affect the induced eddy currents and consequently the induced magnetic field. These variations in the induced magnetic field can be detected by their effect on the coil inducing the currents or by a search coil in the neighbourhood of the specimen.

10. Surface Thickness. Gauging the thickness of material and the thickness of coatings or claddings is done by the use of contact or noncontact gauges. Radiation gauges are probably the best known and most widely used.

Ultrasonic Testing. When certain crystals like 'quartz' are placed in an electric field, a mechanical deformation is produced, which is known as **piezo-electric effect.** The deformations or oscillations, produced in the crystal due to an alternating current, is of the same frequency as the current. The frequency of vibrations used is in the ultrasonic range of 10^5 to 2×10^7 cycles per sec (cps or Hz), whereas the audiable or sonic range is only 16 to 2×10^4 cps (Hz).

The probe containing the crystal (transducer) is placed against the test piece (using an oil film for better transmission), the waves travel like a narrow light beam producing oscillations in the path. Thus, if a 5×10^6 cps (Hz) frequency current is applied for a millionth of a second, five waves are sent to the surface of the object. When the current is put off, the crystal ceases to produce ultrasonic waves and functions as receiver for the reflected waves, see Figure 19.2. The reflected waves generate vibrations in the crystal, producing electrical impulses which are fed into a cathode ray oscilloscope. The oscilloscope is so timed that the beam sweeps to the right the same number of times per second as the electrical impulses. The initial vibration striking the test surface makes a sharp peak at the extreme left side of the oscilloscope screen and the reflected waves from the opposite surface of the test piece produce another sharp peak at the extreme right, a distance depending upon the thickness traveled and the time required, see Figure 19.3 (*a*).

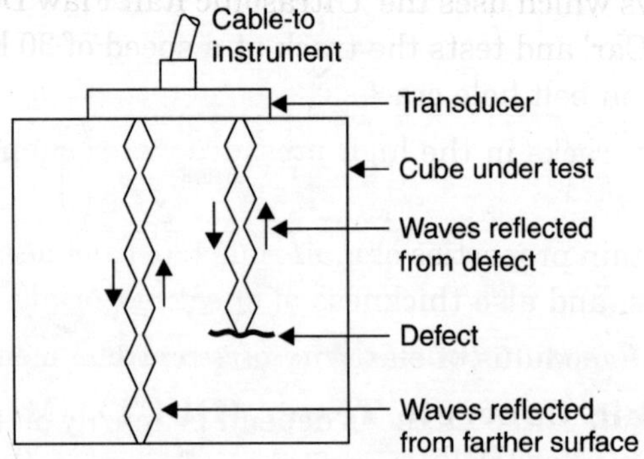

Figure 19.2. Reflection of ultrasonic wave from defect

If the test piece has a defect lying in the path of the beam, an additional peak for the waves reflected from the defect can be noticed, see Figure 19.3 (*b*). A time (or distance) scale in the form of a square wave is shown on the oscilloscope from which the distance to the defect can be known. The distance scale may be varied so that one cycle of the wave may indicate, say 2 cm or 10 cm.

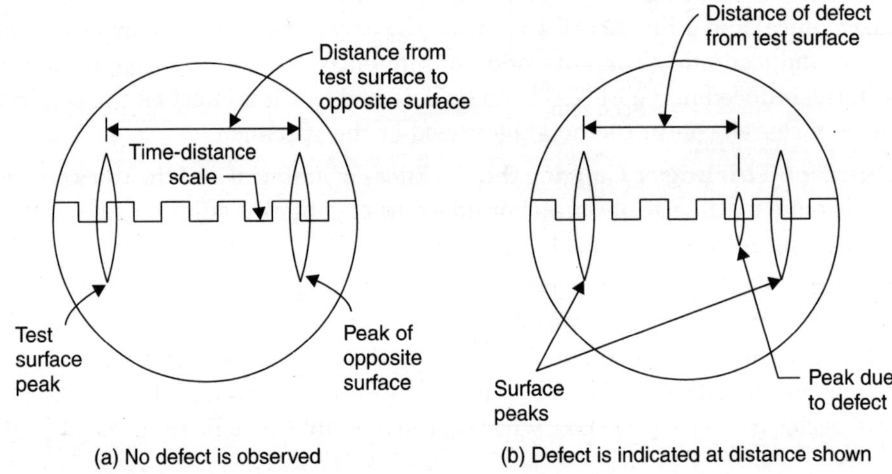

Figure 19.3. Appearance on the oscilloscope screen during ultrasonic test

Since, the beam is very narrow, the whole surface of the test piece has to be covered by progressively moving the probe. Since, some cracks lying parallel to the beam, reflect very little, the test process has to be repeated over the other two surfaces normal to the face tested (so that the beam now is normal to the earlier).

The performance of the ultrasonic equipment can be first checked by using standard blocks with holes of various sizes drilled in them at various distances from the test surface.

Though the initial investment, for the ultrasonic detector is high, it is of great utility for :

(*i*) testing defects in a crank shaft or cam shaft forging without the need of sending them to the machine shop.

(*ii*) routine inspections of locomotive axles and wheel pins for fatigue cracks performed in place.

(*iii*) Indian Railways which uses the 'Ultrasonic Rail Flaw Detector' (URFD) by carrying the equipment in a 'Test Car' and tests the track at a speed of 30 kmph. Axles in position and rails in track are tested and bolt hole cracks can be detected without dismantling.

(*iv*) testing for hair cracks in the high pressure water supply pumping mains from the river.

(*v*) determining certain properties of materials like thickness of enamel paint and nickel coatings on metallic bases, and also thickness of sheet materials.

(*vi*) determining the modulus of elasticity of structural members or parts under load.

Digital Ultrasonic Concrete Tester (DUCT). (Manufactured by Toshniwal Instruments and Engg. Co., New Delhi)

An ultrasonic pulse applied on one face of the concrete block is received by a sensitive unit at the opposite face, see Figure 19.4. The path length and the transit time (in microseconds) are measured and the

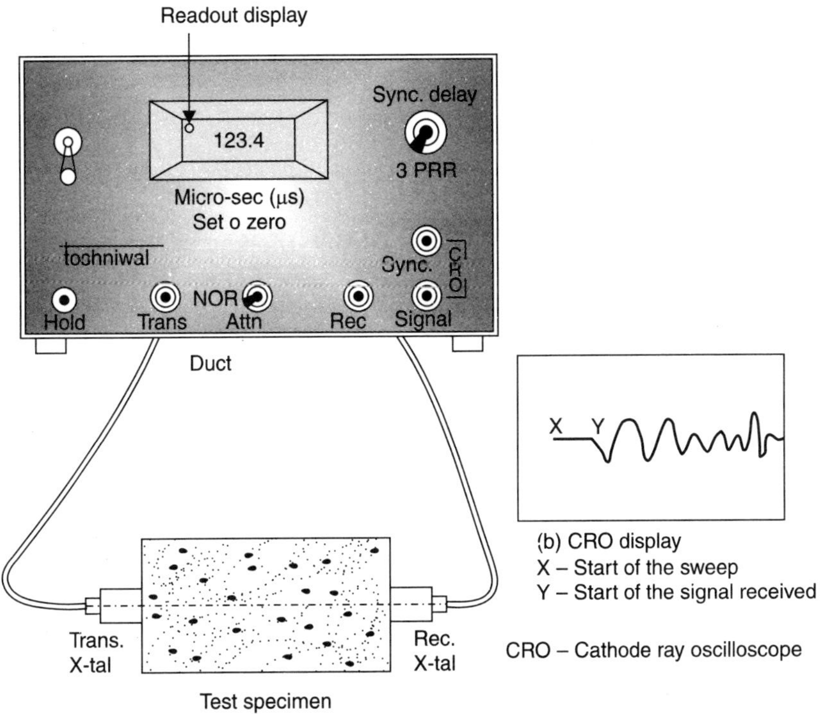

Figure 19.4. Digital ultrasonic concrete tester–Toshniwal

$$\text{Pulse velocity } (V) = \frac{\text{Path length}}{\text{Transit time}} \qquad ...(19.2)$$

The pulse velocity of longitudinal ultrasonic vibrations travelling in an elastic solid is given by

$$V = \frac{E(1-\mu)}{\rho(1+\mu)(1-2\mu)} \qquad ...(19.3)$$

where ρ is the density of the material. The velocity of the pulse depends upon the density and elastic property of the material. The modulus of elasticity obtained from Eq. (19.2) is taken as a measure of the compressive strength.

The least lateral dimension (i.e., dimensions measured normal to the path of pulses) should not be the less than the wavelength of the pulse vibrations. The pulse frequency for testing concrete or timber is much low (20–250 kHz corresponding to wavelengths of 200 – 16 mm, respectively) than that used for metal. The higher the frequency, the narrower the beam of pulse propagation but greater the attenuation of the pulse vibrations.

Pulses are not transmitted through large air voids. The instrument (DUCT) indicates the time taken by the pulse which circumvents the voids by the quickest route. It is thus possible to detect large voids when a grid of pulse velocity measurements is made over a region in which these voids are present.

The pulse velocity method of testing may be used for testing plain, reinforced, and prestressed concrete whether it is precast or cast in situ. The method is used to determine :

(*i*) homogeneity of the concrete.

(*ii*) presence of voids, cracks and other imperfections.

(*iii*) changes in concrete which may occur with time or through the action of fire, frost, or chemical attack.

(*iv*) quality of concrete, generally expressed in terms of compresive strength.

The Toshniwal DUCT uses TTL Integrated Circuits and silicon semiconductors. It has the facility for viewing the received waveform on CRO, see Figure 19.4 (*b*). The display consists of four digit readout with a positioned decimal (123.4 μs, 10 MHz oscillator), completely portable with optional battery operation.

Concrete Test Hammer (Types N and NR). A nondestructive impact test for determining the hardness and quality of concrete in the finished structure (ordinary building or bridge construction) can be made by using the Schmidt (Swiss made) concrete test hammer (Type *N*). The tubular unit is pressed against the finished surface of the concrete to be tested causing a spring loaded hammer inside the tube automatically strike the concrete surface, see Figure 19.5. The Rebound number R, of the hammer after impact is indicated on a scale. The rebound number thus obtained, in combination with calibration curves supplied (based on measurements on a large number of test cubes), can be used to obtain the most likely value of the compressive strength of the concrete. Calibration curves and Tables for cube and cylinder compressive strengths W_m, Z_m (in N/mm^2) against the rebound number R, are supplied from the manufacturer.

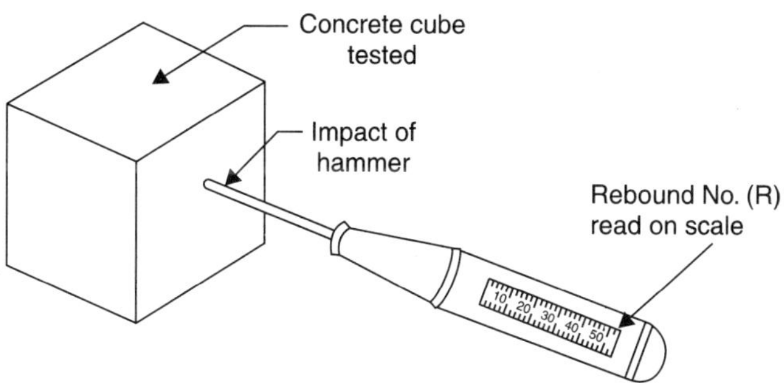

Figure 19.5. Concrete test hammer [Schmidt]

The probable compressive strength of test cylinders cured at the site or of the cores cut from the hardened concrete, can thus be estimated.

20 Strain Gauges and Rosettes

Strain gauges are the devices that sense the change in length, magnify it and indicate it in some form. Strain gauges can be classified broadly into five groups based on the physical principle adopted for the magnification of the change in length as follows:

(*i*) Mechanical strain gauges (extensometers) using a spring and lever, see Figure 2.2.

(*ii*) Optical gauges

(*iii*) Electrical gauges-inductance, capacitance and resistance types

(*iv*) Pneumatic strain gauges

(*v*) Acoustic strain gauges.

Reference may be made to (Ref. 2,13) for their constructional details and working principle.

Whole-Field Methods. The strain gauges mentioned above give strain at a point. Therefore many gauges are needed to investigate a strain field over a finite area. Some important methods developed for the measurement of strain simultaneously at many points over the whole field are:

1. Moire method
2. Brittle-lacquer technique
3. Photoelastic method
4. Holography

Methods 2 and 3 are discussed in this book. Reference may be made to [Ref. 2,13] for all the four methods.

Electrical Resistance Strain Gauges. It was first discovered by Lord Kelvin in 1856 that if a metal wire is stretched, its electrical resistance changes. This property has been used in electrical strain gauges for measuring the surface strain of a structural member and machine parts.

Electrical resistance strain gauges with a metallic-sensing element may be of unbonded or bonded wire, foil or weldable type, and semiconductor gauges.

Unbonded wire strain gauges have excellent stability characteristics and are used in transducers for measurement of quantities such as acceleration, pressure etc.

Bonded wire strain gauges is about the size of a postage stamp and generally much lighter. It consists of a grid or filament of 0.3 to 0.5 mm diameter wire mounted on a paper or plastic, see Figure 20.1.

The foil gauges are made by printing or punching on the metallic foil. The foil grid (of ≈ 2.5 micron strain sensitive coil) is bonded to a backing, see Figure 20.2. Due to greater bonding surface area higher sensitivity can be achieved.

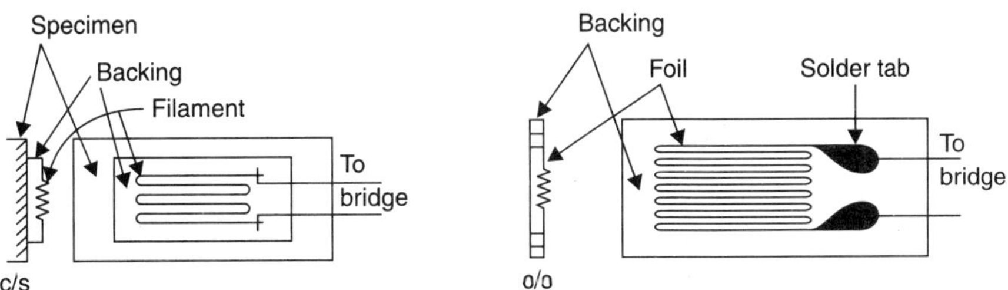

Figure 20.1. Bonded wire strain gauge *Figure 20.2. Foil strain gauge*

To measure the strain in a structural member or machine part, one or more of these strain gauges are cemented to the surface of the part with suitable adhesives like nitro-cellulose, epoxy, cyanoacrylate, ceramic, epoxy-phenolic. The strain gauge is connected to a Wheatstone bridge to measure the small change in resistance (ΔR). The axial sensitivity or **gauge factor** F, of a strain gauge is the fractional change of resistance for a given strain and is defined as

$$F = \frac{\Delta R/R}{\Delta L/L} = \frac{\Delta R/R}{\varepsilon}$$

or

$$\varepsilon = \frac{\Delta R/R}{F} \qquad \text{...(20.1)}$$

This factor F is determined by calibration of samples from a batch of similar gauges. Most commonly used strain gauges have gauge factor ranging from 2 to 3.5. Strain gauges are made with a resistance R, of a few hundred ohms and carry a current around 25 mA although an overload of twice this value is permissible for short periods.

Gauge shape and size are important factors. If the variation of strain on the surface of a component is small, a long narrow gauge 10-25 mm (gauge length L) can be used, say SR-4 strain gauge. In locations of likely stress concentrations and steep strain gradients, a shorter gauge length (1.5-2 mm) must be used, the width now being greater than the length in order to obtain a suitable resistance value. While measuring strains in a two-dimensional (biaxial) stress field with strain gauges, the measured strains have to be corrected for transverse strains (Poisson's effect) by using a transverse sensitivity factor K_t, if the error due to transverse sensitivity effect is significant.

Strain gauge installations are affected by environments containing sea-water, moisture chemical vapours, temperature, humidity, etc. and are given protective coatings of wax or acrylic, epoxy plastics, synthetic rubber, etc.

Dummy gauges may be mounted on a separate piece of the same material for temperature compensation. Self-temperature compensated (STC) strain gauges are commercially available.

Example 20.1. *A change of 0.001 Ω is observed in the contact resistance of a 120 Ω strain gauge. What is the strain induced, if the gauge factor is 2 ?*

Solution. $\varepsilon = \dfrac{\Delta R/R}{F} = \dfrac{0.001/120}{2}$ = **4.2 μm/m**

Measurement of Static Strains. Two electrical circuits—the potentiometer and Wheatstone bridge, are used to measure small changes in resistance, the latter with some variation being common.

Arrangements of strain gauges and bridge circuits for measurement of direct strain (due to axial load-tension or compression), strain due to bending moment M, and torque T, are given below.

***(i)* Direct Strain.** The arrangement of four gauges cemented on a bar with a square cross-section, under axial load P, is shown in Figure 20.3. Gauges A and C are parallel to the axis and gauges B and D transverse to the axis in the Poisson arrangement to obtain automatic temperature compensation.

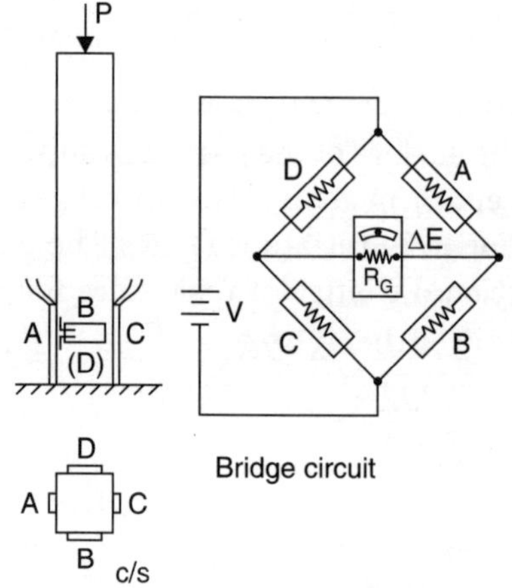

Figure 20.3. Arrangement of strain gauges for direct (axial) strain measurement

Their arrangement on the Wheatstone bridge will eliminate any bending or torsional strain and the bridge output ΔE, is directly proportional to the axial strain ε_a, due to the load P. If V is the applied voltage from the battery

$$\Delta E = \frac{VF}{2}(1 + \mu)\,\varepsilon_a \qquad \text{...(20.2)}$$

***(ii)* Bending Strain.** Two gauges A and B are cemented on the top and bottom surfaces of a member subjected to a bending moment M, and axial load P, see Figure 20.4. Each gauge forms one arm of the bridge and acts as a dummy for the other. Both arms are active and the sensitivity of the bridge is doubled i.e., since the stresses on the top and bottom are opposite in sign, the output voltage ΔE is twice that due to gauge$A$ alone. The direct strain and temperature variations are common to both and cancel each other.

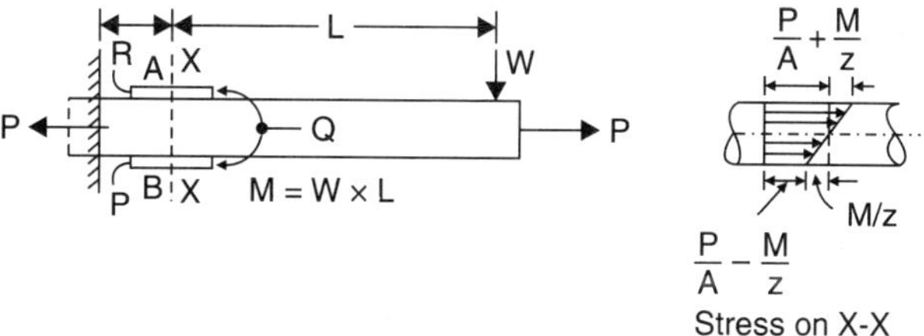

(a) Shank of a crane hook under combined bending and tension

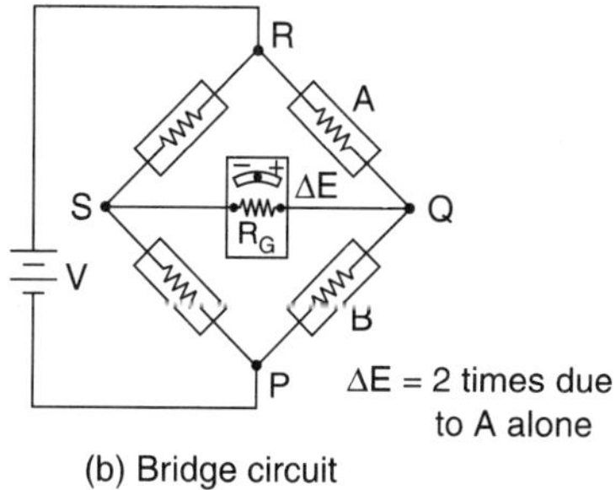

(b) Bridge circuit

Figure 20.4. Strain gauge arrangement for measurement of bending strain

***(iii)* Torsional Shear Strain.** A cylindrical shaft is subjected to a torque T, and an axial load P, see Figure 20.5. The principal stresses and strains on the surface act along directions at 45° to the axis of the shaft, also they will be equal in magnitude and opposite in sign. Four gauges are mounted so that a pair of them (1 and 2) are along the principal directions at the required point on the surface of the shaft, the other pair of gauges (3 and 4) are mounted diametrically opposite to the first pair (1 and 2).

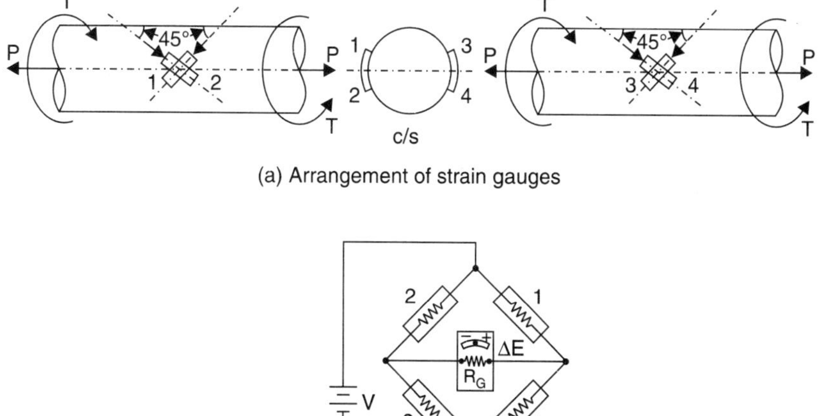

(a) Arrangement of strain gauges

(b) Bridge circuit: ΔE = 4 times due to 1 alone

Figure 20.5. Arrangement of strain gauges for torsional strain measurement

The gauges from the four arms of the bridge, act as dummies for each other and give a combined output ΔE, which is four times as sensitive as for a single gauge (mounted along one principal direction). Strains due to axial load bending and temperature variations are eliminated since they cancel each other. Thus the output of the bridge is directly proportional to the torque T, applied to the shaft.

$$\Delta E = \frac{16(1+\mu)}{E\pi D^3} \text{ F.V.T} \qquad ...(20.3)$$

Slip Rings. Strain gauges are often mounted on rotating equipment for measurement of quantities such as vibratory stresses, torque and thrust. For connecting strain gauges on rotating equipment to a stationary measuring system, slip rings and brushes are used, see Figure 20.6. The main problem in the use of slip rings is that the change in contact resistance at the brushes during rotation is of the same order as the signal from the strain gauge i.e., the reduction of noise/signal ratio. The choice of contact materials is important. Silver wire or silver graphite and stainless steel brushes are commonly used. Slip ring noise can be reduced by the careful design of the slip ring/brush assembly with a fairly small contact area.

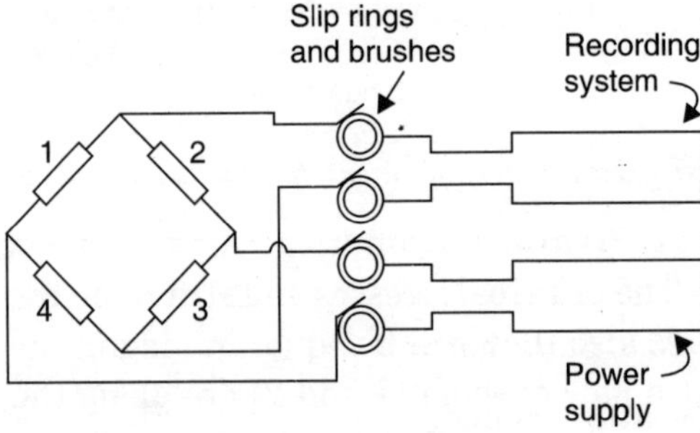

Figure 20.6. Strain measuring system connected to bridge through slip rings

Load Cells and Torque Bars. In same testing machines and structures, the forces transmitted are determined by an elastically strained block of metal to which strain gauges are attached, which is in series with the specimen or structural member, see Figure 20.7. The strain gauge block of metal is termed a **load cell**.

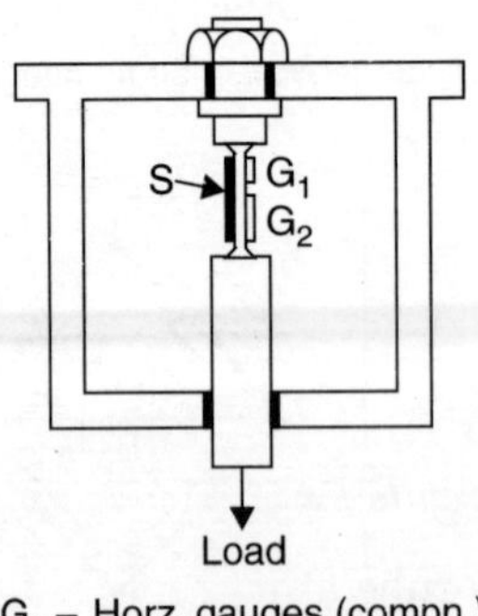

G_1 – Horz. gauges (compn.)
G_2 – Vert. gauges (tension)
S – Test specimen

Figure 20.7. Strain gauge load cell

A simple variation of the principal of load cell for measuring torque is an elastically twisted bar to which strain gauges are fixed, arranged in series with the member or specimen subjected to torsion.

The circuits described earlier for direct strain, bending and torsion measurement can be used for load cells and torque bars and are known as **strain gauge transducers**.

Dynamic Strains. Strain which varies appreciably in magnitude over a short time interval is termed **dynamic strain**, for example strain in a spring under forced vibration, see Figure 20.8 (*b*). Idealized patterns of static and dynamic strains are shown in Figure 20.8. In actual practice, the magnitude, frequency and the wave shape of the dynamic strains may vary with time. The dynamic strains are measured by Wheatstone bridge and potentiometer circuits. Because of the short time interval (0.1 μs) during which most dynamic and transient

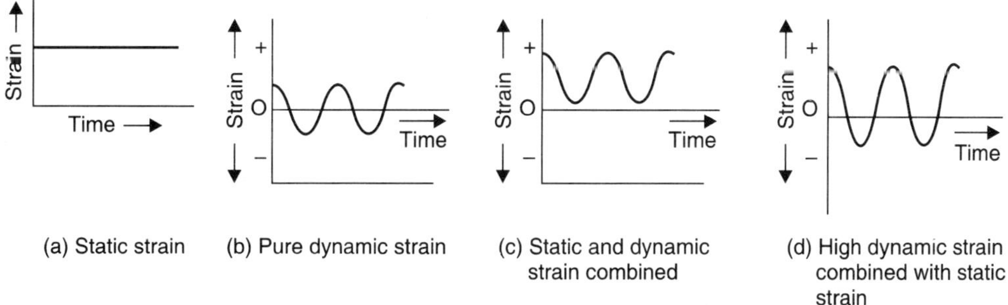

Figure 20.8. Static and dynamic strain pattern idealised

strains are measured, temperature compensation is not necessary. It is necessary to apply the output voltage of the bridge to an oscillograph or cathode-ray oscilloscope for indications. D.C. amplifier is used for measuring dynamic strains of low frequency and A.C. circuit is employed for high frequency measurement, see Figure 20.9.

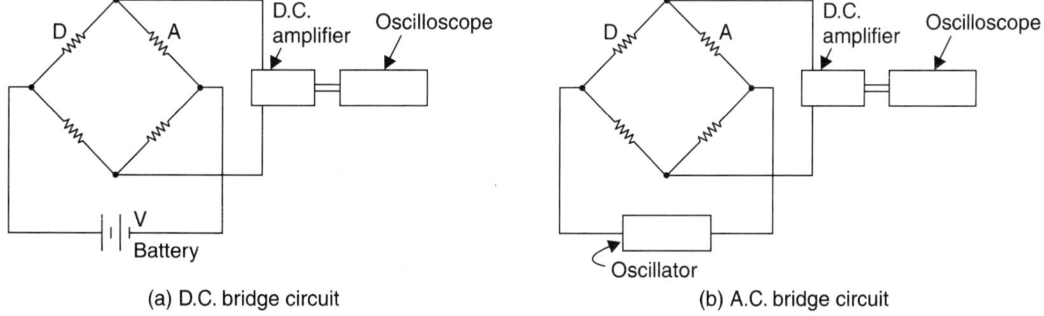

Figure 20.9. Circuits for dynamic strain measurement

The image on the cathode ray tube appears as a waveform of frequency determined by the oscillator and enveloped or modulated by the signal wave from the strain gauge termed **carrier wave system**, see Figure 20.10. This can be used equally well for static or high frequency strain measurement upto about 1/5 of the carrier wave frequency.

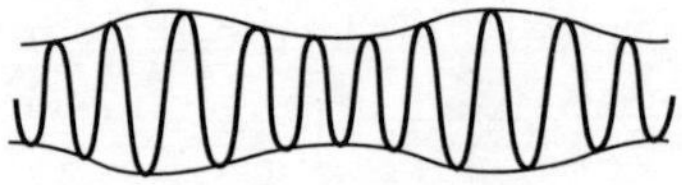

Figure 20.10. Oscilloscope trace in a carrier wave system

Strain Gauge Rosettes. Strain gauges record the component of strain parallel to their length. Hence, at a point on the surface of a structural member where the directions of strain are completely unknown, it becomes necessary for three gauges to be cemented on a small area at the point and set at known relative angles. From the strain components measured in these three directions, the principal strains and stresses can be completely determined either analytically or by construction of Mohr's strain circle. The set of gauges is termed as a **rosette**. Four arrangements are shown in Figure 20.11. Manufacturers can supply made-up rosette gauges of these forms.

(a) Rectangular rosettes

(b) Delta rosette

(c) T-delta rosette

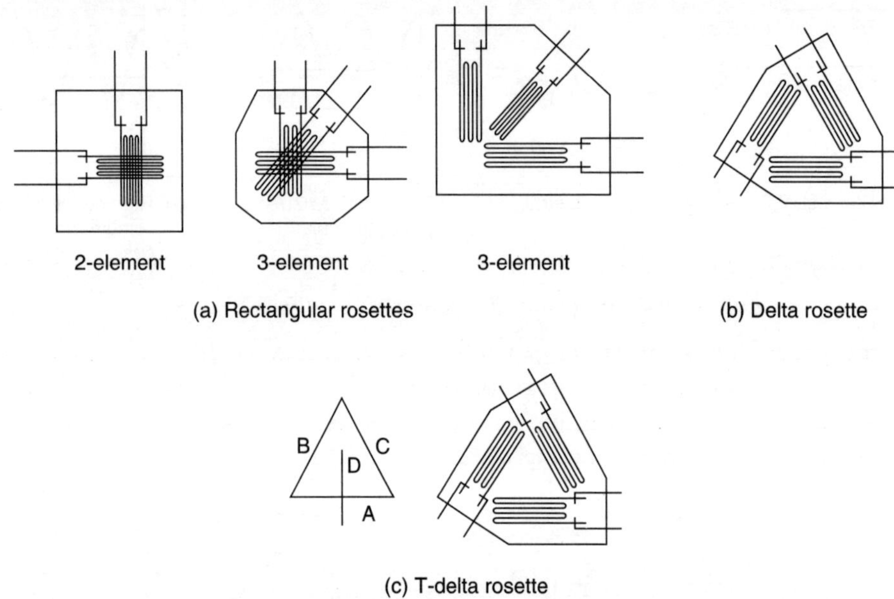

Figure 20.11. Rosette arrangements

If the principal directions are known, or in some cases determined by the 'brittle-lacquer' coating method, a **2-element rectangular rosette** can be used for determination of principal strains and stresses, see Figure 20.11 (*a*).

Analytical Approach. From elementary mechanics, the strain in a direction at an angle ϕ to the x-axis is related to the strains ε_x, ε_y, γ_{xy} in the *XOY*-plane as (see Figure 20.12),

$$\varepsilon_\phi = \frac{\varepsilon_x + \varepsilon_y}{2} + \frac{\varepsilon_x - \varepsilon_y}{2} \cos 2\phi + \frac{\gamma_{xy}}{2} \sin 2\phi \qquad ...(20.4)$$

It is customary to take one arm of the gauge *OA*, as the x-axis, see Figure 20.13.

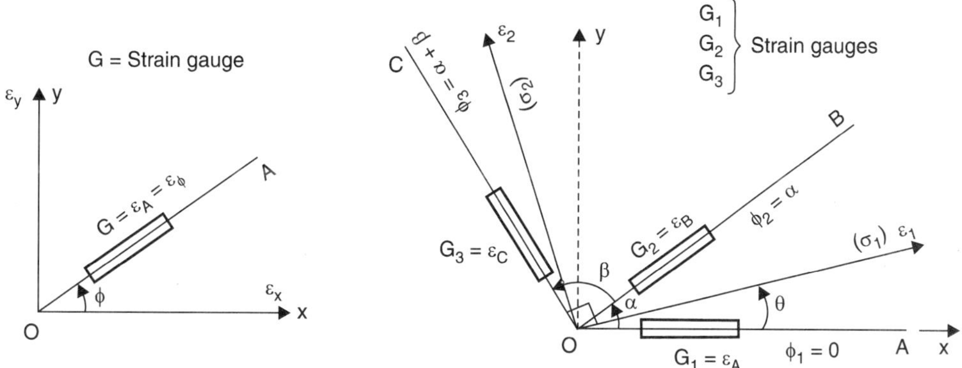

Figure 20.12. x-and y-strain components of ε_A　　*Figure 20.13. Principal strains due to ε_A, ε_B, and ε_C*

For a **3-element rectangular rosette,** putting $\phi = 0°$, 45° and 90° in Eqn. (20.4), three equations can be obtained and their solutions are given as

$$\varepsilon_x = \varepsilon_A, \quad \varepsilon_y = \varepsilon_C, \quad \gamma_{xy} = 2\,\varepsilon_B - (\varepsilon_A + \varepsilon_C) \qquad ...(20.5)$$

Similarly, for the **equiangular or delta rosette,** putting $\phi = 0°$, 60° and 120° in Eqn. (20.4), and solving yields

$$\varepsilon_x = \varepsilon_A, \quad \varepsilon_y = -\frac{\varepsilon_A + 2\,\varepsilon_B + 2\,\varepsilon_C}{3}, \quad \gamma_{xy} = \frac{2}{\sqrt{3}}(\varepsilon_B - \varepsilon_C) \qquad ..(20.6)$$

The principal strains and their directions with x-axis (arm OA) can be obtained as

$$\varepsilon_{1,2} = \frac{\varepsilon_x + \varepsilon_y}{2} \pm \frac{1}{2}\sqrt{(\varepsilon_x - \varepsilon_y)^2 + \gamma_{xy}^2} \qquad ...(20.7)$$

$$\tan 2\theta = \frac{\gamma_{xy}}{\varepsilon_x - \varepsilon_y} \qquad ...(20.8)$$

$$\gamma_{max} = \sqrt{(\varepsilon_x - \varepsilon_y)^2 + \gamma_{xy}^2} \qquad ...(20.9)$$

at 45° to $\varepsilon_{1,2}$ directions.

Then the principal stresses can be determined

$$\sigma_1 = \frac{E}{1-\mu^2}(\varepsilon_1 + \mu\varepsilon_2) \qquad ...(20.10)$$

Similarly,

$$\sigma_2 = \frac{E}{1-\mu^2}(\varepsilon_2 - \mu\varepsilon_1) \qquad ...(20.11)$$

$$\tau_{max} = \frac{E}{2(1+\mu)}\gamma_{max} \qquad ...(20.12)$$

at 45° to $\sigma_{1,2}$ directions.

In a 'tee-delta' (T-Δ) rosette (see Figure 20.11 (c)), three gauges are arranged as in the delta rosette such that the included angle between any two arms is 60°. The fourth gauge D is placed in a direction normal to any one of the three arms of the Δ. The fourth gauge will give an extra strain reading to check the values of ε_x, ε_y and γ_{xy} determined by the solution of the first three strain gauge readings.

The theoretical computations are tedious and the following values for the trigonometric functions of 2ϕ are helpful:

ϕ	0°	45°	60°	90°	120°
$\cos 2\phi$	1	0	$-\frac{1}{2}$	$-\frac{1}{2}$	$-\frac{1}{2}$
$\sin 2\phi$	0	1	$\frac{\sqrt{3}}{2}$	0	$-\frac{\sqrt{3}}{2}$

Hewson (1946) has constructed 'nomographs' for obtaining principal stresses from strain rosette data [Ref. 9].

Solutions can be easily made by construction of Mohr's circle for strain and are illustrated in the following examples.

Example 20.2. *In a 3-element rectangular rosette fixed to the fuselage of a jet airliner (E = 70,000 N/mm², μ = 0.33), the strain indicator readings when cruising at an altitude of 12,000 m were*

$$0° : 485\ \mu m/m, \quad 45° : 300\ \mu m/m, \quad 90° : -85\ \mu m/m$$

Determine the magnitude and direction of the principal stresses (i) analytically, (ii) by construction of Mohr's strain circle.

Solution. (*i*) **Analytically : For the 3-element rectangular rosette, from Eqn. (20.5),**

$$\varepsilon_x = \varepsilon_A = 485, \varepsilon_y = \varepsilon_C = -85$$

$$\gamma_{xy} = 2\,\varepsilon_B - (\varepsilon_A + \varepsilon_C) = 2 \times 300 - (485 - 85) = 200$$

$$\therefore \quad \varepsilon_1 = 502\ \mu m/m, \varepsilon_2 = -102\ \mu m/m$$

From Eqn. (20.9), $\gamma_{max} = 604 \times 10^{-6}$ radian

From Eqn. (20.8), $\tan 2\theta = \dfrac{200}{485 + 85} = 0.351$, $2\theta = 19.34°$

$$\theta_1 = 9.67°, \theta_2 = 99.67°$$

(*ii*) **Construction of Mohr's strain circle, see Figure 20.14**

At any point O as origin, draw horizontal (ε) and vertical $\left(\dfrac{\gamma}{2}\right)$ lines. Measure

$$\varepsilon_A, \varepsilon_B, \varepsilon_C \text{ and } OD = \frac{\varepsilon_A + \varepsilon_C}{2}.$$

At ε_A draw offset $\dfrac{\gamma}{2} = \varepsilon_B - \dfrac{\varepsilon_A + \varepsilon_C}{2}$ and obtain A.

With D as centre and DA as radius, draw the Mohr's strain circle. Draw $DB \perp AC$. DA, DB and DC represent the three arms of the rosette, the angle at the centre of the circle being double (= 2 × 45° = 90°). ε_B = 300 μm/m serves as a check. Measure $\varepsilon_{1,2}$, $2\theta_1$ and $\left(\dfrac{\gamma}{2}\right)_{max}$.

$$\varepsilon_1 = 502\ \mu m/m, \varepsilon_2 = -102\ \mu m/m$$

$$2\theta_1 = 19.34°, \quad \theta_1 = 9.67°, \theta_2 = 99.67°$$

$$\left(\frac{\gamma}{2}\right)_{max} = 302 \times 10^{-6} \text{ radian}, \; \gamma_{max} = 604 \times 10^{-6} \text{ radian}$$

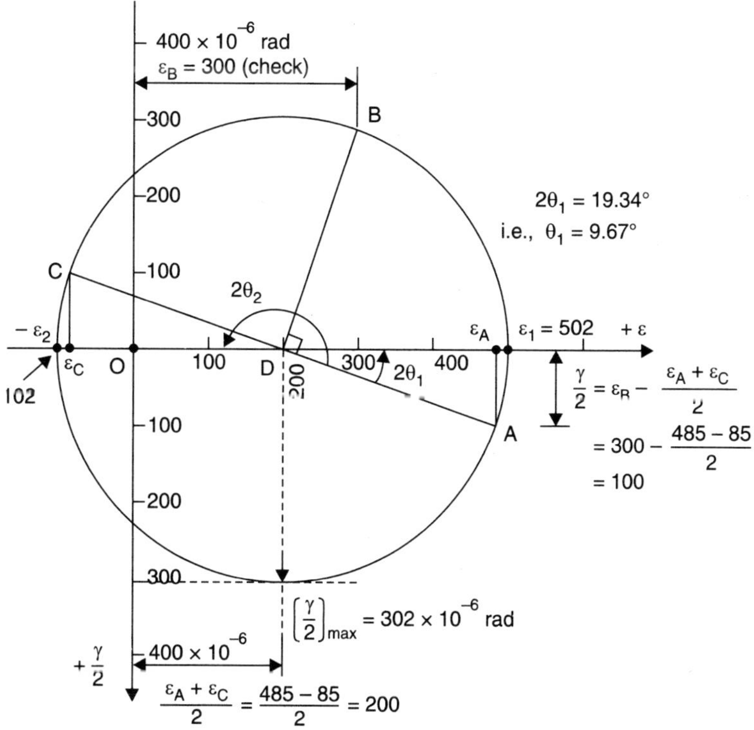

Figure 20.14. Mohr's strain circle for 3-element rectangular rosette

Principal and shear stresses

$$\sigma_1 = \frac{E}{1-\mu^2}(\varepsilon_1 + \mu\varepsilon_2)$$

$$= \frac{70{,}000}{1-(0.33)^2} \times (502 + 0.33 \times -102) \times 10^{-6} = \mathbf{36.8\ N/mm^2}$$

$$\sigma_2 = \frac{E}{1-\mu^2}(\varepsilon_2 + \mu\varepsilon_1) = \frac{70{,}000}{1-(0.33)^2} \times (-102 + 0.33 \times 502) \times 10^{-6}$$

$$= \mathbf{5\ N/mm^2}$$

$$\tau_{max} = \frac{E}{1+\mu} \times \frac{\gamma_{max}}{2} = \frac{70{,}000}{1+0.33} \times (302 \times 10^{-6})$$

$= \mathbf{15.9\ N/mm^2}$ at 45° to the principal directions.

Check : $\tau_{max} = \dfrac{\sigma_1 - \sigma_2}{2} = \dfrac{36.8 - 5}{2} = \mathbf{15.9\ N/mm^2}$,

It can be shown that the angle θ_1 is

$$0° < \theta_1 < 90°, \text{ when } \varepsilon_B > \frac{\varepsilon_A + \varepsilon_C}{2}$$

$$-90° < \theta_1 < 0°, \quad \text{when } \varepsilon_B < \frac{\varepsilon_A + \varepsilon_C}{2}$$

$$\theta_1 = 0°, \text{ when } \varepsilon_A > \varepsilon_C \text{ and } \varepsilon_A = \varepsilon_1$$

$$\theta_1 = \pm 90°, \text{ when } \varepsilon_A < \varepsilon_C \text{ and } \varepsilon_A = \varepsilon_2$$

In the given example $\varepsilon_B = 300$ μm/m

$$\frac{\varepsilon_A + \varepsilon_C}{2} = \frac{485 - 85}{2} = 200 \text{ μm/m} < \varepsilon_B,$$

$$\therefore \qquad 0° < \theta_1 < 90°.$$

The disposition of strain gauges and the directions of principal and shear stresses are shown in Figure 20.15.

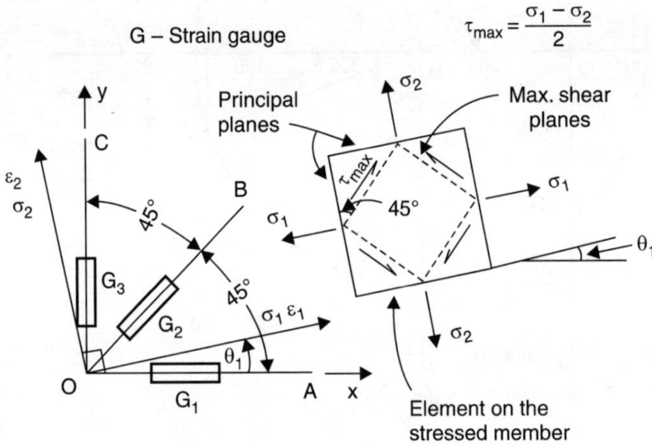

Figure 20.15. Disposition of strain gauges and principal-and shear stresses

Example 20.3. *A 3-element delta rosette is mounted at a point O on the surface of a stressed steel component (E = 2 × 10⁵ N/mm² and μ = 0.3), see Figure 20.16 (a). The strains measured by the three gauges are 0°: 550 μm/m, 120° : 150 μm/m, 240° : –100 μm/m. Determine the principal strains and their directions in the plane of the three gauges. If there is no stress perpendicular to the given plane, determine the principal stresses at the point.*

Solution. (*i*) **Analytically : For the Δ-rosette, from Eqn. (20.6),**

$$\varepsilon_x = \varepsilon_A = 550, \quad \varepsilon_y = \frac{-550 + 2x - 100 + 2 \times 150}{3} = -150$$

$$\gamma_{xy} = \frac{2}{\sqrt{3}}(-100 - 150) = -288.7$$

From Eqn. (20.7),

$$\varepsilon_{1,2} = \frac{550 - 150}{2} \pm \frac{1}{2}\sqrt{(550 + 150)^2 + (288.7)^2}$$

$$= 200 \pm \frac{1}{2}(757.2)$$

$$\varepsilon_1 = 578.6 \text{ μm/m}, \quad \varepsilon_2 = -178.6 \text{ μm/m}, \quad \gamma_{max} = 757.2 \times 10^{-6} \text{ radian}$$

From Eqn. (20.8), $\tan 2\theta = \dfrac{-288.7}{550 + 150} = -0.4124, \quad 2\theta = -22.4°$

$$\theta_1 = \mathbf{-11.2°},\ \theta_2 = -11.2° + 90° = \mathbf{78.8°}$$

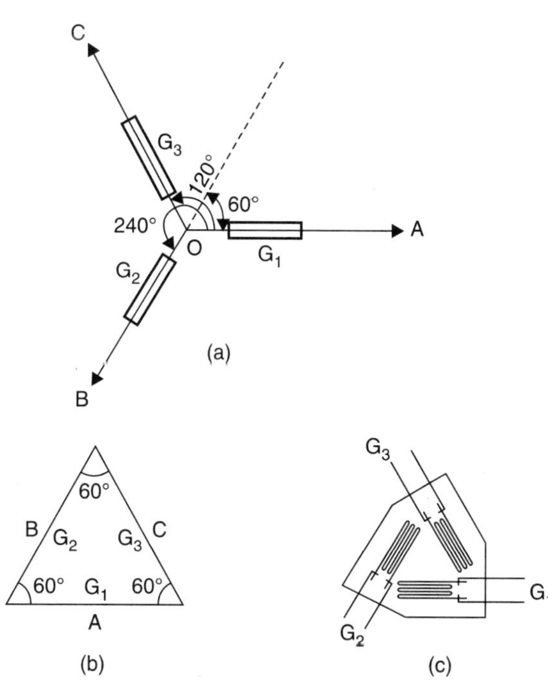

Figure 20.16. Disposition of strain gauges in delta rosette (Included angles = 60°)

(*ii*) **Construction of Mohr's Strain Circle, see Figure 20.17.** From an arbitrary point *O*, draw verticals at horizontal distances of ε_A, ε_B and ε_C. From a point *P* on the middle vertical, draw lines *PA* and *PB* at 60° to the vertical at *P* meeting the other two verticals at *A* and *B*. The intersection of the perpendicular bisectors of *PA* and *PB* gives the centre *G* of the required strain circle of radius *GA* (or *GB*). The strain circle intersects the vertical through *P* at *C*. Join *GC*. *GA*, *GB* and *GC* will give the disposition of the three strain gauges, the angle at the centre of the circle being double (i.e., 2 × 60° = 120°). A horizontal through *G*, intersects the vertical at *O* to give the origin *O* for ε and γ/2 axes. $OG = \dfrac{\varepsilon_A + \varepsilon_B + \varepsilon_C}{3}$ serves as a check.

Measure ε_1, ε_2, $2\theta_1$ and $\left(\dfrac{\gamma}{2}\right)_{max}$:

$$\varepsilon_1 = 578\ \mu\text{m/m},\ \varepsilon_2 = -178\ \mu\text{m/m},\ 2\theta_1 = -22.5°$$

$$\theta_1 = -11.2°,\ \theta_2 = -11.2° + 90° = 78.8°\ ;$$

$$\left(\frac{\gamma}{2}\right)_{max} = 375 \times 10^{-6}\ \text{radian},\ \gamma_{max} = 750 \times 10^{-6}\ \text{radian}$$

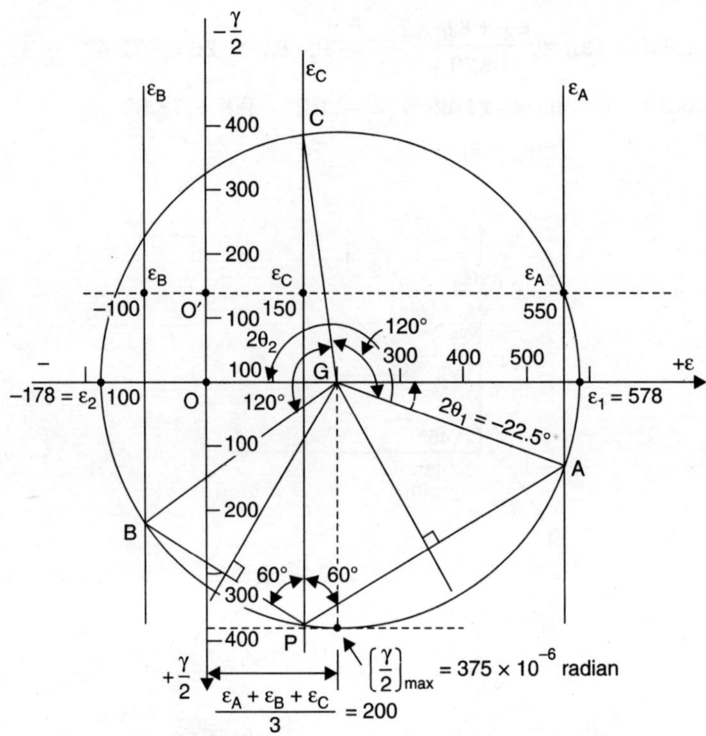

Figure 20.17. Mohr's strain circle for delta rosette

Principal and shear stresses

$$\sigma_1 = \frac{E}{1-\mu^2}(\varepsilon_1 + \mu\varepsilon_1)$$

$$= \frac{2\times 10^5}{1-0.32}\{578.6 + 0.3\,(-178.6)\} \times 10^{-6} = \mathbf{115.4\ N/mm^2}$$

$$\sigma_2 = \frac{E}{1-\mu^2}(\varepsilon_2 + \mu\varepsilon_1)$$

$$= \frac{(2\times 10^5)}{1-0.3^2} \times (-178.6 + 0.3 \times 578.6) \times 10^{-6} = \mathbf{1.1\ N/mm^2}$$

$$\tau_{max} = \frac{E}{1+\mu} \times \frac{\gamma_{max}}{2}$$

$$= \frac{2\times 10^5}{1+0.3} \times \left(\frac{757.2}{2} \times 10^{-6}\right) = \mathbf{58.2\ N/mm^2}$$

Check : $\tau_{max} = \frac{\sigma_1 - \sigma_2}{2} = \frac{115.4 - 1.1}{2}$ = **57.15 N/mm²**

Shear-Strain Gauges. By orienting a 2-element rectangular rosette such that the x-axis bisects the 90°angle and connecting the gauge elements to the adjacent arms of the Wheatstone bridge (see Figure 20.18), an output from the rosette equal to the shear strain γ_{xy}, can be obtained directly

$$\gamma_{xy} = \frac{\varepsilon_A + \varepsilon_B}{\sin 2\theta}$$

If the angle between the two gauges, $2\theta = 90°$

$$\gamma_{xy} = \varepsilon_A - \varepsilon_B \qquad \text{...(20.13)}$$

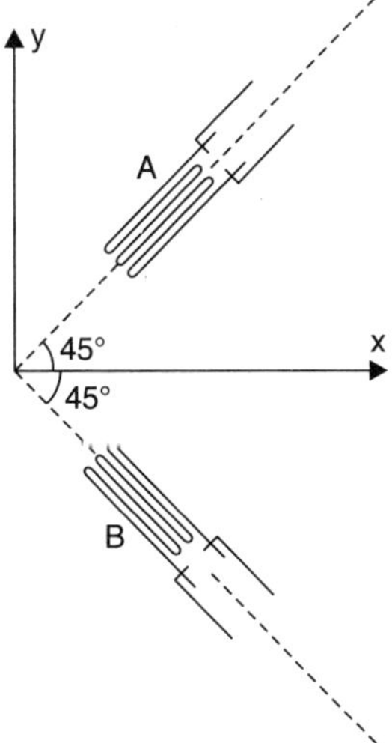

Figure 20.18. 2-Element rectangular rosette for measuring shear strain

Stress Gauge. In a biaxial stress field, the maximum principal stress is generally of primary interest. The principal stress directions can very often the determined from a **stress coat study**, from the shape of the member and mode of loading, or sometimes from the nature of a fracture.

In such cases (where the principal stress directions are known), it is only necessary to mount a strain gauge at an angle θ to the σ_1-axis (see Figure 20.19), such that

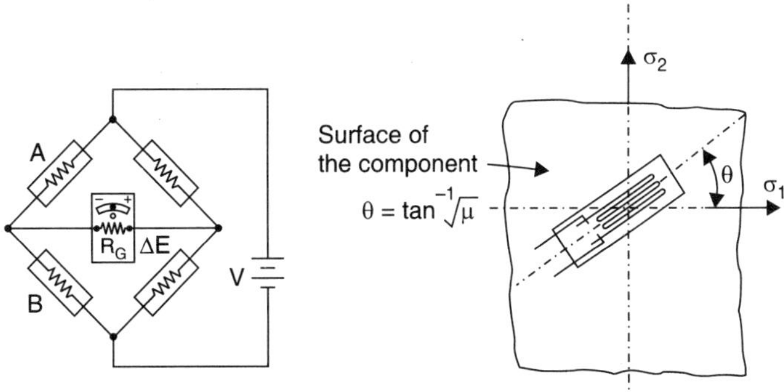

Figure 20.19. Stress gauge

$$\theta = \tan^{-1} \mu \qquad \text{...(20.14)}$$

and if the strain gauge reading is ε_θ,

$$\sigma_1 = \frac{E}{1-\mu}\,\varepsilon_\theta. \qquad ...(20.15)$$

Brittle Lacquer Coatings. The technique consists in applying a coating of brittle lacquer to the structural member to be tested, loading the member and analysing the resulting lacquer cracks for information about surface strains and direction of the principal stresses on the surface. The lacquer will crack perpendicular to the direction of maximum principal strain (or stress). This information will enable strain gauges to be cemented in the appropriate directions. The regions of strain or stress concentration are immediately revealed and the cracks form an accurate pattern of stress trajectories or isostatics. The strain is estimated by the spacing density of the cracks as compared with a **calibration bar** known as loading.

For the solution of dynamic problems (which are more complicated than the static ones) the coating should be properly calibrated. One of the difficulties to be overcome is the 'closing up' of cracks at stress levels much lower than the levels at which the cracks close under static load.

Basically, the stress coat lacquers are made of certain wood resins dissolved in carbon-disulphide. The stress coat is sensitive to temperature and humidity and this has been overcome by the development of a ceramic brittle coating under the trade name **Stress Coat All Temp.**

Lacquers are generally of brittle nature and are applied by brushing or spraying on the surface giving relatively good adhesion, the thickness ranging from 0.05 to 0.2 mm. It is allowed to dry for one or more days when the strain sensitivity will increase considerably.

Lacquers exhibit the property of creep at room temperature and in a lengthy test the results have to be corrected to allow for this.

When compared to photoelastic coatings, brittle coatings have the important advantage of being **much thinner**.

21 Photoelasticity

The photoelastic method is based on the principle that models made of certain transparent materials like, Araldite, Bakelite, Fosterite, Kriston, etc. When subjected to stress and placed in a field of polarised light exhibit the property of double refraction or 'birefringence' but return to normal refraction when the stress (load) is released. The two refracted components of light are plane polarised in perpendicular planes. The relative retardation between the two components is proportional to the difference of principal stresses ($\sigma_1 - \sigma_2$) at every point in the material. The number of wavelengths of relative retardation is given by

$$n = \frac{\alpha}{2\pi} = \frac{Ct}{\lambda}(\sigma_1 - \sigma_2) \qquad \text{...(21.1)}$$

where α = relative phase difference or phase angle between the two ray paths,

t = thickness of the model,

σ_1, σ_2 = major and minor principal stresses at the point considered in the model,

λ = wavelength of light used,

c = an optical coefficient (having dimensions $\frac{\text{m}^2}{\text{N}}$ or $\frac{1}{\text{Pa}}$ which depends on the wavelength, temperature and model material,

n = number of multiples of wavelength referred to as 'fringe order' (order of interference),

One wavelength of relative path difference ($\alpha = 2\pi$ radians) is called a **fringe**. The model fringe value f, is defined as

$$f = \frac{\lambda}{Ct} \qquad \text{...(21.2)}$$

and expressed as 'stress per fringe' i.e., N/mm^2 per fringe or Pa per fringe. Since, the retardation is proportional to the thickness of the model t, the material fringe constant (value) is defined as $F = ft$, which is the stress required to produce one fringe order in a unit thickness of the material and has the dimensions N/mm or kN/m. Thus, the Eqn. (21.1) can be written as

$$\sigma_1 - \sigma_2 = fn = \frac{F}{t}n \qquad \text{...(21.3)}$$

Further, the value of the maximum shear stress (at 45° to the principal stresses)

$$\tau_{max} = \frac{\sigma_1 - \sigma_2}{2} = \frac{f}{2} n \quad \text{or,} \quad f' n \qquad ...(21.4)$$

$$= \frac{F}{2t} n \quad \text{or} \quad \frac{F'}{t} n \qquad ...(21.4a)$$

where f' and F' are the model and material fringe values (constants) in shear, respectively, which are half the corresponding values in tension.

Photoelastic Bench. A typical arrangement of observing a loaded model in a beam of polarised light is shown in Figure 21.1. It consists of

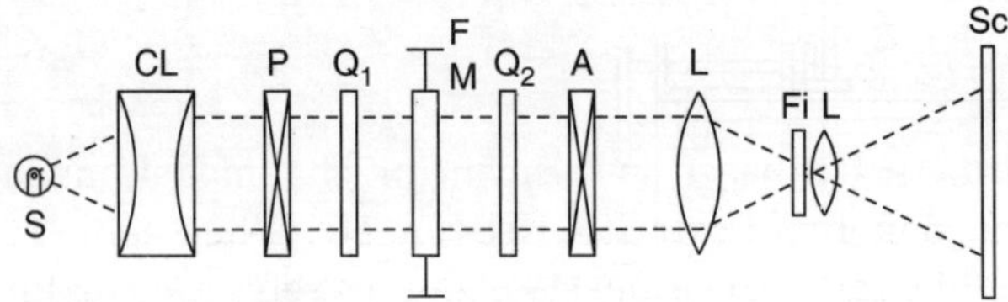

Figure 21.1. Photoelastic bench

1. A Light source (*S*)–either white light or monochromatic light (a mercury vapour lamp with a suitable filter usually green, or a sodium vapour lamp as commonly used)
2. A condenser lens (*CL*) to give a parallel beam over the working section.
3. Polariser (*P*)–natural crystals, or 'poloroid' disc made of mica or nicol to obtain plain-polarised beam.
4. Quarter wave plate (Q_1) made of mica, the thickness being related to the wave-length of light source.
5. Stressed transparent model (*M*) kept in a loading frame (*F*)
6. Quarter wave plate (Q_2) made of mica
7. Analyser (*A*)–another polaroid sheet
8. Collimating lens (*L*)
9. A filter (*Fi*) usually green (to make the light as monochromatic as possible) and projection lens (*L*)
10. A viewing screen (*Sc*) or camera

Diffused Light Polariscope, see Figure 21.2. It is quite simple and compact and uses diffused light source. The only lens is that of the camera having a large focal length or it can be viewed directly by the observer. It occupies much less length of the bench and is inexpensive. The field depends upon the size of the polaroid and quarter wave plates and hence can be made very large, for research projects.

Universal loading frame is shown in Figure 21.3.

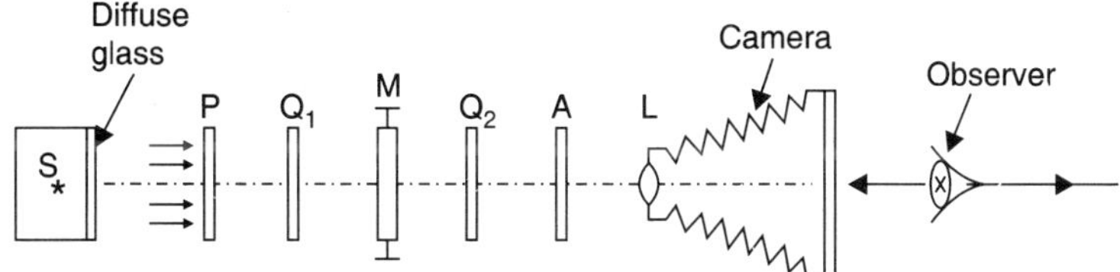

Figure 21.2. Diffused light polariscope

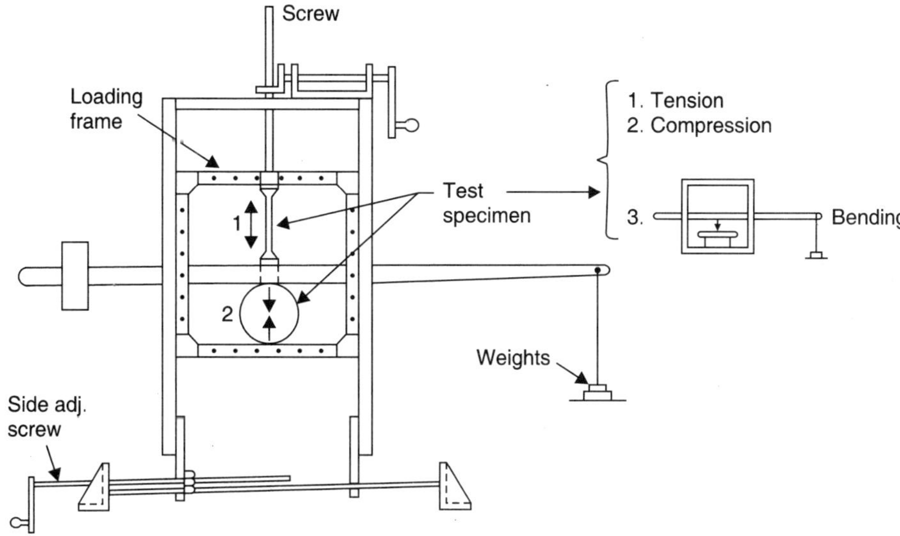

Figure 21.3. Loading frame

A. Plane Polariscope (Crossed polariser and Analyser and no Quater wave plates), see Figure 21.4.

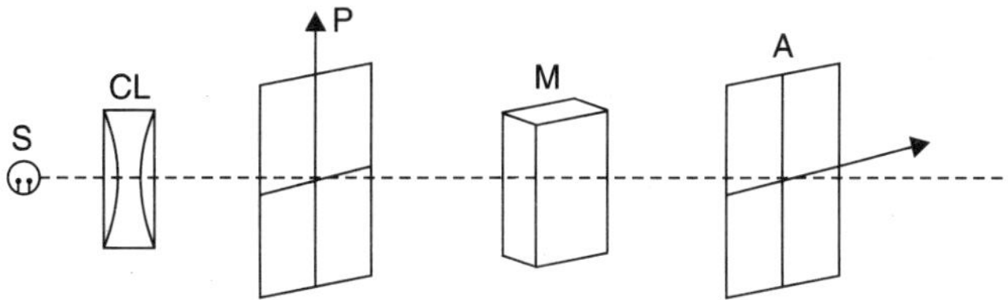

Figure 21.4. Plane polariscope

The intensity of light transmitted through a plane polariscope can be shown as

$$I = Ca^2 \sin^2 2\theta \sin^2 \left(\frac{\alpha}{2}\right) \qquad ...(21.5)$$

where θ = angle which σ_1 and σ_2 make with the axes of polariser P, and analyser A, see Figure 21.5,

a = amplitude of the light wave.

The intensity of light coming out of the analyser is zero under two conditions:

(*i*) when, $\theta = 0°$ or $\dfrac{\pi}{2}$

(*ii*) when, $\alpha = 2m\pi$, $m = 0, 1, 2, \ldots\ldots$

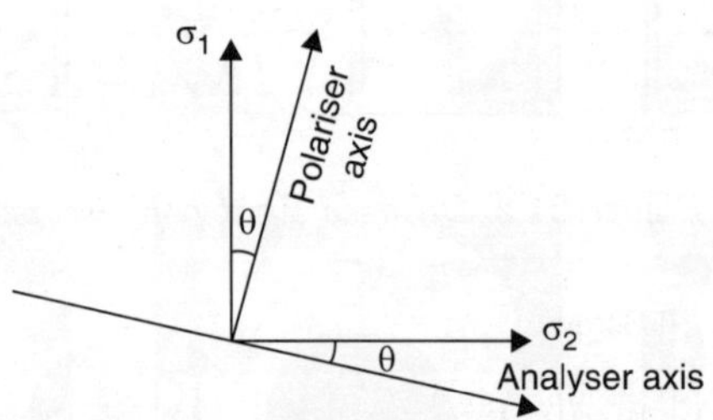

Figure 21.5. Isoclinic parameter

Condition (*i*) implies that extinction occurs at all points in the model where the direction of the principal stresses are parallel to the axes of polariser *P*, and analyser *A*, and on the image of the model (on screen) 'dark bands' will be formed which are termed '**isoclinics**' i.e., the loci of all points at which the principal stresses are in the same directions (as *P* and *A*). If the crossed *P* and *A* combination (see Figure 21.6), is now rotated new isoclinics will appear. Thus, a complete set of isoclinics will appear.

Thus a complete set of isoclinics of successive parameters (angular positions of *P* and *A*: 0°, 15°, 30°, ...) may be obtained and photographs taken (see Figure 21.7), which represent the direction of principal stresses at all points in the model, see Figure 21.8, (and so the maximum shear stress which is at 45° to the direction of the principal stresses) and they are independent of the load applied on the model.

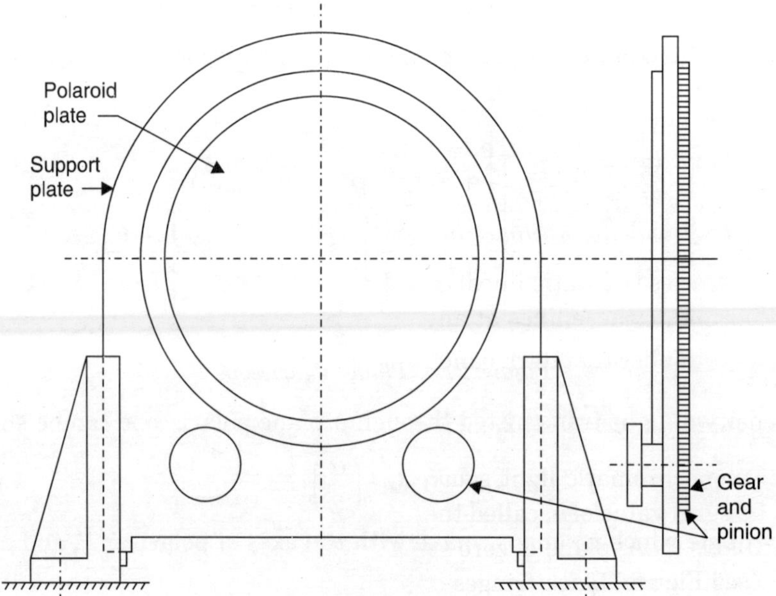

Figure 21.6.

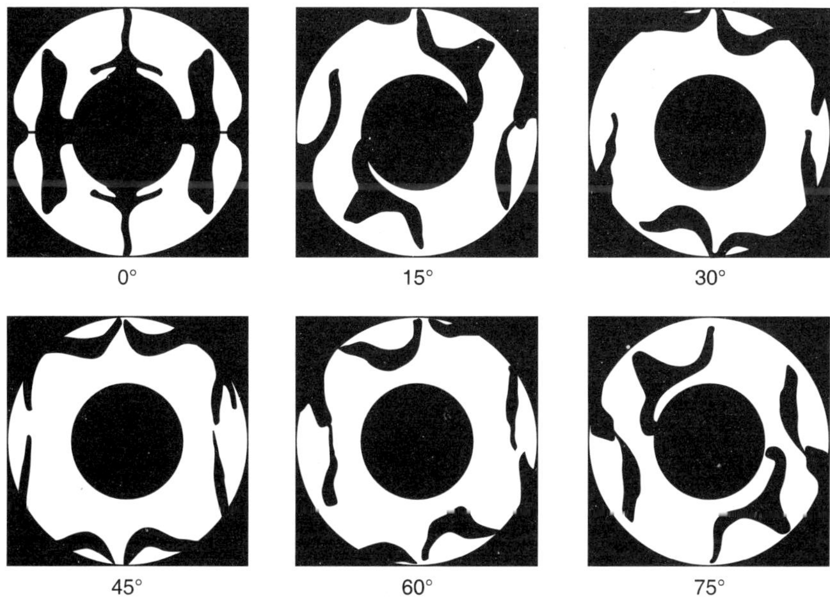

Figure 21.7. Isoclinics in a circular ring subjected to diametral compression

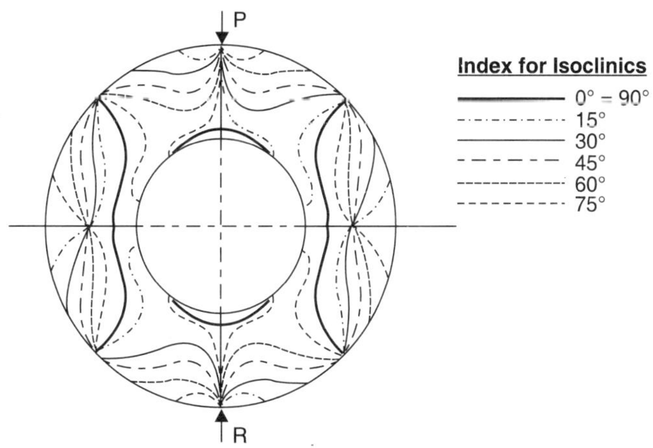

Figure 21.8. Family of isoclinics in a circular ring under diametral compression

Stress trajectories (isostatic lines) can be constructed so that the principal stresses are tangential and normal to these lines at any point, see Figure 21.9.

Condition (*ii*) implies that dark bands appear on the image of the model when the relative phase difference $\alpha = 2m\pi$, $m = 0, 1, 2, \ldots\ldots$, $n = \dfrac{\alpha}{2\pi} = 0, 1, 2, 3, \ldots\ldots$, i.e., an integral number of wavelengths. If monochromatic light source is used black lines or fringes are formed on the image for each integral value of n, called the 'fringe order', the order of the fringe being counted from the neutral axis (N.A.-plane of zero stress). This is called 'dark-field setup' for isoclinics. Here, both the isoclinic and black fringes appear as black bands and the resulting pattern is

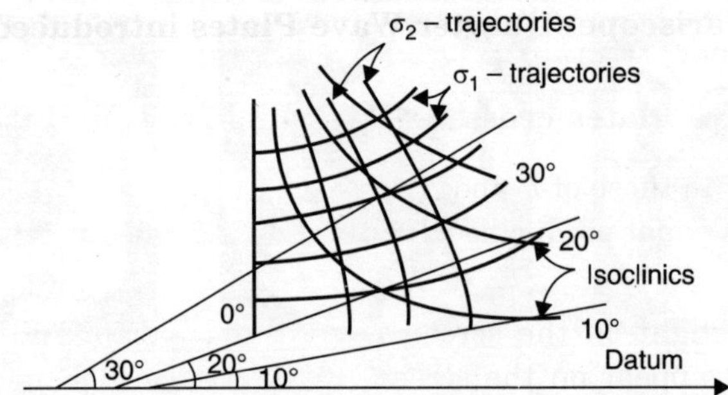

Figure 21.9. Construction of stress trajectories

mixed and ambiguous, see Figure 21.10. If white light source is used, the isoclinics still appear as dark bands but the black fringes are replaced by coloured bands called **isochromatics**. In the fringe pattern on the screen only the isoclinics and the zero-order fringe appear black (dark). The other fringes appear coloured. If the *P* and *A* combination is rotated, the black (dark) fringe corresponding to the isoclinic moves while the isochromatics including the black zero-order fringe do not change their positions. The orientations of the *P* and *A* with a reference frame (*x-y* axes) give the directions of the principal stresses at the point of interest. A photograph taken on a panchromatic film gives the isoclinics clearly, see Figure 21.11.

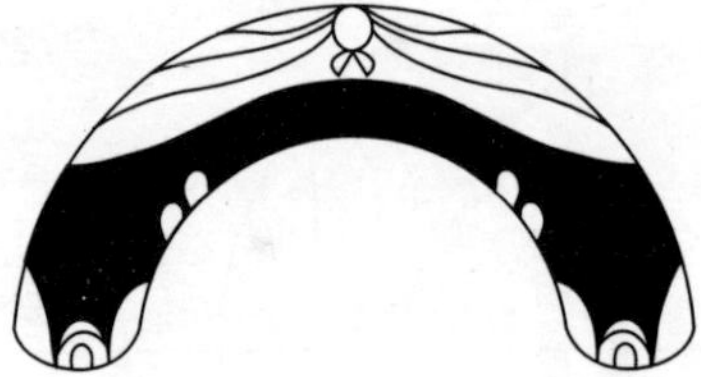

Figure 21.10. 0°-Isoclinic and dark fringes (monochromatic light)

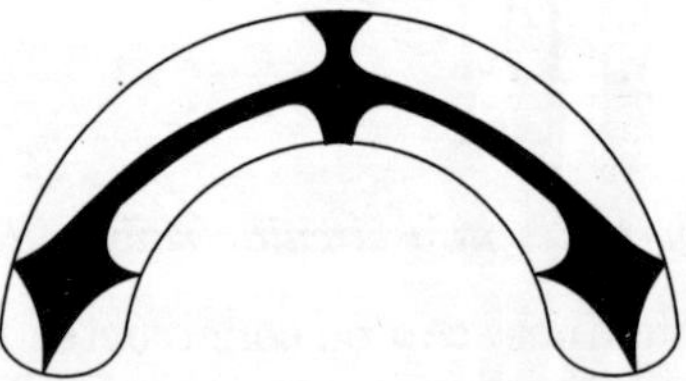

Figure 21.11. 0°-Isoclinic on panchromatic film (white light)

A fringe pattern represents points having equal principal stress (or maximum shear stress) difference in the model. If the isochromatic fringe pattern is restricted to the position of purple bands between red and blue, known as tint of passage, the fringes can be ascertained with a higher degree of precision.

B. Circular Polariscope (Quarter Wave Plates introduced)

(*a*) **Quarter Wave Plates crossed—Dark-Field Setup.** If the axes of the $\frac{\lambda}{4}$ plates are kept inclined at 45° to those of *P* and *A*, (i.e., crossed), Figure 21.12, the amplitude of the two components will be equal producing **circularly polarised light** and serve to eliminate the isoclinics.

The intensity of light on the screen is zero and is called a dark-field setup. Only **isochromatic fringes** appear on the screen. If the source is monochromatic then all the isochromatic fringes appear as **dark bands** which represent integral number of wavelengths of relative retardation (n = 0, 1, 2, ...). The fringe pattern of a circular disc under diametral compression is shown in see Figure 21.13.

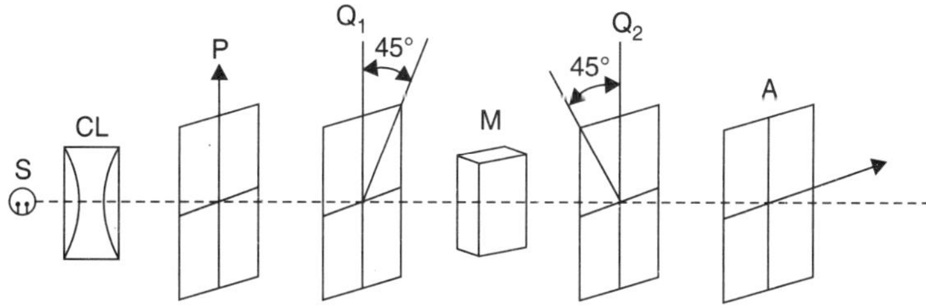

Figure 21.12. Circular polariscope

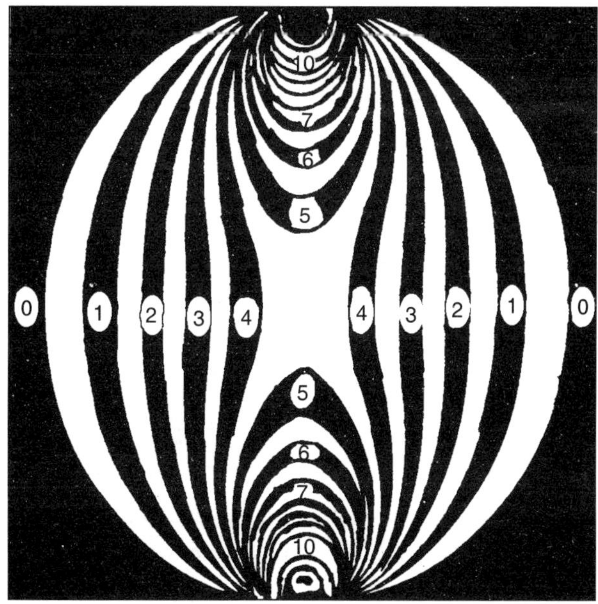

Figure 21.13. Isochromatics (dark fringes) for a circular disc under diametral compression [Dark-field set up]

(*b*) **Quarter Wave Plates parallel—Light-Field Setup.** If one of the $\frac{\lambda}{4}$ plates is rotated by 90°, so that they are now parallel and cancel each other, bright lines appear on the screen due to an overall change of $\frac{1}{2}\lambda$ in the polariscope. $I = \cos^2 \alpha/2$ and extinction or a fringe (dark) now occur when the condition of stress in the model are such that the relative phase difference $\alpha = (2m + 1)\pi$, $m = 0, 1, 2, \ldots\ldots$ when the relative retardation $n = \frac{\alpha}{2\pi} = \frac{1}{2}, \frac{3}{2}, \frac{5}{2}, \ldots\ldots$ wavelengths. Thus, in the bright-field setup, the dark fringes are seen for $\frac{1}{2}\lambda$, $1\frac{1}{2}\lambda$, $2\frac{1}{2}\lambda$, ..., etc., retardation i.e., one-half wavelength or its odd integer multiple. If white light is used, isochromatic fringes (spectral bands) can be seen.

A fringe represents the locus of all points having the same principal stress difference or maximum shear stress. The fringe pattern for a bright and dark-field setups for a circular ring under diametral compression is shown in Figure 21.14. If two of these photographs are superimposed, the lines of the one photograph would fall in between the lines of the other. The use of the two photographs increases the accuracy and sensitivity of the determinations.

Figure. 21.14. Isochromatics in a circular ring under diametral compression [Dark-field and Light-field patterns superimposed]

The combined knowledge of dark-field and bright field fringe patterns will enable to determine the relative retardation at a given point to an accuracy of half a wavelength. Also, if a plot of fringe order versus distance along a suitable chosen line, is prepared, one can determine fairly accurately the integral and fractional fringe order at the point of interest.

The difference in value between two fringes (lines or bands) is obtained by calibration (i.e., stress per fringe). By knowing the fringe order (with reference to a plane of zero stress), the principal stress difference and the maximum shear stress can be determined at a desired point as

$$\text{Stress} = \text{Fringe order} \times \text{Stress per fringe}$$

The various polariscope setups for stress analysis are given in Table 21.1.

Table 21.1 Polariscope Set Up

Arrangement	Light source	Field	Present	Fringe order	Determination	Remarks
(*a*) Plane polariscope *P* & *A* (crossed)	White light (usually)	Dark bands	Isoclinics (move when *P* and *A* rotated)	Integral (0, 1, 2, 3, ...)	Directions (θ°) of σ_1 & σ_2	Max. or min. intensity of light depending on θ°
		Coloured bands	Isochromatics (stationary)	Integral	$(\sigma_1 - \sigma_2)$ & τ_{max}	
	Monochromatic light	Dark bands	Isoclinics (by rotating *P* & *A*)	Integral	σ_1, σ_2 & τ_{max}	Difficult to distinguish isoclinics from the fringe pattern
(*b*) Circular polariscope						
1. *P* & *A* (crossed) Q_1 & Q_2 (crossed)	Monochromatic light (usually)	Dark	Isochromatics (dark fringes)	Integral (0, 1, 2, 3, ...)	$(\sigma_1 - \sigma_2)$ & τ_{max}	
2. *P* & *A* (crossed) Q_1 & Q_2 (parallel)	Monochromatic light (usually)	Light	Isochromatics (dark fringes)	Odd multiple of half $(\frac{1}{2}, 1\frac{1}{2}, 2\frac{1}{2}, 3\frac{1}{2}, ..)$	$(\sigma_2 - \sigma_2)$ & τ_{max}	Fringes can be read to an accuracy of $\frac{1}{2}\lambda$ by combining photographs of (1) & (2)
3. *P* and *A* (crossed) Q_1 & Q_2 (parallel)	White light	Light	Isochromatics (spectral bands)	Odd multiple of half	$(\sigma_1 - \sigma_2)$ & τ_{max}	

Calibration. $\sigma_1 - \sigma_2 = fn = \frac{F}{t}n, \quad \tau_{max} = \frac{\sigma_1 - \sigma_2}{2} f'n$

or, $(\sigma_1 - \sigma_2) \sim n, \quad \tau_{max} \sim n$

It is necessary to determine the constant of proportionality f, so that the principal stress difference or the maximum shear stress at any point can be obtained from a record of fringe pattern. This optical constant or calibration value for the model material is expressed as

$$f = \frac{\sigma_1 - \sigma_2}{n} \text{ or } \frac{\tau_{max}}{n} \qquad ...(21.6)$$

i.e., 'stress per fringe' or 'load per fringe × a constant'.

This constant f (or F) can be determined by calibration from a known stress field such as simple tension, pure bending or circular disc under diametral compression on a specimen cut from the same sheet of material as the model.

(*a*) Calibration from Tension Specimen

$$\sigma_1 = \sigma_y = \frac{P}{bt}, \quad \sigma_2 = 0, \text{ (see Figure 21.15 (a))}$$

$$\sigma_1 - \sigma_2 = fn$$

$$\frac{P}{bt} = fn$$

$$\therefore \quad f = \left(\frac{P}{n}\right)\frac{1}{bt} = \frac{P^*}{bt} = \text{stress per fringe} \qquad ...(21.7)$$

$$F = ft = \frac{P^*}{b}$$

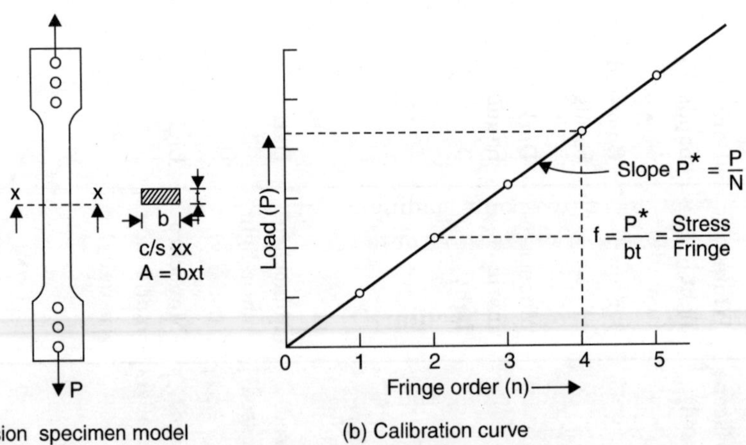

(a) Tension specimen model

(b) Calibration curve

Figure 21.15. Calibration of model in tension

The tension specimen is loaded (see Figure 21.15 (*a*)), and the fringe order is observed in a circular polariscope. As the load is increased or decreased, the corresponding fringe order is noted and a curve of P *versus* n, is plotted which yields a straight line, see Figure 21.15 (*b*). The slope of this line

$$= \frac{P}{N} = P^* = \text{Load per fringe}$$

(*b*) Calibration from a Rectangular Beam under Pure Bending

The beam specimen is subjected to a two-point loading so that it is under pure bending, see Figure 21.16 (*a*),

$$\sigma_1 - \sigma_2 = fn, \quad \sigma_1 = \sigma_x, \quad \sigma_2 = 0$$

$$\frac{M}{I} = \frac{\sigma_x}{Y}, \quad \sigma_x = M \times \frac{Y}{I}, \quad M = \frac{P}{2} \times a, \text{ and } I = \frac{bd^3}{12}$$

The beam is subjected to a load *P*, and the horizontal fringes of the stressed beam are observed in a circular polariscope set for dark back ground with monochromatic light, see Figure 21.16 (*a*). The trace of the neutral surface is represented by the zero-order fringe. If y_1, y_2, are the distances of the 1, 2, ... order fringes from the zero-order fringe, a curve 'σ_x versus *n*' can be plotted which yields a straight line (see Figure 21.16 (b)), whose slope gives

$$f = \frac{\sigma_x}{n} = \text{Stress per fringe} \qquad ...(21.8)$$

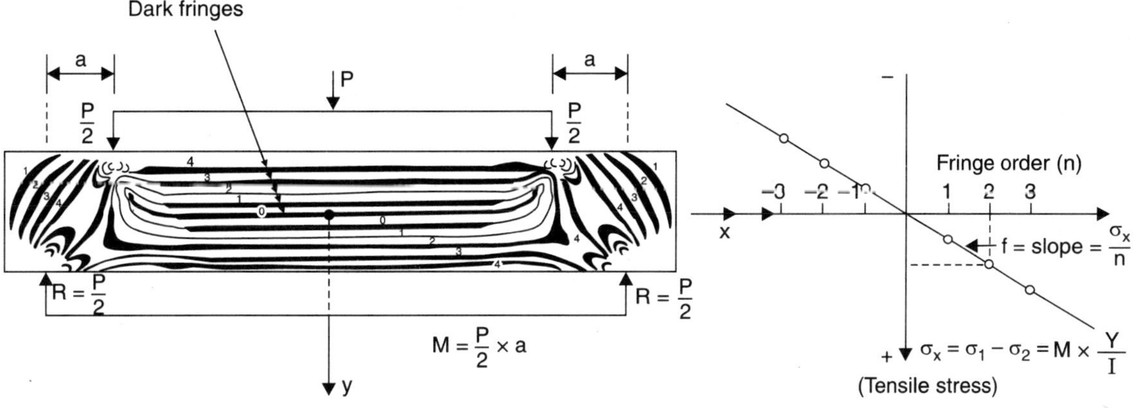

(a) Rectangular beam with two-point loading

(b) Calibration for bending stress σ_x

Figure 21.16. Calibration for a beam under pure bending

The advantage of two-point loading is that calibration plot can be drawn with a single loading unlike the case (*a*) where three or four different loads are required to obtain a calibration plot.

(*c*) Calibration from a Circular Disc subjected to Diametral Compression, see Figure 21.17.

The stress distribution along the horizontal diameter is given by

$$\sigma_x = \sigma_1 = \frac{2P}{\pi t D}\left[\frac{D^2 - 4x^2}{D^2 + 4x^2}\right]^2$$

$$\sigma_y = \sigma_2 = -\frac{2P}{\pi t D}\left[\frac{4D^4}{(D^2 + 4x^2)^2} - 1\right]$$

At the centre O of the disc, $x = 0, y = 0$

$$\sigma_1 - \sigma_2 = \frac{8P}{\pi t D} = fn$$

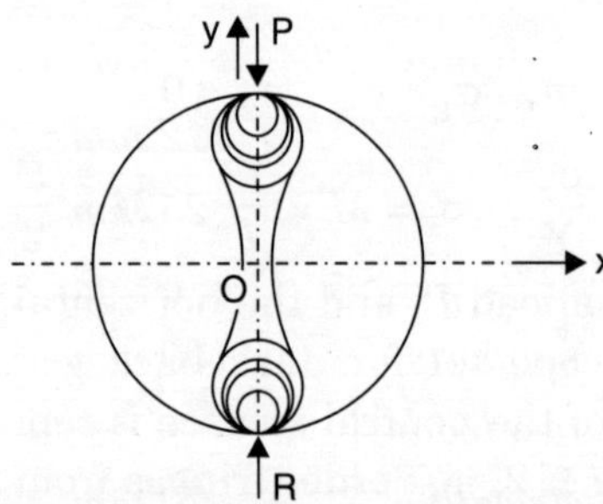

Figure 21.17. Circular disc under diametral compression

The formation of fringes at the centre for the load P applied can be observed in a circular polariscope. Different loads are applied and a curve 'P versus n' is plotted which yields a straight line. The slope of this line

$$\frac{P}{n} = P^* = \text{Load per finge}$$

$$f = \frac{\sigma_1 - \sigma_2}{n} = \frac{8}{\pi t D}\left(\frac{P}{n}\right) = \frac{8P^*}{\pi t D} \qquad ...(21.9)$$

$$F = ft = \frac{8P^*}{\pi D} \qquad ...(21.9a)$$

Example 21.1. *The fringe order observed at a point in a stressed model is 3.5 with mercury light (λ = 548.1 nm). The material fringe constant in tension is 20 kN/m. If the model has a thickness of 6 mm, calculate the maximum shear at the point.*

What fringe order will be observed if sodium light (λ = 589.3 nm) is used ? Assume that C is independent of λ.

Solution. From Eqn. (21.1),

$$\sigma_1 - \sigma_2 = \frac{\lambda}{Ct} n = fn \text{ , } ft = F$$

$$\sigma_1 - \sigma_2 = \frac{F}{t} n, \text{ where } F = 20 \text{ kN/m} = 20 \text{ N/mm}$$

$$= \frac{20}{6} \times 3.5 = \frac{70}{6} \text{ N/mm}^2$$

$$\tau_{max} = \frac{\sigma_1 - \sigma_2}{2} = \frac{70}{6 \times 2} = 5.83 \text{ N/mm}^2$$

$$\frac{\lambda n}{c} = (\sigma_1 - \sigma_2)\, t = \text{constant}$$

Assuming that C is invariant with λ (actually C is a function of) and remains constant,

$$\lambda n = a \text{ constant i.e., } \lambda_1 n_1 = \lambda_2 n_2$$

$$\therefore \qquad n_2 = \frac{\lambda_1 n_1}{\lambda_2} = \frac{548.1}{589.3} \times 3.5 = \mathbf{3.25}$$

Hence, fringe pattern changes with the wavelength of light used.

Note. 1 nm = 10^{-9} m

Example 21.2. *Calculate the material fringe constant given the following data:*

Length of beam	*112 mm*
Depth of beam	*25 mm*
Width of beam	*6 mm*
Distance between supports	*100 mm*
Distance between two loads 400 N each symmetrically placed on either side of the centre of span	*70 mm*
Number of fringes observed on the whole depth of beam	*7*

Solution. For a beam subjected to pure bending (under 2-point symmetrical loading)

$$\frac{M}{I} = \frac{\sigma_x}{Y}, \quad \sigma_x = M \times \frac{Y}{I}$$

$$M = \frac{P}{2} \times a = 400 \times \frac{100 - 70}{2} = 6000 \text{ N·mm}$$

At top or bottom fibre of the beam

$$Y = \frac{d}{2}, \quad n = \frac{7}{2} = 3.5$$

$$I = \frac{bd^3}{12}, \; \sigma_x = M \times \frac{d/2}{bd^3/12} = \frac{6\,M}{bd^2} = \frac{6 \times 6000}{6 \times 25^2} = 9.6 \text{ N/mm}^2$$

$$\sigma_1 - \sigma_2 = fn, \quad \sigma_1 = \sigma_x, \quad \sigma_2 = 0$$

$$9.6 = f \times 3.5$$

$$f = 2.743 \text{ N/mm}^2 \text{ per fringe}$$

$$F = ft = 2.743 \times 6 = \mathbf{16.46 \text{ N/mm or } 16.46 \text{ kN/m.}}$$

Example 21.3. *Calculate the fringe value of the material of the model given the following data:*

Concentrated load on a big plate supported completely at bottom	*= 1 kN*
Thickness of plate	*8 mm*
Fringe order at a distance of 4 mm from the load	*14*
Fringe order at a distance of 16 mm from the load	*3*

Solution. For a concentrated load P, on a semi-infinite plate, at a distance r, from the point of loading (Ref. 15),

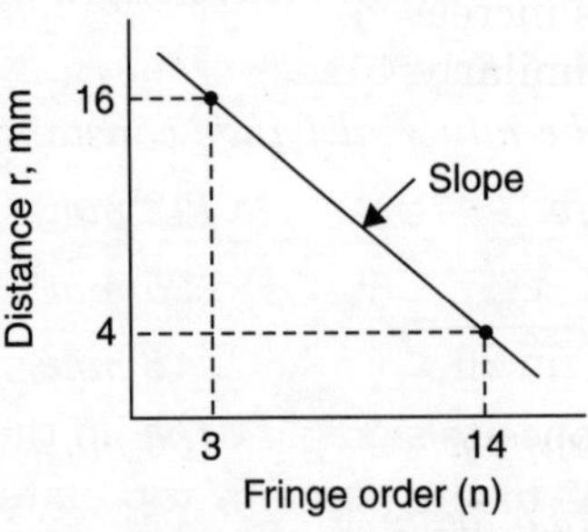

Figure 21.18.

$$\sigma_1 - \sigma_2 = \frac{2P}{\pi r} = \frac{2(1 \times 1000)}{\pi r} = \frac{636.62}{r}$$

r (mm)	$\sigma_1 - \sigma_2$ (N/mm^2)	n
4	159.16	14
16	39.8	3
	119.36	11

$$\sigma_1 - \sigma_2 = fn$$

From the above table, $f = \frac{\sigma_1 - \sigma_2}{n} = \frac{119.36}{11} = 10.851$ N/mm^2 per fringe

$F = ft = 10.851 \times 8 =$ **86.8 N/mm or 86.8 kN/m**

Stress Analysis. Photoelastic methods serve as an invaluable tool in determination of the following:

(*i*) The state of stress at a desired point.

(*ii*) Evaluation of the magnitude and direction of the principal stresses and maximum shear stress at the desired point.

(*iii*) Magnitude and distribution of stresses in a structural member or a machine element.

(*iv*) Zones of stress concentration and magnitudes of maximum principal and shear stresses.

The model is carefully prepared from a suitable material and annealed to remove initial strains induced by shaping. It is placed in a frame by means of which loads may be applied as desired, see Figure 21.3. At a given point in the model for a particular load two observations are made.

(*i*) Direction of the Principal Stresses (PPS). The $\frac{\lambda}{4}$ plates are removed and the P and A crossed combination is rotated until a black spot appears at the point. The direction of PPS corresponds to the direction of the axes of P and A.

***(ii)* Maximum Shear Stress (τ_{max}).** The $\frac{\lambda}{4}$-plates are introduced. As the load is applied gradually, a black band or fringe will appear at the point or zone of maximum shear stress. It will gradually move as the load is increased. When the shear stress is doubled, a second band or 2nd order fringe will appear. Similarly, higher order fringes appear as the load is increased. From calibration value

$$\tau_{max} = fn$$

Further, $$\tau_{max} = \frac{\tau_{xy}}{\sin 2\theta} = \frac{\sigma_1 - \sigma_2}{2} \qquad ...(21.10)$$

where $\theta = 45°$, and σ_1, σ_2 are of opposite signs; if σ_1, σ_2 in the plane are of the same sign, there is a larger τ_{max} on a plane at 45° to the plane of the plate. This shear has no effect on the isochromatics.

***(iii)* Principal Stress Difference ($\sigma_1 - \sigma_2$).** From the fringe or interference order n, at a given point and the model fringe value

$$\sigma_1 - \sigma_2 = fn$$

In Figure 21.16, for the case of a beam under pure bending, the trace of the neutral surface is represented by the zero-order fringe and the other fringes are equally spaced above or below the zero-order fringe. The closeness of the fringes at the support and at the two load points indicates that high stress concentrations exist there. In other cases, starting from a point where the fringe order n, is known to be zero, if a plot of fringe order versus distance along a suitable chosen line is prepared, one can determine fairly accurately the integral (n = 0, 1, 2, ...) or fractional fringe order ($n = \frac{1}{2}, 1\frac{1}{2}, 2\frac{1}{2}$, ...) at the point of interest.

From the use of **compensation techniques** the fractional fringe order at a point can be accurately determined. The method of compensation consists in bringing the existing fringe order at a point to an integral value in the case of a dark-field setup or to an odd multiple of half order fringes in the case of bright field setup.

In the Tardy's method of compensation, the $\frac{\lambda}{4}$ plates are kept crossed at 45°, the analyser is then rotated for extinction and the angle rotation $\phi°$ of the analyser with the horizontal, is noted. Then $\alpha = \pm 2\phi$, and the fractional fringe order.

$$n = \frac{\alpha}{2\pi} = \frac{2\phi}{2\pi} = \pm \frac{\phi°}{\pi} \qquad ...(21.11)$$

The sign (±) indicates that the compensation is by increasing the existing fringe order to the next higher integral value or decreasing it to the next lower integral value.

***(iv)* Magnitude of Principal Stresses.** On a free (unloaded) boundary, one of the principal stress (normal to the boundary) is zero, (see Figure 21.19), and therefore the boundary stresses can be obtained directly from the fringe pattern ; this may be of sufficient indication of stress concentrations. For example, the tenth order fringe on the inner boundary, (see Figure 21.14), gives the maximum tensile stress and its position on the body. Like this, the stress patterns on geometries such as fillets, shoulders, gear teeth, holes, etc. can be determined which are valuable for design studies.

Individual principal stresses, normal or shear stresses in the model can be obtained by several methods listed below, one usual approach being to find the sum of the principal stresses $(\sigma_1 + \sigma_2)$ at a point and combine this with the difference $(\sigma_1 - \sigma_2)$.

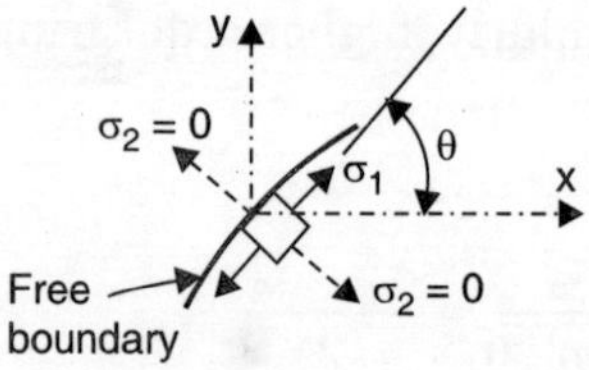

Figure 21.19. Stress at a free boundary

Separation Techniques (to evaluate σ_1 and σ_2)

(*i*) Numerical determination of $(\sigma_1 + \sigma_2)$ by **relaxation or iteration method**, values at the free boundaries being known.

(*ii*) Numerical integration along a line starting from a free boundary by **shear difference method**.

(*iii*) Integration along a stress trajectory making use of Lame-Maxwell equations of equilibrium.

(*iv*) Lateral strain method by the use of a highly sensitive **lateral extensometer** or by the use of an **interferometer** (by an 'interference fringe' pattern).

(*v*) From the fringe orders obtained from normal and oblique incidence. For the details of these methods reference may be made to [Ref. 13,17]

Three-Dimensional Photoelasticity (Frozen Stress Model)

In two-dimensional photoelasticity models subjected to plane states of stress are studied. A great majority of structural components and machine elements are subjected to three-dimensional states of stress.

If a three-dimensional model is cast or machined and loaded in the same manner as the component and polarised light is passed, the fringe pattern would not give a direct relationship for the difference of principal stresses, since it is most unlikely that all along the path of light through the model, two of the principal stresses would be in the plane of the wavefront and the third coincident with the axis of propagation.

It was observed by Solakian (1935), Brewster, Cleark Maxwell and later investigators that certain gelatinous materials such as isinglass and resins like bakelite and fosterite when loaded while hot and allowed to cool under load, the stress becomes 'frozen'. The model could be sliced without disturbing the fringe pattern and the thin slice could be analysed optically as if it were a two-dimensional model. The components τ_{xz}, τ_{yz}, σ_z are known to have no effect on a ray in the *z*-direction i.e., normal to the slice. Isoclinics or the direction of stress at a point can be determined by the plane poloriscope and the oblique incidence technique can be used for separation of the principal stresses.

Photoelastic Coating Method. A thin sheet of photoelastic material is bonded on to the metal surface of the component. When the component is loaded, the surface strain is transmitted to the coating thus inducing birefringence. If a polariscope is setup the incident polarised light on the coating is reflected and a fringe pattern appears proportional to the principal stress difference in the metal. This technique enables the measurement of plastic strains in the metal in a complicated structure.

PART C

STRENGTH OF MATERIALS

(Theory and Problems)

1 Simple Stresses and Strains

Structural members or machine components are subjected to loads (forces) such as tension, compression, shear, bending, torsion, or a combination of these. Internal forces (per unit area) developed over the cross-section to resist these external forces are called **stresses,** resulting in deformations or **strains.** The cross-sectional area (or dimensions) of the member or component should be sufficient such that the stresses developed are within the safe allowable limits (to avoid fracture or collapse).

Direct (axial) Stress due to Tension or Compression, see Figure 1.1

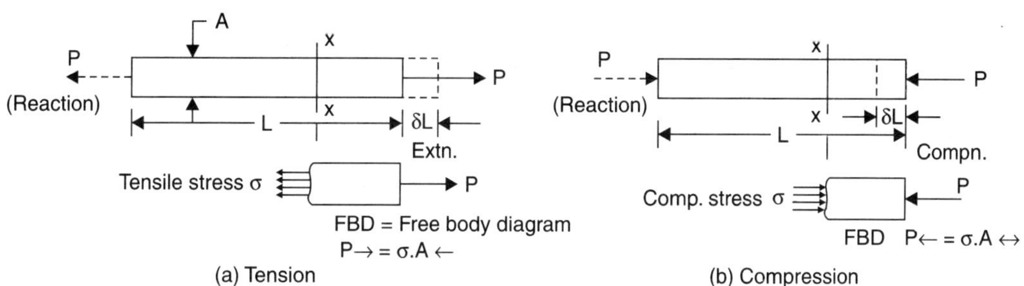

Figure 1.1

Stress (σ) is the load per unit area.

$$\therefore \quad \text{Stress } (\sigma) = \frac{\text{Load}\,(P)}{\text{Area}\,(A)} \text{ N/mm}^2$$

Deformation = Extension or compression = ΔL

Strain (ε) is deformation per unit length (original) and has no dimensions.

$$\therefore \quad \text{Strain } (\varepsilon) = \frac{\Delta L}{L}$$

Within elastic limit (more precisely proportional limit),

$$\text{Young's modulus of elasticity } (E) = \frac{\text{Stress } (\sigma)}{\text{Strain } (\varepsilon)}$$

E has dimensions of stress in N/mm², higher the value of E of the material (or the beam), higher the **stiffness** and less is the deflection.

A typical **stress-strain curve** for mild steel (M.S.) in tension, with salient points marked, is shown in Figure 1.2. The slope of the initial straight portion of the stress-strain curve (i.e.,within elastic or proportional limit) gives the value of E.

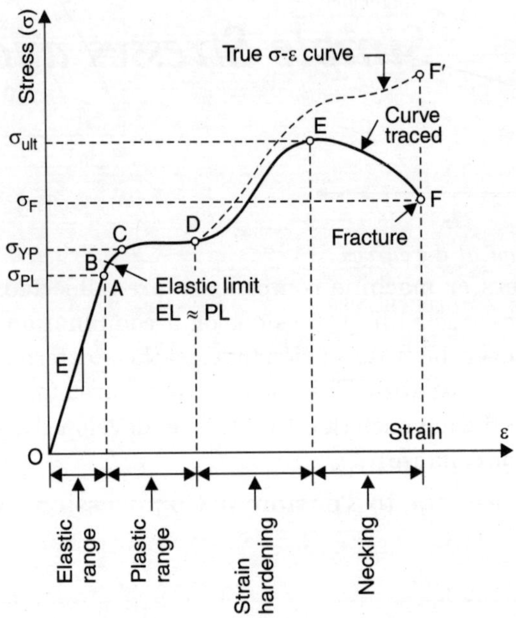

Figure 1.2. Tension test on M.S. specimen

OA—Hooke's law is obeyed; linear

A—Linearity ceases : Proportional limit (PL)

B—Elasticity ceases : Elastic limit (EL)

C—Plastic flow commences; the material yields at constant load : Yield point (YP)

D—Strain hardening begins : metal undergoes change in its atomic and crystalline structure resulting in the increased resistance of the material to further deformation.

E—Necking starts resulting in the decrease of load.

F—Fracture or collapse.

Typical stress-strain curve for some other materials are shown in Figure 1.3.

Proof Stress. Materials like some alloy steels and light alloys of *Al* and *Mg* will not show a definite yield. For such materials a **proof stress** is used to determine the onset of plastic range.

Proof stress is the stress at which if the material is unloaded, a specified percentage of strain is permanently left in it; this permanent set is usually taken as 0.1 or 0.2% offset, see Figure 1.4.

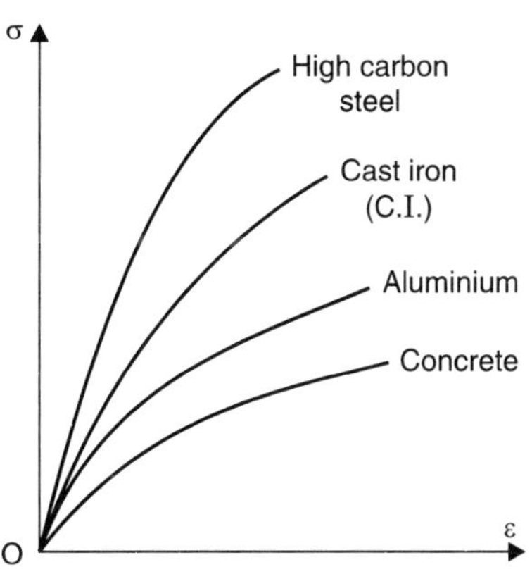

Figure 1.3. Typical σ-ε curves

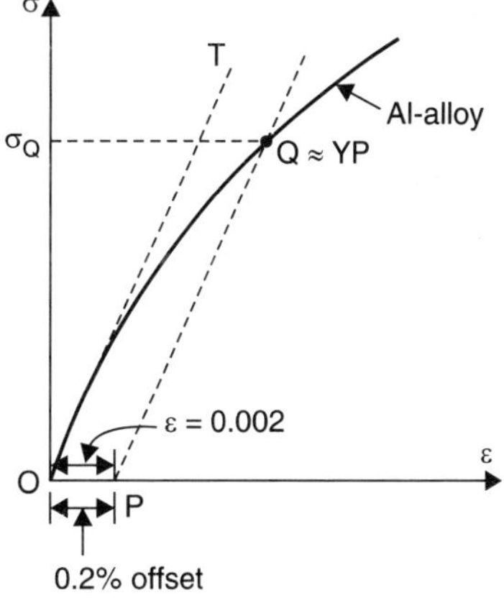

Figure 1.4. 0.2% proof stress

To determine, say 0.2% proof stress, a tangent to the curve at the origin (OT) is drawn which is the initial straight portion of the curve. A 0.2% offset is measured such that $OP = 0.002$.

Now a line $PQ \parallel OT$ is drawn to intersect the 'σ-ε curve' at Q; then

$$\sigma_Q = 0.2\% \text{ Proof stress} \approx \text{Yield stress} \approx \text{YP}$$

Allowable stress or working stress

$$\sigma_{\text{allow}} = \frac{\text{Yield stress}}{\text{Factor of safety } (\approx 2)}$$

$$\sigma_{\text{allow}} = \frac{\sigma_{\text{ult}}}{\text{F.S. } (\approx 3)}$$

$$\text{Load factor} = \frac{\text{Load at fracture (collapse)}}{\text{Service or working load}}$$

Strain Energy and Resilience

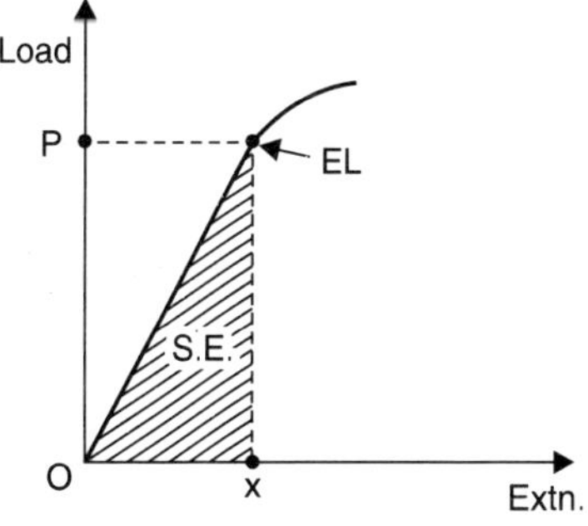

Figure 1.5. Load-extn. curve

Strain energy (S.E.) of the bar is the work done by the load P, in straining it. For gradually applied or 'static' load,

$$\text{S.E.} = \text{Shaded area in Figure 1.5}$$

$$= \frac{1}{2} Px$$

But $P = \sigma A$ and $\varepsilon = \dfrac{x}{L} = \dfrac{\sigma}{E}$

$$= \frac{1}{2}(\sigma A)\left(\frac{\sigma}{E}L\right)$$

$$= \left(\frac{\sigma^2}{2E}\right) AL$$

where, AL is volume of bar.

$$\textbf{S.E.} = \frac{\sigma^2}{2E} \times \textbf{Volume}$$

Resilience U, is the strain energy per unit volume,

$$U = \frac{\sigma^2}{2E}$$

in simple tension or compression.

Proof resilience is the value at the elastic limit or at the proof stress for non-ferrous materials.

Poisson's Ratio. In an elastic material, the axial strain due to direct stress (tension or compression) is associated with a lateral strain (which is negative for tension and positive for compression), see Figure 1.6.

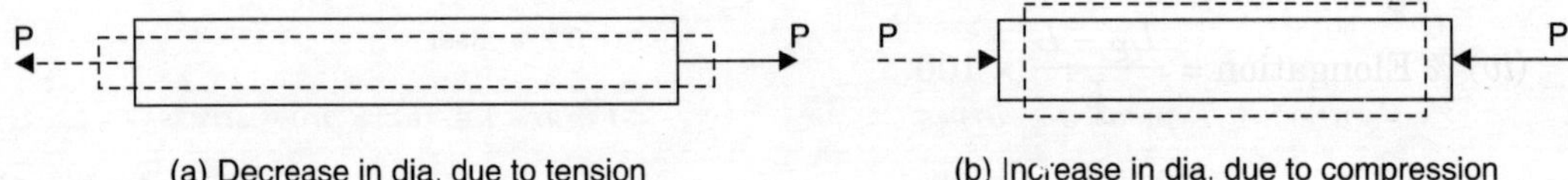

Figure 1.6. Poisson's ratio

$$\text{Poisson's ratio } (\nu) = -\frac{\text{Lateral strain}}{\text{Axial strain}}$$

For most metals, ν lies in the range of 0.25 to 0.35 (i.e., $\nu < 1$)

$$\nu = \frac{1}{m}, \text{ where } m = 2 \text{ to } 4 \ (m \text{ always} > 2)$$

It can be shown that due to the axial stress (σ),

$$\text{Volumetric strain } (\varepsilon_v) = \frac{\Delta V}{V} = \varepsilon\,(1 - 2\nu) = \frac{\sigma}{E}\,(1 - 2\nu).$$

Example 1.1. *A M.S. specimen of gauge length 200 mm and dia 25 mm was tested into destruction and the following observations were made:*

Extn. of 0.16 mm under a load of 80 kN

Load at EL = 160 kN

Maximum load = 180 kN

Gauge length and diameter at fracture = 256 mm and 18 mm, respectively.

Determine: σ_{EL}, E, σ_{ult}, *% elongation, % reduction in area, and proof resilience.*

Solution. (*i*) $\sigma_{EL} = \dfrac{P_{EL}}{A} = \dfrac{160 \times 10^3}{\frac{\pi}{4} \times 25^2} = \dfrac{160{,}000}{490.87} = \mathbf{326\ N/mm^2}$

(*ii*) At P = 80 kN within EL

$$\sigma = \frac{P}{A} = \frac{80 \times 10^3}{490.87} = 163 \text{ N/mm}^2$$

$$\varepsilon = \frac{x}{L} = \frac{0.16}{200} = 0.0008$$

$$E = \frac{\sigma}{\varepsilon} = \frac{163}{0.0008} = \mathbf{203{,}718\ N/mm^2}$$

Note. $E = \dfrac{PL}{Ax}$ = Slope of load-extn. curve $\dfrac{P}{x} \times \dfrac{L}{A}$, see Figure 1.5 ; $L = L_0$, $A = A_0$

$$= \left(\frac{80 \times 10^3}{0.16}\right) \times \frac{200}{\frac{\pi}{4} \times 25^2} = \mathbf{203{,}718\ N/mm^2}$$

(*iii*) $\sigma_{ult} = \dfrac{P_{ult}}{A_0}$, where $P_{ult} = P_{max}$

$$= \frac{180 \times 10^3}{490.87} = \mathbf{367\ N/mm^2}$$

σ_{ult} = Ultimate tensile strength (UTS)

(*iv*) % Elongation $= \dfrac{L_F - L}{L} \times 100$

$$= \frac{256 - 200}{200} \times 100 = \mathbf{28\%}$$

(*v*) % Reduction in area $= \dfrac{D^2 - {D_F}^2}{D^2} \times 100$

$$= \frac{25^2 - 18^2}{25^2} \times 100 = \mathbf{48.16\%}$$

(*vi*) Proof resilience (U) $= \dfrac{\sigma_{EL}^2}{2E} = \dfrac{326^2}{2 \times 203{,}718}$ = **0.26 N.mm/mm³** or **0.26 N/mm²**

Example 1.2. *Find the stresses developed in the bar and the total elongation, see Figure 1.7. Take E = 80,000 N/mm².*

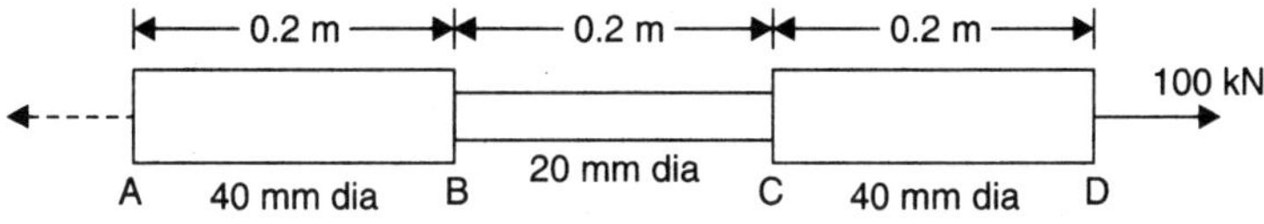

Figure 1.7

Solution.

$$\sigma_{AB} = \sigma_{CD} = \frac{P}{A} = \frac{100 \times 10^3}{\frac{\pi}{4} \times 40^2} = \mathbf{79.58\ N/mm^2}$$

$$\varepsilon_{AB} = \varepsilon_{CD} = \frac{\sigma}{E} = \frac{79.58}{80{,}000} = 9.95 \times 10^{-4}$$

$$\sigma_{BC} = \frac{P}{A} = \frac{100 \times 10^3}{\frac{\pi}{4} \times 20^2} = \mathbf{318.3\ N/mm^2}$$

$$\varepsilon_{BC} = \frac{\sigma}{E} = \frac{318.3}{80{,}000} = 3.98 \times 10^{-3}$$

Total elongation $(\Delta L) = 2(\varepsilon_{AB} \times 0.2) + (\varepsilon_{BC} \times 0.2)$ $\quad \because \varepsilon = \dfrac{\Delta L}{L}, \Delta L = \varepsilon.L$

$$\therefore \quad \Delta L = 2(9.95 \times 10^{-4} \times 0.2) + (3.98 \times 10^{-3} \times 0.2)$$
$$= \mathbf{1.194 \times 10^{-3}\ m} \quad \text{or } \mathbf{1.2\ mm}$$

Example 1.3. *What should be the value of P in Figure 1.8, if the overall axial deformation of the bar should not exceed 2 mm ?*

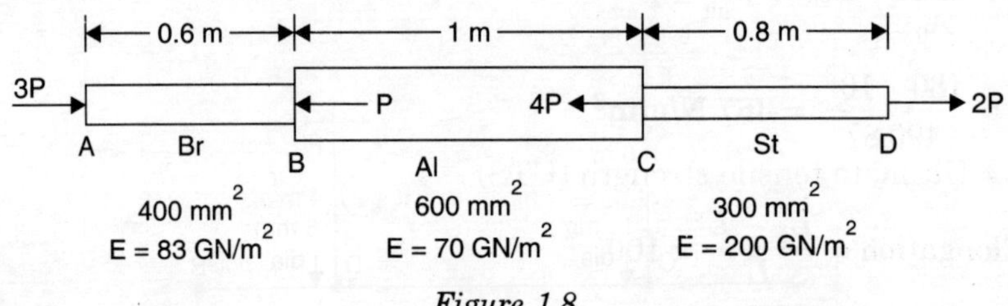

Figure 1.8

Solution.

FBD – AB (sec. in)

FBD – BC (sec. in)

FBD – CD (sec. in)

FBD – AB : $\quad -\Delta L = \varepsilon_L = \dfrac{\sigma}{E} \times L = \dfrac{3P/(400 \times 10^{-6})}{83 \times 10^9} \times 0.6, \quad \text{GN/m}^2 = 10^9 \text{ N/m}^2$

FBD – BC : $\quad -\Delta L = \dfrac{2P/(600 \times 10^{-6})}{70 \times 10^9} \times 1$

FBD – CD : $\quad +\Delta L = \dfrac{2P/(300 \times 10^{-6})}{200 \times 10^9} \times 0.8$

$$\Sigma \Delta L = \frac{P}{10^3}\left(-\frac{3}{83 \times 400} \times 0.6 - \frac{2}{70 \times 600} \times 1 + \frac{2}{200 \times 300} \times 0.8\right)$$
$$= 2 \times 10^{-3} \text{ m}$$

$$\therefore \quad P = \mathbf{26{,}599.74\ N} \text{ or } \mathbf{26.6\ kN}$$

Note. AB and BC are in compression and CD is in tension, the resultant deformation being 2 mm shortened.

Example 1.4. *A rigid beam AB 4 m long is hinged at A and is suspended by wires CE and DF as shown in Figure 1.9. If a load of 5 kN is applied at B, find the stresses developed in the wires and the deflection of the end B. E-wires = 2 × 10⁵ N/mm².*

Solution. Since, AB is a rigid beam, from similar triangles,

$$\frac{x}{1} = \frac{y}{3}, \quad y = 3x \qquad ...(i)$$

The elongations $x = \left(\frac{PL}{AE}\right)CE, \quad y = \left(\frac{PL}{AE}\right)DF$

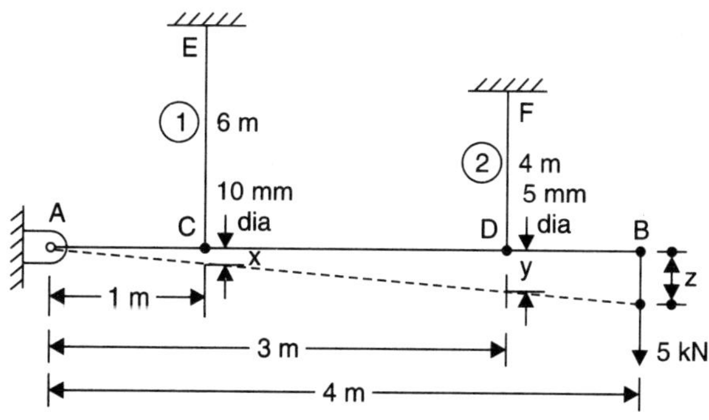

Figure 1.9

From (i), $\frac{P_2(4 \times 1000)}{\left(\frac{\pi}{4} \times 5^2\right)E} = 3 \times \frac{P_1(6 \times 1000)}{\left(\frac{\pi}{4} \times 10^2\right)E}$

$$\therefore \quad P_2 = \frac{9}{8} P_1 \qquad ...(ii)$$

Taking moments about A

$$5 \times 4 = P_1 \times 1 + P_2 \times 3$$

$$20 = P_1 \times 1 + \left(\frac{9}{8} P_1\right) 3$$

$$P_1 = 4.57 \text{ kN}, \quad P_2 = 5.14 \text{ kN}$$

$$\sigma_{CE} = \frac{4570 \text{ N}}{\frac{\pi}{4} \times 10^2} = \mathbf{58.19 \ N/mm^2}$$

$$\sigma_{DF} = \frac{5140 \text{ N}}{\frac{\pi}{4} \times 5^2} = \mathbf{232.75 \ N/mm^2}$$

Deflection of the end B

$$\frac{x}{1} = \frac{z}{4}, z = 4x = 4\left(\frac{PL}{AE}\right)CE$$

$$\therefore \quad z = 4 \times \frac{4570 \times 6000}{\left(\frac{\pi}{4} \times 10^2\right) 2 \times 10^5} = \textbf{6.982, say 7 mm}$$

Example 1.5. *For a conical bar shown in Figure 1.10, obtain an expression for elongation under an axial pull P.*

Solution. $d_1 = d + \dfrac{D-d}{L}\,x$

The extension of this short length (dx),

$$= \frac{P \cdot dx}{A_1 E}$$

$$= \frac{P \cdot dx}{\dfrac{\pi}{4}\left[1 + (D-d)\dfrac{x}{L}\right]^2 E}$$

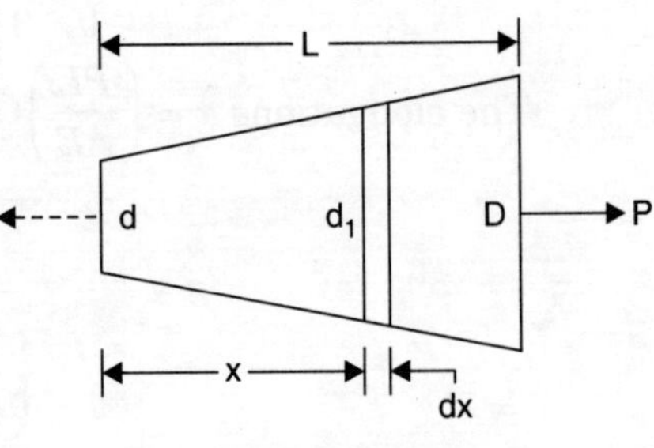

Figure 1.10

and for the whole conical bar

$$\Delta L = \int_0^L \frac{4P \cdot dx}{\pi\left[d + (D-d)\dfrac{x}{L}\right]^2 E}, \quad | \text{ Put } u = d + (D-d)\frac{x}{L}$$

$$\frac{du}{dx} = \frac{D-d}{L} \qquad dx = \frac{L}{D-d} \times du$$

$$= \frac{4P}{\pi E} \cdot \frac{L}{D-d} \int_{x=0}^{L} \frac{du}{u^2}$$

$$= -\frac{4\,PL}{\pi E(D-d)} \cdot \left.\frac{1}{d + (D-d)\dfrac{x}{L}}\right]_0^L$$

$$\Delta L = \frac{4\,PL}{\pi E(D-d)} \cdot \left[\frac{1}{d} - \frac{1}{D}\right]$$

$$\Delta L = \frac{\mathbf{4\,PL}}{\boldsymbol{\pi}\,\mathbf{E\,d\,D}}$$

Example 1.6. *For a tapered bar of uniform thickness (t) shown in Figure 1.11, obtain an expression for elongation under an axial pull P.*

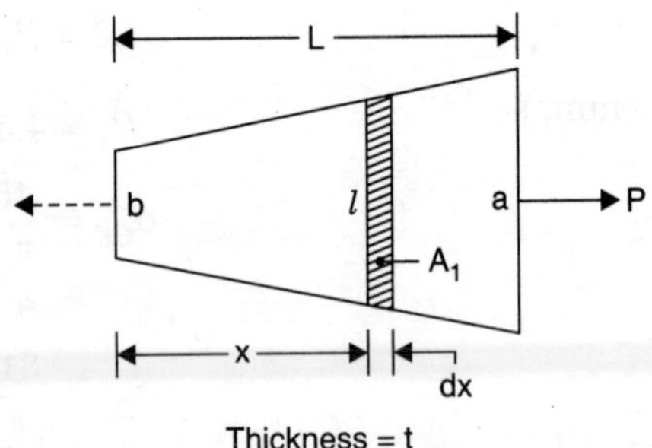

Figure 1.11

Solution. $l = a + (a-b)\,\dfrac{x}{L}$

Extn. of length (dx) $= \dfrac{P \cdot dx}{A_1\, E}$

$$= \frac{P \cdot dx}{\left[a + (a-b)\dfrac{x}{L}\right] t \cdot E}$$

and for the whole bar

$$\Delta L = \int_0^L \frac{P \cdot dx}{\left[a + (a-b)\dfrac{x}{L}\right] t \cdot E}, \qquad \text{Put } a + (a-b)\,\frac{x}{L} = u$$

Then, $\dfrac{du}{dx} = \dfrac{a-b}{L}$, $\quad dx = \dfrac{L}{a-b}\,du$

$$\Delta L = \frac{P}{tE} \cdot \frac{L}{a-b} \int_{x=0}^{L} \frac{du}{u}$$

$$\therefore \qquad \Delta L = \frac{P}{tE} \frac{L}{a-b} \ln \frac{a}{b}$$

Note. For a steel plate 8 mm thick, taking $a = 80$ mm, $b = 40$ mm, $L = 0.5$ m

$P = 30$ kN, $E = 200$ GN/m^2

$$\Delta L = \frac{30 \times 10^3}{0.008\,(200 \times 10^9)} \cdot \frac{0.5}{(0.08 - 0.04)} \ln \frac{80}{40} = 0.162 \times 10^{-3} \text{ m}$$

$$\Rightarrow \qquad \Delta L = \mathbf{0.162 \text{ mm}}$$

Example 1.7. *A bar 5 m long is made up of two materials. The first 1.7 m of its length is of brass and 750 mm² in cross-section and the remaining of its length is of steel and 600 mm² in cross-section. The bar is subjected to axial tension and the total elongation of bar is 1.2 mm. Determine the load and the work done in elongating the bar E-steel = 2.1 × 10⁵ N/mm², E-Brass = 8.4 × 10⁴ N/mm².*

Solution. (*i*) Total elongation (δ) = $\sum \frac{PL}{AE}$, where P = axial load

$$1.2 = \frac{P(1.7 \times 1000)}{750\,(8.4 \times 10^4)} + \frac{P(3.3 \times 1000)}{600\,(2.1 \times 10^5)} \quad \Rightarrow \quad P = \mathbf{22.6 \text{ kN}}$$

(*ii*) Workdone or strain energy (S.E.) = 2 E × Volume

$$\text{S.E.} = \frac{\{(22.6 \times 10^3)/750)\}^2}{2\,(8.4 \times 10^4)} \times (750 \times 1700) + \frac{\{(22.6 \times 10^3)/600\}^2}{2(2.1 \times 10^5)} \times (600 \times 3300)$$

= **13,579.73 N·mm** or **13.6 N·m**.

Example 1.8. *A M.S. bar 0.2 m long is 10 mm square over part of its length. The remainder is circular and alone has a volume of 24 × 10⁻⁶ m³. The bar is subjected to a pull of 20 kN. Find the dimensions of the circular portion so that the total strain energy is minimum and find its value if E = 200 GN/m².*

Solution. If the length of the circular portion is x mm, the remaining length is $(200 - x)$ mm, see Figure 1.12.

$$\text{S.E.} = \sum \frac{\sigma^2}{2E} \times \text{Volume}$$

$$\text{S.E.} = \frac{1}{2E}\left[\left(\frac{20 \times 10^3}{10 \times 10}\right)^2 \times 10^2\,(200 - x) + \left(\frac{20 \times 10^3}{(24 \times 10^3)/x}\right)^2 24 \times 10^3\right]$$

For S.E. to be minimum, $d\,\frac{(\text{S.E.})}{dx} = 0$

$$\frac{1}{10^2}(-1) + \frac{2x}{24 \times 10^3} = 0, \quad x = \mathbf{120 \text{ mm}}$$

$$\text{Min. S.E.} = \frac{(20 \times 10^3)^2}{2(2 \times 10^5)}\left[\frac{200 - 120}{10^2} + \frac{120^2}{24 \times 10^3}\right]$$

= **1400 N·mm** or **1.4 N·m**

To find the radius of the circular portion,

$$\pi r^2 \times 120 = 24 \times 10^3 \quad \text{mm}^3$$

$r = 7.9788$ mm, say 8 mm or **16 mm dia.**

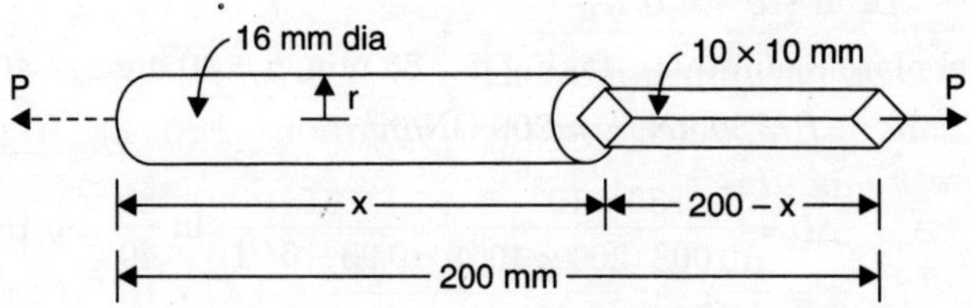

Figure 1.12. M.S. bar

Volumetric Strain. Consider a prism *xyz* subjected to stress in the principal directions, *Figure 1.11.*

$$\varepsilon_x = \frac{\sigma_1}{E} - \nu\left(\frac{\sigma_2}{E} + \frac{\sigma_3}{E}\right) \quad ...(i)$$

$$\varepsilon_y = \frac{\sigma_2}{E} - \nu\left(\frac{\sigma_1}{E} + \frac{\sigma_3}{E}\right) \quad ...(ii)$$

$$\varepsilon_z = \frac{\sigma_3}{E} - \left(\frac{\sigma_1}{E} + \frac{\sigma_2}{E}\right) \quad ...(iii)$$

$$\delta_x = x\varepsilon_x, \quad \delta_y = x\,\varepsilon_y, \quad \delta_z = x\varepsilon_z$$

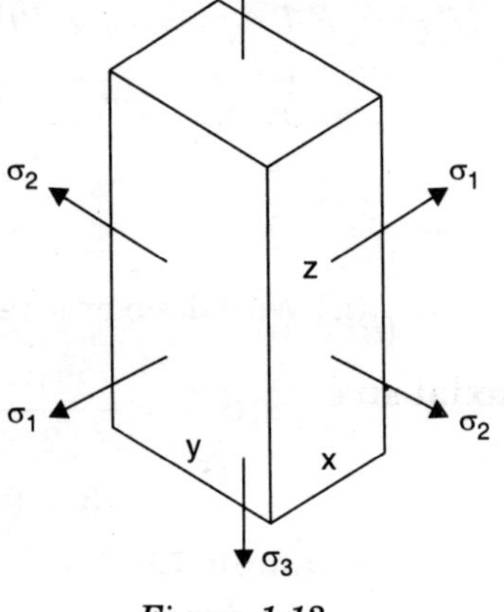

Figure 1.13

$$\text{Volumetric strain } (\varepsilon_v) = \frac{\Delta V}{V} = \frac{\text{Change in volume}}{\text{Original volume}}$$

$$\varepsilon_v = \frac{x(1+\varepsilon_x)\cdot y(1+\varepsilon_y)\cdot z(1+\varepsilon_z) - xyz}{xyz}$$

$$= (1+\varepsilon_x)(1+\varepsilon_y)(1+\varepsilon_z) - 1$$

$$= 1 + \varepsilon_x + \varepsilon_y + \varepsilon_z + \varepsilon_x\varepsilon_y + \varepsilon_y\varepsilon_z + \varepsilon_z\varepsilon_x - \varepsilon_x\varepsilon_y\varepsilon_z - 1$$

Neglecting multiples of ε, $\varepsilon_v = \varepsilon_x + \varepsilon_y + \varepsilon_z$

Volumetric strain = Algebric sum of the three principal strains

Substituting for ε_x, ε_y and ε_z from (*i*), (*ii*) and (*iii*),

$$\text{Volumetric strain } \varepsilon_v = (\sigma_1 + \sigma_2 + \sigma_3)\,\frac{(1-2\nu)}{E}$$

If $\sigma_1 = \sigma_2 = \sigma_3 = \sigma$, as in fluid pressure ($\sigma = p = \gamma_w h$),

$$\varepsilon_v = 3\sigma\cdot\frac{(1-2\nu)}{E} = \textbf{3 times principal strains}$$

Then the Bulk modulus *K*, is defined as the ratio of the triaxial stress σ, or fluid pressure *p*, to the volumetric strain ε_v.

$$K = \frac{\sigma}{\Delta V/V} = \frac{\sigma}{3(1-2\nu)\,\sigma/E} = \frac{E}{3(1-2\nu)}$$

$$\therefore \quad \boldsymbol{E = 3K\,(1-2\nu)},$$

the relation between E & K; also see page 19, 164, 225

Note: $E = K$, when $\nu = \frac{1}{3}$, say $\frac{1}{m}$, $m = 3$

Strain energy per unit volume $= \frac{\sigma^2}{\mathbf{2K}}$

Example 1.9. *A 30 mm aluminium rod 3 m long is subjected to an axial pull of 100 kN. Taking E = 70 GN/m² and ν = 1/3, determine the elongation, change in diameter and volume of the rod. Estimate the bulk modulus.*

Solution. (*i*) $\delta L = \varepsilon L$, $\varepsilon = \frac{\sigma}{E}$, $\sigma = \frac{P}{A} = \frac{100 \times 10^3}{\frac{\pi}{4} \times 0.03^2} = 1.41 \times 10^8 \text{ N/m}^2$

$$\varepsilon = \frac{1.41 \times 10^8}{70 \times 10^9} = 0.00201 = 2.01 \times 10^{-3}$$

$$\delta L = \varepsilon L = 2.01 \times 10^{-3} \times 3 = 6.03 \times 10^{-3} \text{ or } 6 \text{ mm}$$

(*ii*) Poisson's ratio $= \frac{\varepsilon_{lateral}}{\varepsilon_{axial}}$

$$\varepsilon_{lat} = -(2.01 \times 10^{-3})\ 1/3 = -6.71 \times 10^{-4}$$

$$\delta D = \varepsilon D = (-6.71 \times 10^{-4}) \times 30 = -\mathbf{0.02\ mm}$$

(*iii*) Volumetric strain $(\varepsilon_v) = \frac{\Delta V}{V} = \frac{\sigma(1-2\nu)}{E} = \frac{\sigma}{3K}$, since the rod is subjected to only **axial stress (σ).**

$$\Delta V = V\varepsilon\,(1 - 2\nu), \quad \varepsilon = \frac{\sigma}{E} = \text{axial strain} = 2.01 \times 10^{-3}$$

$$= \left(\frac{\pi}{4} \times 0.03^2 \times 3\right) 2.01 \times 10^{-3} \times \left(1 - 2 \times \frac{1}{3}\right)$$

$$= 1.42 \times 10^{-6} \text{ m}^3 \text{ or } \mathbf{1420\ mm^3}$$

(*iv*) $K = \frac{E}{3(1-2\nu)} = \frac{70 \times 10^9}{3\left(1 - \frac{2}{3}\right)} = \mathbf{70\ GN/m^2}$

Note. $E = K$ when, $\nu = \frac{1}{3}$ (or $m = 3$).

Example 1.10. *A 10 mm steel rod 1 m long is inserted inside an aluminium tube 20 mm external dia., 2 mm thick and of the same length, and rigidly connected at their ends such that they are coaxial.*

$$E_{St} = 2 \times 10^5 \text{ N/mm}^2 = 3E_{Al}\,;\ \alpha_{St} = 11 \times 10^{-6}/°\text{C},\ \alpha_{Al} = 2\alpha_{St}$$

Calculate the stresses in steel and aluminium and change in their length when they are subjected to

(*a*) *axial load of 20 kN*

(*b*) *a temperature increases by 50°C.*

Solution. Case (*a*). Compound bar subjected to axial load *P*, see Figure 1.14.

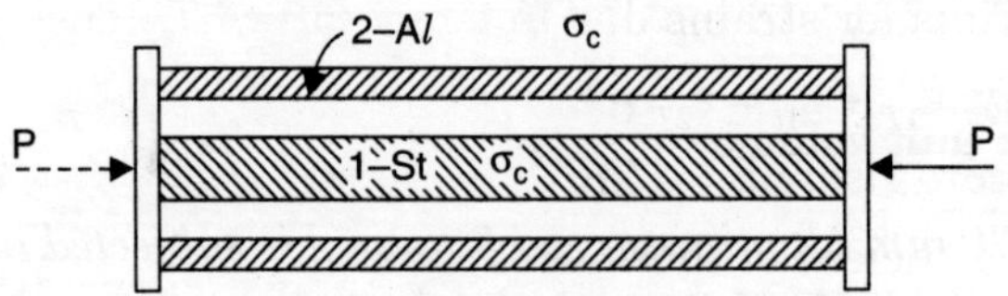

Figure 1.14

Strain is same:

$$\frac{\sigma_1}{E_1} = \frac{\sigma_2}{E_2} \quad \text{...(i)}$$

Load *P*, is shared by *St* and *Al*:

$$\sigma_1 A_1 + \sigma_2 A_2 = P \quad \text{...(ii)}$$

σ_1 and σ_2 are **similar stresses** proportional to the moduli of elasticity; solve (*i*) and (*ii*) for σ_1 and σ_2

From (*i*), $\sigma_1 = \sigma_2 \dfrac{E_1}{E_2} = 3\sigma_2, \because \; E_{St} = 3E_{Al}$

From (*ii*), $3\sigma_2 \left(\dfrac{\pi}{4} \times 10^2\right) + \sigma_2 \times \dfrac{\pi}{4}\,(20^2 - 16^2) = 20 \times 10^3$

$\therefore \quad$ $\sigma_2 = \mathbf{57.35\ N/mm^2}$, $\sigma_1 = 57.35 \times 3 = \mathbf{172\ N/mm^2}$

(*iii*) $\Delta L = \varepsilon L = \dfrac{\sigma_1}{E_1} L = \dfrac{172}{2 \times 10^5} \times 1000 = \mathbf{0.86\ mm}$

Case (*b*). Compound bar subjected to temperature increase $(\Delta T) = 50°C$, see Figure 1.15.

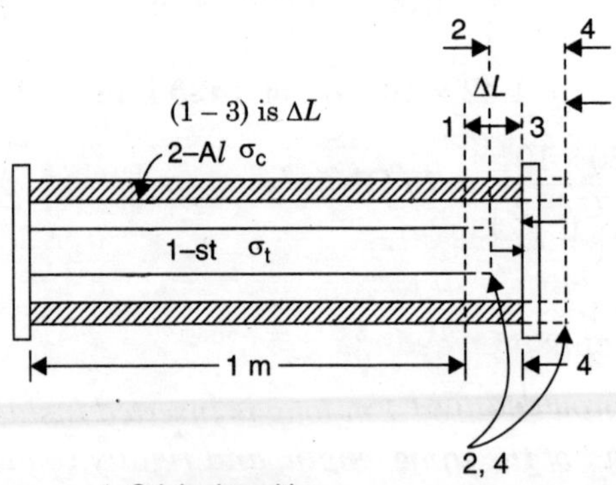

Figure 1.15

If the materials were free to expand, they would have reached positions (2) & (4) since $\alpha_{Al} = 2\,\alpha_{St}$. Since, they are rigidly connected at ends, their final position is (3), i.e., steel rod is under tension and *Al*-tube under compression.

For equilibrium pull = Push: $\sigma_1 A_1 = \sigma_2 A_2$ ...(*i*)

Denoting subscripts for strains due to temperature T, tension t, compression c,

For compatibility: $\varepsilon_{1T} + \varepsilon_{1t} = \varepsilon_{2T} - \varepsilon_{2c}$ ...(*ii*)

Due to temperature rise, $\Delta L = L\,\alpha\,\Delta T$

where α is the coefficient of thermal expansion.

$$\text{Temperature strain } (\varepsilon_T) = \frac{\Delta L}{L} = \alpha\,\Delta T$$

From (*ii*), $\varepsilon_{1t} + \varepsilon_{2c} = \varepsilon_{2T} - \varepsilon_{1T}$

$$\frac{\sigma_1}{E_1} + \frac{\sigma_2}{E_2} = (\alpha_2 - \alpha_1)\,\Delta T \quad ...(iii)$$

i.e., Sum of strains due to pull and push = Diff. in temperature strains

Solve for σ_1 and σ_2 from (*i*) and (*iii*); σ_1 and σ_2 are **dissimilar stresses** inversely proportional to the areas.

$$\text{From } (i),\ \sigma_1 = \sigma_2 \frac{A_2}{A_1} = \sigma_2 \times \frac{20^2 - 16^2}{10^2} = 1.44\,\sigma_2$$

$$\text{From } (iii),\ \frac{1.44\,\sigma_2}{E_1} + \frac{\sigma_2}{E_1/3} = (2\alpha_1 - \alpha_1)\,50$$

$$\sigma_2 \left(\frac{1.44}{2\times 10^5} + \frac{3}{2\times 10^5}\right) = (11 \times 10^{-6})\,50$$

$\therefore$ σ_2 = **24.77 N/mm²** (compn.), $\sigma_1 = 24.77 \times 1.44$ = **35.67 N/mm²** (tension)

Resulting expansion $\Delta L = (\varepsilon_{1T} + \varepsilon_{1t})L$

$$= \left(\alpha_1\,\Delta T + \frac{\sigma_1}{E_1}\right) L$$

$$= \left(11\times 10^{-6} \times 50 + \frac{35.67}{2\times 10^5}\right) \times 1000 = \mathbf{0.7283\ mm}$$

Example 1.11. *A steel rail is 30 m long and is at a temperature of 24°C. Calculate the thermal stresses developed in the rail when the temperature rises to 44°C if*

(*i*) *no expansion joint is provided*

(*ii*) *a 6 mm gap is provided for expansion*

$$E_{St} = 2 \times 10^5\ \text{N/mm}^2,\ \alpha_{St} = 11 \times 10^{-6}/°\text{C}.$$

Solution. If the rail were free to expand, AC is the final length of rail,

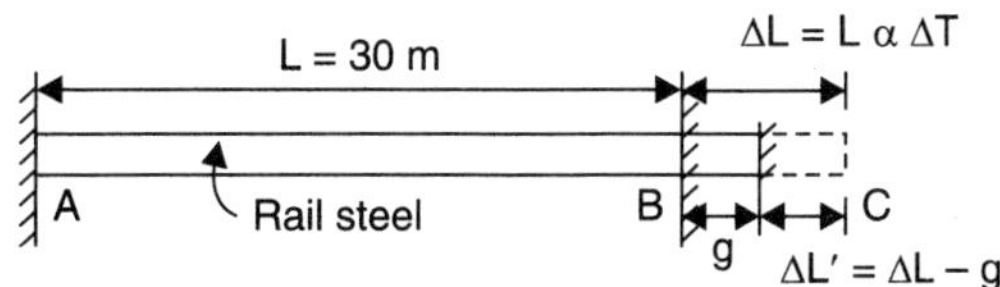

Figure 1.16

In that case, elongation due to temperature rise

$$\Delta L = L\,\alpha\,\Delta T\,,\quad \Delta L = 30 \times 10^3\,(11 \times 10^{-6})\,(44 - 24) = 6.6 \text{ mm}$$

(*i*) If no expansion gap is provided (end supports are rigid), 6.6 mm of expansion is prevented, resulting in thermal stresses in the rail (σ_T)

$$\sigma_T = \varepsilon_T E = \frac{6.6}{30{,}000}\,(2 \times 10^5) = \mathbf{44\ N/mm^2} \text{ (compn.)}$$

(*ii*) If a 6 mm expansion gap *g*, is provided or in some other case, the rigid end supports yield by amount *g*, the free expansion prevented becomes only

$$\Delta L' = \Delta L - g = 6.6 - 6 = 0.6 \text{ mm}$$

Thermal stress due to prevention of 0.6 mm expansion

$$\sigma_T = \varepsilon_T E = \frac{0.6}{30{,}000} \times (2 \times 10^5) = \mathbf{4\ N/mm^2}$$

Example 1.12. *A composite shaft is fixed between two rigid supports as shown in Figure 1.17. If the temperature is raised by 100°C, determine:*

(*i*) *the force P exerted on the supports*

(*ii*) *the relative moment of the junction B.*

Take E-Brass = 85 kN/mm², E-steel = 210 kN/mm²

$$\alpha_{Br} = 2 \times 10^{-5}/°C,\ \alpha_{St} = 1.1 \times 10^{-5}/°C.$$

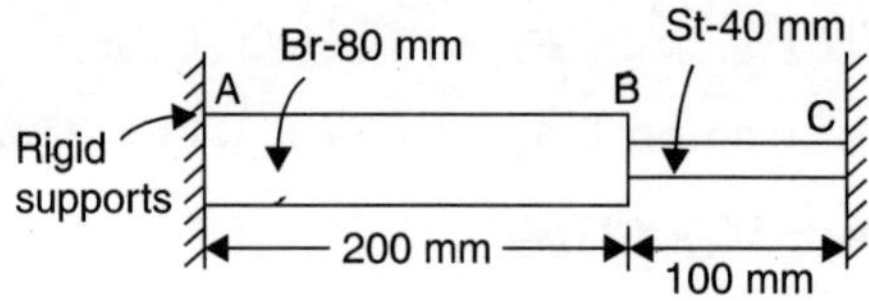

Figure 1.17

Solution. (*i*) $\sum$ ΔL due to temperature = $\sum$ ΔL due to force *P*, since ΔL is prevented due to rigid supports

$$\sum L\,\alpha\,\Delta T = \sum \frac{PL}{AE}$$

$$\{200\,(2 \times 10^{-5}) \times 100° + 100\,(1.1 \times 10^{-5}) \times 100°\} = P\left\{\frac{200}{\frac{\pi}{4}(80^2) \times 85} + \frac{100}{\frac{\pi}{4}(40^2) \times 210}\right\}$$

$$0.51 \text{ mm} = 8.47 \times 10^{-4} P$$

$$\Rightarrow \qquad \mathbf{P = 602.1\ kN}$$

(*ii*) Actual ΔL of $AB = \Delta L_T - \Delta L_P = L\,\alpha\,\Delta T - \dfrac{PL}{AE}$, negative sign to compn. *P*.

$$= 200 \times (2 \times 10^{-5}) \times 100 - \frac{602.1 \times 200}{\frac{\pi}{4}(80^2)\,85}$$

$$= 0.400 - 0.282 = 0.118$$

$$\simeq 0.12 \text{ mm extn.} = \text{contraction of } CB.$$

Example 1.13. *A thin tyre is to be shrunk on to a wheel of 1.2 m dia. Assuming the wheel to be rigid what should be the inside dia. of the tyre, if after shrinking the hoop stress, should not exceed 80 N/mm². What is the least temperature to which the tyre must be heated above that of the wheel before it could be slipped on ?*

Take α-tyre = 1 × 10^{-5}/°C, E-tyre = 2 × 10^5 N/mm²

Solution. (*i*) Suppose I.D. of tyre = d and after *heating to T°*, final dia (D) = 1.2 m

$$\text{Strain } (\varepsilon) = \frac{\sigma}{E} = \frac{\pi D - \pi d}{\pi d} = \frac{D-d}{d} = \frac{D}{d} - 1 \qquad \text{...from}(i)$$

$$\frac{80}{2 \times 10^5} + 1 = \frac{D}{d}, \quad \frac{D}{d} = 1.0004$$

$$d = \frac{1.2}{1.0004} = \mathbf{1.19952\ m}$$

(*ii*) $\varepsilon_T = \alpha\,\Delta T = 1.0004 - 1 = 0.0004$ ...from(*i*)

$$\therefore \qquad T = \frac{0.0004}{1 \times 10^{-5}} = \mathbf{40°C}$$

Shear. When two equal and opposite parallel forces act so as to make one part of the body to slide over the other at a section are called shear forces F. The shear stress acts parallel or tangential to the shear plane or section of shear failure. Though the shear stress distribution varies over a cross-section, an average shear stress τ, is taken in design.

$$\text{Shear stress } (\tau) = \frac{\text{Shear force } (F)}{\text{Area of } c/s\ (A)} \quad \text{N/mm}^2$$

On a rectangular element in a stressed material, a balancing couple of forces exists equal and opposite to the couple of forces causing shear, see Figure 1.18, and the shear stresses are called **complimentary shear**. Pure (direct) shear occurs commonly in riveted and cottered joints.

The shear deformation is measured by the angle of slide (or rotation) from the rectangular position due to shear force F. In Figure 1.18, the rectangular element $ABCD$ has been distorted to $A'B'CD$.

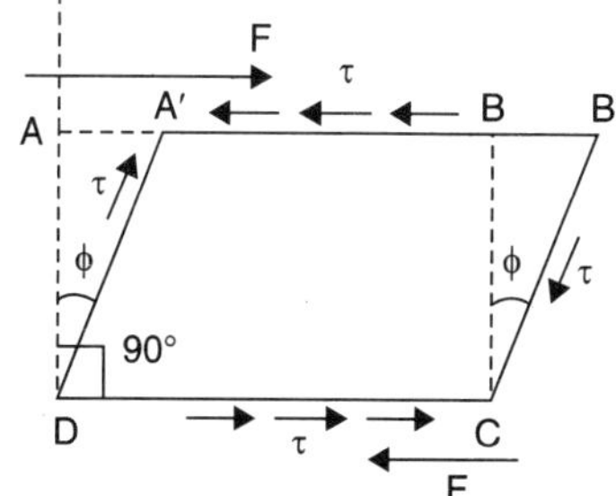

Figure 1.18

$$\text{Shear strain } (\gamma) = \tan \phi = \frac{AA'}{AD} = \phi \text{ radian}$$

Since, the angle ϕ is very small.

The shear stresses τ, acting on the pairs of opposite sides form complimentary shear.

In elastic materials, the shear strain is proportional to the shear stress producing it and the ratio of shear stress to shear strain is called rigidity modulus G.

$$G = \frac{\text{Shear stress } (\tau)}{\text{Shear strain } (\gamma)} \quad \text{N/mm}^2$$

$$\text{Strain energy S.E. due to shear, S.E.} = \frac{\tau}{2G} \times \text{Volume}$$

The relation among the three moduli of elasticity

$$E = \frac{\sigma}{\varepsilon},\ G = \frac{\tau}{\gamma} \text{ and } K = \frac{\sigma}{\Delta V/V} \text{ is given by}$$

$$\mathbf{E = 2G\,(1 + \nu) = 3K\,(1 - 2\nu)}, \text{ see page 158 \& 225.}$$

Example 1.14. *A 50 mm shaft is keyed to a pulley by means of a key, of 10 mm × 10 mm × 70 mm long, see Figure 1.19. If the torque is 1 kN·m, what is the shear stress in the key ?*

Solution. See Figure 1.19

Torque $(T) = F \cdot r$

where F is the shear force on the key 40 kN.

$$F = \frac{T}{r} = \frac{1000 \times 10^3}{25} \text{ N·mm/mm}$$

$F = \tau A$, where A is the area of shear of key and τ is the shear stress in the key.

$$\tau = \frac{F}{A} = \frac{40 \times 10^3}{10 \times 75} = \mathbf{53.5\ N/mm^2}$$

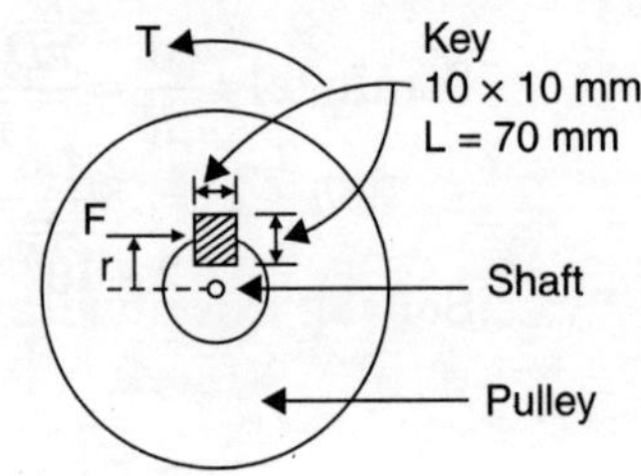

Figure 1.19

Example 1.15. *During a tensile test on a M.S. rod of gauge length 400 mm and dia. 20 mm, the elongation under a load of 32 kN was measured as 0.2 mm, and the lateral contraction 0.003 mm. If the load was well within elastic limit, compute the three moduli of elasticity.*

Solution. Poisson's ratio $= \frac{\varepsilon_{lateral}}{\varepsilon_{axial}} = \frac{0.003/20}{0.2/400} = 0.3$

$$E = \frac{\sigma}{\varepsilon} = \frac{(32 \times 10^3)/\frac{\pi}{4}(20^2)}{0.2/400} = \mathbf{203{,}718\ N/mm^2}$$

$$E = 2G\,(1 + \nu) = 3K\,(1 - 2\nu)$$

$$G = \frac{E}{2(1+\nu)} = \frac{203{,}718}{2(1+0.3)} = \mathbf{78{,}353\ N/mm^2}$$

$$K = \frac{E}{3(1-2\nu)} = \frac{203{,}718}{3(1-2\times 0.3)} = \mathbf{169{,}765\ N/mm^2.}$$

Example 1.16. *A steel punch can be worked to a compressive stress of 700 N/mm². What is the least diameter of the hole that can be punched through a steel plate 8 mm thick, if its ultimate shear strength is 300 N/mm².*

Solution. See Figure 1.20.

Compressive punching force = Force shearing the plate

$$\frac{\pi d^2}{4} \cdot \sigma_c = (\pi\, d \cdot t)\,\tau$$

$$\therefore \quad d = \frac{4t \cdot \tau}{\sigma_c} = 4 \times 8 \times \frac{300}{700} = \mathbf{13.7\ mm}$$

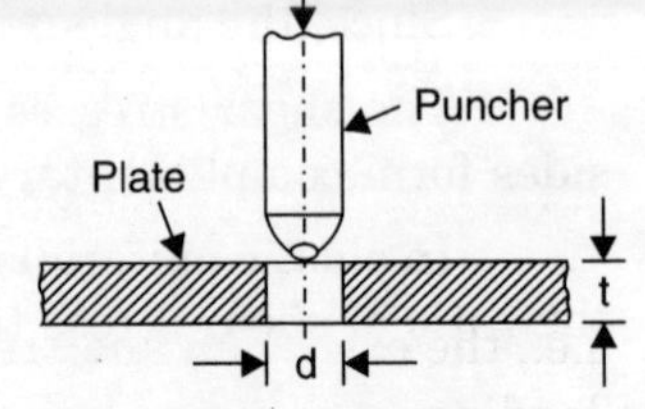

Figure 1.20

Dynamic (Impact) Loading. In Figure 1.21, a sliding weight W, drops on the collar securely fixed at the bottom of the vertically hanging bar and due to impact, the bar extends by δ. From the principle of conservation of energy,

Loss of P.E. = Gain of S.E. of bar

$$W(h + \delta) = \frac{\sigma^2}{2E} \times \text{Volume}$$

$$= \frac{\left(\frac{\delta}{L} E\right)^2}{2E} \times AL$$

$$W(h + \delta) = \frac{EA\,\delta^2}{2L}$$

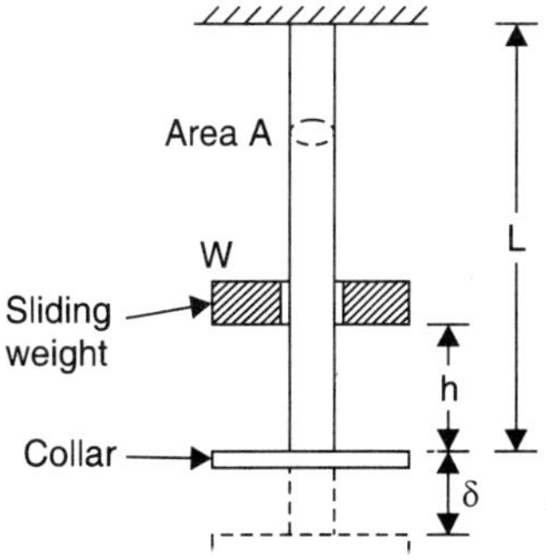

Figure 1.21

Solving this quadratic in δ and taking the positive root.

$$\delta = \frac{WL}{EA} + \left[\left(\frac{WL}{EA}\right)^2 + \frac{2\,WLh}{EA}\right]^{1/2} \qquad ...(i)$$

Denoting $\delta_{St} = \dfrac{WL}{EA}$ (due to static load)

$$\delta = \delta_{St} + [\delta_{St}^2 + 2h\ \delta_{St}]^{1/2}$$

Since $\delta_{St} << h, \quad \delta \approx \sqrt{2h\delta_{St}}, \quad \dfrac{\delta}{\delta_{St}} = \sqrt{2h/\delta_{St}}$

From (i), $\sigma_{max} = \varepsilon E = \dfrac{\delta}{L} E = \dfrac{W}{A} + \left[\left(\dfrac{W}{A}\right)^2 + \dfrac{2WhE}{AL}\right]^{1/2}$

$$\sigma_{max} = \frac{W}{A}\left(1 + \sqrt{1 + \frac{2h}{WL/AE}}\right)$$

$$\sigma_{max} = \frac{W}{A}\left(1 + \sqrt{1 + \frac{2h}{\delta_{St}}}\right) \qquad \left[\because\ \delta_{St} = \frac{WL}{EA}\right]$$

$$P_{max} = \sigma_{max} A = W\left(1 + \sqrt{1 + \frac{2h}{\delta_{St}}}\right)$$

Denoting $\sigma_{St} = \dfrac{W}{A}$, $\quad \sigma_{max} = \sigma_{St}\left(1 + \sqrt{1 + \dfrac{2h}{\delta_{St}}}\right) \qquad ...(ii)$

If $h \to 0$, i.e., for a suddenly applied load

$$\delta = 2\delta_{St}, \quad \sigma_{max} = 2\sigma_{St} \quad \text{i.e.,} \quad \frac{\delta}{\delta_{St}} = \frac{\sigma}{\sigma_S} = 2 \qquad ...(iii)$$

i.e., the extension and stress due to a suddenly applied load are twice of those caused by static loading.

Further, $\delta = \sqrt{2h\delta_{St}}$, where $\delta_{St} = \dfrac{WL}{AE}$ and $W = Mg$

$$\delta = \sqrt{\frac{2L}{AE} \cdot Mgh}$$

The shock stress caused by sudden stopping of mass M, that is moving with a velocity v, when it strikes the collar, replacing the P.E. by the K.E.,

$$Mgh = \frac{1}{2} Mv^2, \quad v = \sqrt{2gh}$$

$$\sigma = \frac{\delta}{L} E = \sqrt{\frac{2L}{AE} \cdot Mgh} \times \frac{E}{L} = \sqrt{\frac{2E}{AL} \frac{Mv^2}{2}}$$

$$\sigma = \sqrt{\frac{EW}{ALg} v^2} \qquad ...(iv)$$

Note. $$\sigma = \sqrt{\frac{2hE}{L} \cdot \sigma_{St}} \qquad ...(v)$$

The ratio of maximum instantaneous deflection to the static deflection is called the **'impact factor I_f'**

$$I_f = \frac{\delta}{\delta_{St}} = \frac{\sigma}{\sigma_{St}} = 1 + \sqrt{1 + \frac{2h}{\delta_{St}}} \approx \sqrt{\frac{2h}{\delta_{St}}} \qquad ...(vi)$$

Work done due to the suddenly applied load $(W) = W \cdot \delta$ ...(vii)

Example 1.17. *A 15 mm thick rod 2 m long hangs vertically with a collar securely fixed at its bottom. A sliding weight of 200 N is dropped from a height of 60 mm on the collar. Determine the elongation and the stress developed in the rod due to impact.*
[Take E = 200 kN/mm².]

Solution.
$$\delta_{St} = \frac{WL}{EA} = \frac{200 \times 2000}{(200 \times 10^3) \times \frac{\pi}{4} \times 15^2} = 0.0113 \text{ mm}$$

$$\delta \approx \sqrt{2h\delta_{St}} = \sqrt{2 \times 60 \times 0.0113} = 1.165 \text{ mm}$$

$$\sigma_{max} = \frac{W}{A}\left(1 + \sqrt{1 + \frac{2h}{\delta_{St}}}\right)$$

$$= \frac{200}{\frac{\pi}{4} \times 15^2}\left(1 + \sqrt{1 + \frac{2 \times 60}{0.0113}}\right) = \mathbf{117.76 \text{ N/mm}^2}$$

Alternatively,
$$\sigma = \varepsilon E = \frac{\delta}{L} E = \frac{1.165}{2000} \times (200 \times 10^3) = 116.5 \text{ N/mm}^2$$

Note. Impact factor
$$I_f = \frac{\delta}{\delta_{St}} = \frac{1.165}{0.0113} = \mathbf{103}$$

$$\sigma_{St} = \frac{W}{A} = \frac{200}{\frac{\pi}{4} \times 15^2} = 1.131 \text{ N/mm}^2$$

$$\frac{\sigma}{\sigma_{St}} = \frac{116.5}{1.131} = \mathbf{103}$$

which is the same as the corresponding ratio of deflections.

Example 1.18. *A bar of certain metal 40 mm dia and 1.2 m long has a collar securely fixed at the bottom. A load of 20 kN is gradually applied to the collar causing extension of the bar of 0.2 mm. Find the height from which the load can be dropped on to the collar so that the maximum stress in the bar will not exceed 120 N/mm².*

Solution.

$$\sigma_{max} = \frac{W}{A}\left(1 + \sqrt{1 + \frac{2h}{\delta_{St}}}\right)$$

$$120 = \frac{20 \times 10^3}{\frac{\pi}{4} \times 10^2}\left(1 + \sqrt{1 + \frac{2h}{0.2}}\right)$$

$$\Rightarrow \qquad h = \mathbf{4.176 \text{ mm}}$$

Note :

$$\sigma_{St} = \frac{W}{A} = 15.915 \text{ N/mm}^2$$

$$\frac{\sigma}{\sigma_{St}} = \frac{120}{15.915} = \mathbf{7.54}$$

$$\delta \approx \sqrt{2h\delta_{St}} = \sqrt{2 \times 4.176 \times 0.2} = 1.292 \text{ mm}$$

$$I_f = \frac{\delta}{\delta_{St}} = \frac{1.292}{0.2} = \mathbf{6.46}$$

Also see **Example 6.12** on page 242.

Example 1.19. *A cable supported mine elevator travels downwards at a speed of 6 km/hr with a 20 kN load including its own weight. When it is 300 m above the ground level, the gear mechanism at the surface jams. If the stress in the cable should not exceed 200 N/mm², what should be its cross-sectional area ? [Take, E = 2 × 10⁵ N/mm².]*

Solution. From Eqn. (*iv*), $\sigma = \sqrt{\frac{EW}{AL\,g}v^2}$

$$200^2 = \frac{(2 \times 10^5)(20 \times 10^3)}{A(300 \times 1000)(9.81 \times 1000)} \times \left(\frac{6 \times 10^6}{60 \times 60}\right)^2$$

$$A = \mathbf{94.4 \text{ mm}^2}$$

Stress Concentration. When an abrupt change in geometry or cross-section occurs in a member (component) subjected to tension, compression or bending, or due to some discontinuity such as a notch, hole, keyway, groove, edge fillet, etc., the stress will no longer be uniform and the maximum stress occurs at the edge of the hole, fillet or groove. The maximum stress may be several times greater than the average stress over the net cross-sectional area at the section of discontinuity or hole. The stress distribution is uniform over sections at a considerable distance away from the region of discontinuity. The abrupt increase in stress in regions of discontinuity is called **stress concentration**, and the ratio of the maximum stress to the average stress over the cross-sectional area in the region of discontinuity is called the **Stress concentration factor**,

$$k = \frac{\sigma_{max}}{\sigma_{ave}}$$

and if K is known, $\sigma_{max} = k \cdot \sigma_{ave} = k \cdot \frac{P}{A_{net}}$.

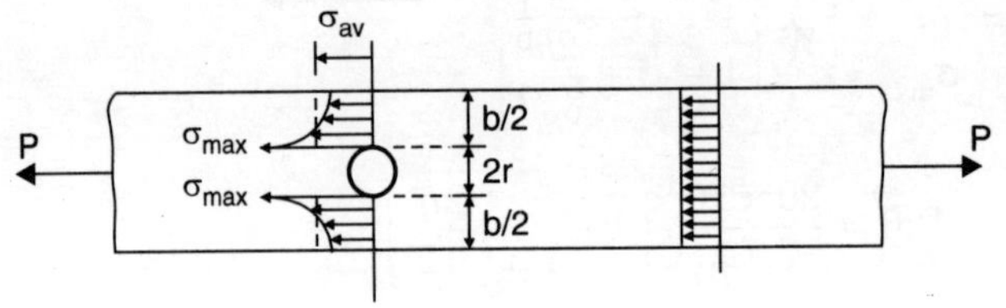

(a) Circular hole

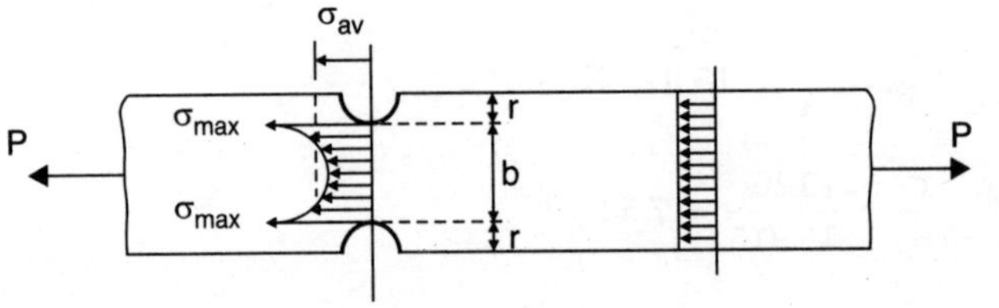

(b) Semi-circular notches

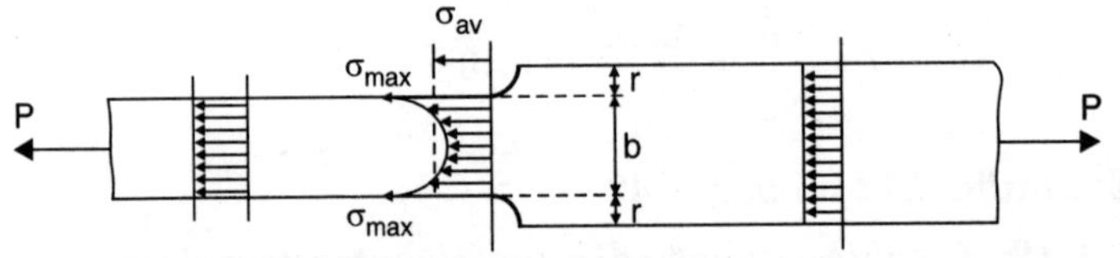

(c) Quarter-circular fillets

Figure 1.22 Stress concentration

The variation of k with the ratio $\frac{r}{b}$ is shown in Figure 1.23 and it can be seen that a smaller ratio of $\frac{r}{b}$ gives a larger value of k, and a higher stress concentration. Hence, the radius of holes notches and fillets should be reasonably large to avoid high stress concentrations.

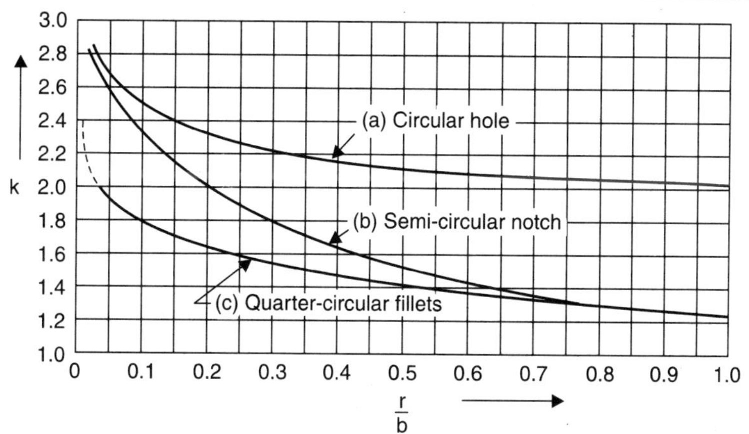

Figure 1.23. Stress concentration factor

In the case of ductile materials (such as M.S.) yielding commences at the point of maximum stress σ_y, and as the load is increased yielding occurs at other points while the maximum stress remains at σ_y, thereby distributing the load more evenly over the net cross-section.

In brittle materials, due to the lack of yield point, the increase in load causes a continuous increase in maximum stress till the material cracks at the point of maximum stress. Therefore, for brittle materials the stress concentration is an important factor, while it is not for ductile materials subjected to static load.

Machine components subjected to fatigue (cyclic stress variations or repetitive reversals of stress) progressive cracks are likely to develop from the points of stress concentration for both ductile and brittle materials.

Also see page 80, Figs. 14.6, 14.7 & 14.8.

Example 1.20. *Determine the maximum load that can be applied to a 10 mm thick plate shown in Figure 1.24, so that the stress at any point does not exceed 100 N/mm².*

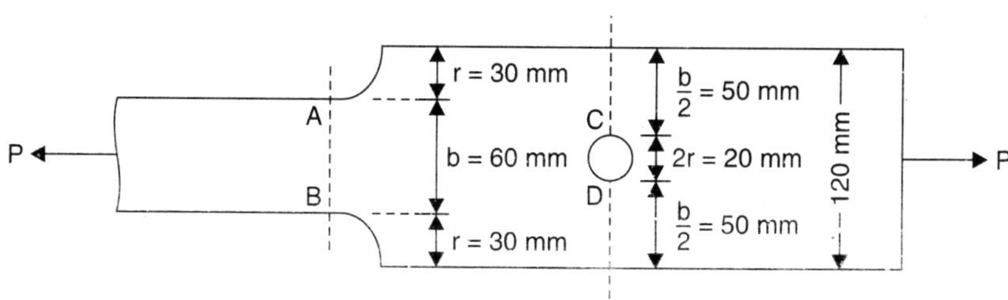

Figure 1.24

Solution. (*i*) Consider stress concentration at the fillets:

From Figure 1.23, for $\frac{r}{b} = \frac{30 \text{ mm}}{60 \text{ mm}} = 0.5, k = 1.42$

At A and B, $\sigma_{max} = k\,\sigma_{av} = k.\,\frac{P}{bt} = 1.42 \times \frac{P}{60 \times 10} \leq 100 \text{ N/mm}^2$

$P \leq 4.23 \times 10^4$ N, or **42.3 kN** ...(*i*)

(*ii*) Consider stress concentration at the circular hole:

From Figure 1.23, for $\frac{r}{b} = \frac{10 \text{ mm}}{100 \text{ mm}} = 0.1, k = 2.5$

At C and D, $\sigma_{max} = k\,\sigma_{av} = k \cdot \frac{P}{bt} = 2.5 \times \frac{P}{100 \times 10} \leq 100 \text{ N/mm}^2$

$P \leq 4 \times 10^4$ N, or **40 kN** ...(*ii*)

From (*i*) and (*ii*), the maximum load $P \not> $ **40 kN.**

PROBLEMS

1.1. A M.S. specimen was tested in tension and the following results were obtained:

Dia of specimen	=	20 mm
Gauge length	=	200 mm
Extn. under a load of 10 kN	=	0.032 mm
Load at yield point	=	82 kN
Maximum load	=	133 kN
Gauge length after fracture	=	2.52 mm
Dia of neck	=	12.6 mm

Determine (*i*) Young's modulus of elasticity, (*ii*) Yield stress, (*iii*) Ultimate stress, (*iv*) % elongation (*v*) Percentage reduction in area, and (*vi*) Working stress if the factor of safety is 2.

1.2. A circular rod 250 mm long and 60 mm in dia was subjected to an axial pull of 300 kN. The increase in length 0.126 mm and decrease in dia was 0.012 mm. Calculate the values of ν, E, G and K for the material.

1.3. For a given material, $E = 110$ kN/mm^2 and $G = 43$ kN/mm^2. Find the bulk modulus and the lateral contraction of a round bar of 40 mm dia and 2.5 m long when stretched by 2.5 mm. What is the change in its volume ?

1.4. An axial compressive load of 500 kN is applied to a metal bar of square section 50 × 50 mm. The contraction on a 200 mm gauge length is found to be 0.55 mm, and the increase in thickness was 0.045 mm. Find the value of Young's modulus and Poisson's ratio.

1.5. A circular rod of 1 m length has 10 mm dia. over the first 500 mm length and 20 mm dia. over the remaining length. If $E = 2 \times 10^5$ N/mm^2, calculate the strain energy stored in the rod when it is subjected to an axial tensile force of 10 kN.

1.6. A rigid cross bar is supported horizontally by two vertical bars A and B of equal length and hanging from their tops. The bars A and B are 0.6 m apart. The cross bar stays horizontal even after application of load of 6 kN (vertical force) at a point 0.4 m from B. If the stress in A is 200 MN/m^2, Find the stress in B and the area of cross-section of the two rods. [Take E_A = 200 GN/m^2 and E_B = 130 GN/m^2.]

1.7. (*a*) Derive an expression for the extension of a bar subjected to an axial tensile load P.

(*b*) A load of 10 kN is to be raised by a steel wire. If the stress in the wire is not to exceed 140 N/mm^2, what should be the dia of the wire ? What will be the extension of a 5 m length of wire in this case. [Take, $E = 2 \times 10^5$ N/mm^2.]

1.8. Two rods one of copper 40 mm dia and another of steel 30 mm dia are fixed at their top ends 0.6 m apart and hang vertically downwards. They are connected at their bottom ends by a horizontal cross bar on which a 15 kN load is to be placed. Determine the location of this load so that the cross bar may remain horizontal. Take E_{Cu} = 1.1 × 10^5 N/mm^2 and E_{St} = 2.2 × 10^5 N/mm^2.

1.9. A steel tube 35 mm internal dia and 2.5 mm thick, 5 m long is covered and lined throughout with copper tubes 2.5 mm thick. The two tubes are rigidly connected at their ends. This compound tube is subjected to tension and the stress produced in steel is 80 N/mm^2. Determine (*i*) the elongation of the tube, (*ii*) stress produced in the copper tube, and (*iii*) the load carried by the combined tube. Take E_{St} 2.1 × 10^5 N/mm^2, E_{Cu} = 1.1 × 10^5 N/mm^2.

1.10. A circular rod of steel tapers uniformly from 60 mm dia to 30 mm dia. Find the elongation in a length of 0.3 m, when subjected to an axial pull of 100 kN. Find also the extension of the rod if the rod has a uniform dia of 40 mm. Take, E = 2.1 × 10^5 N/mm^2.

1.11. A M.S. rod 20 mm dia. passes centrally through a copper tube whose I.D. is 22 mm and thickness 3 mm. The compound tube is 0.6 m long and their ends are rigidly connected. If a load of 50 kN acts axially, determine the stresses in the two metals and their expansions. Take, E_{St} = 205 GN/m^2, and E_{Cu} = 102.5 GN/m^2.

1.12. A compound bar 0.8 m long made of steel for 0.3 m length and of brass for the remaining. If the steel portion has a dia. of 30 mm, what should be the dia of the brass portion if the elongation should be same in each portion under an axial tension of 50 kN. What are the stresses developed in steel and brass. Take E_{St} = 2 × 10^5 N/mm^2 and E_{Br} = 1.2 × 10^5 N/mm^2.

1.13. A hollow copper tube 60 mm I.D. and thickness 5 mm is fitted inside a steel tube of the same thickness and 70 mm I.D. Both tubes are 0.5 m long and rigidly connected at the two ends. If the temperature is raised by 63°C, determine the stresses developed in each tube. Take E_{St} = 2 × 10^5 N/mm^2, α_{St} = 1.2 × 10^{-5}/°C, E_{Cu} = 1 × 10^5 N/mm^2, α_{Cu} = 1.85 × 10^{-5}/°C.

1.14. In tests on a sample of steel bar of 25 mm dia, it is found that a tensile load of 50 kN results in an extension of 0.1 mm on a gauge length of 200 mm, and a torque of 250 N·m produces an angle of twist of 0.9° in a length of 250 mm. Deduce the value of Poisson's ratio for steel and prove the formula you use.

Hint. See Chapter–4: Torsion ; $\dfrac{T}{J} = \dfrac{G\theta}{L}$, $J = \dfrac{\pi d^4}{32}$, θ in radians.

$$E = 2\,G(1 + \nu)$$

1.15. An unknown weight falls through 15 mm on to a collar rigidly attached to the lower end of a vertical bar 3 m long and 600 mm^2 in section. If the instantaneous extension was observed as 1.3 mm, what is the stress induced and the value of the unknown weight Take, E = 2 ×10^5 N/mm^2.

1.16. A 6 mm thick plate shown in Figure P.1.16, is subjected to a tensile load of 50 kN. Sketch the stress distribution indicating the maximum stress.

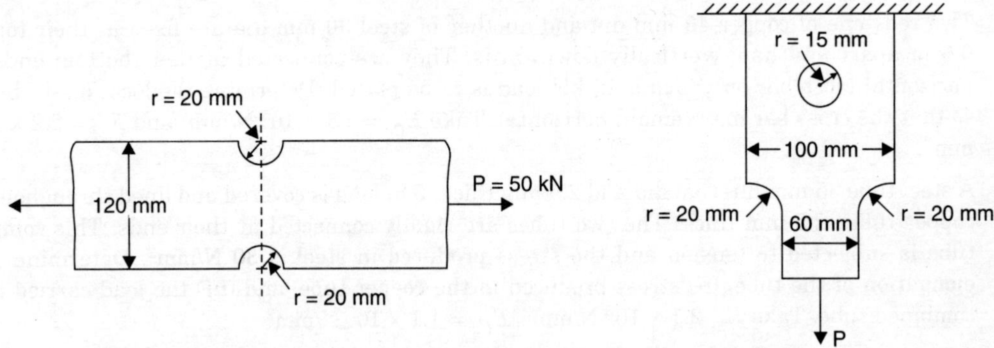

Figure P.1.16

Figure P.1.17

1.17. **Determine the maximum static axial load *P*, that may be applied to the 10 mm thick plate shown in Figure P.1.17, so that the maximum stress does not exceed 150 N/mm².**

2 B.M. and S.F. Diagrams

The shearing force (S. F. or F) at a section is the algebraic sum of all the vertical forces either to the left or to the right of the section i.e., $F_{XX} = \sum V$. If a force is inclined to the beam, the vertical component of the force should be considered. A shearing force diagram (SFD) shows the variation of shearing force along the length of the beam.

The bending moment (B.M. or M) at a section is the algebraic sum of the moments of the forces acting towards the left or the right of the section i.e., $M_{XX} = \sum F_1x_1$. A bending moment diagram (BMD) shows the variation of bending moment along the length of the beam.

Sign Conventions, see Figure 2.1.

S.F. is positive when $\sum V$ towards the left of the section is upward, or towards the right downward; and negative otherwise.

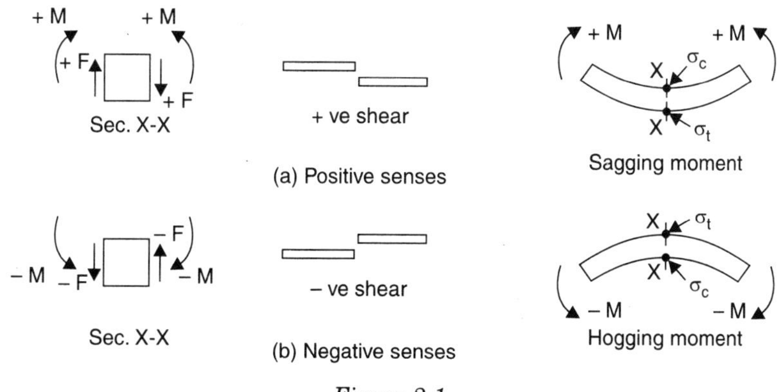

Figure 2.1.

B.M. is positive when $\sum M$ towards the left of the section X-X is clockwise, or anticlockwise towards the right of the section – XX, i.e., the B.M. on the portion of the beam is such as to cause bending of the beam concave upwards (called **sagging moment**).

B.M. is negative when M towards the left of the section X-X is anticlockwise and towards the right clockwise, i.e., the B.M. on the portion of the beam causes bending of the beam convex upwards (called **hogging moment**).

Relation among w, F and M, see Figure 2.2.

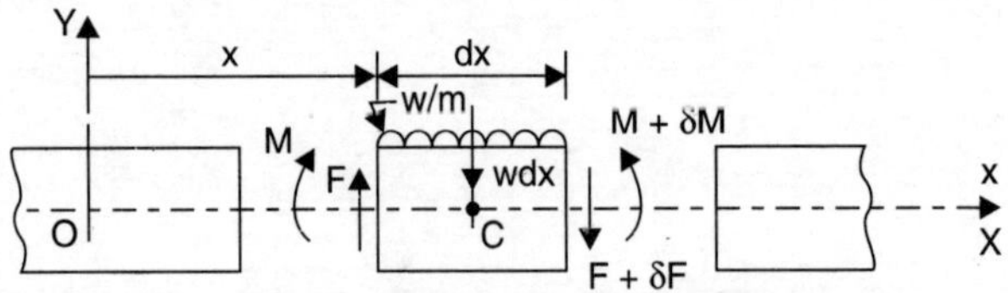

Figure 2.2

$$\sum M_C = 0: \quad M + F \cdot \frac{dx}{2} + (F + \delta F)\frac{dx}{2} - (M + \delta M) = 0$$

Neglecting the product $F \cdot dx$, in the limits, $\mathbf{F} = \dfrac{\mathbf{dM}}{\mathbf{dx}}$ ...(*i*)

$$\sum Y = 0: \quad F - wdx - (F + \delta F) = 0, \quad \mathbf{w} = -\frac{\mathbf{dF}}{\mathbf{dx}} = -\frac{\mathbf{d^2M}}{\mathbf{dx^2}} \qquad ...(ii)$$

S.F. and B.M. Diagrams

1. Cantilever with concentrated load W, at the free end, see Figure 2.3.

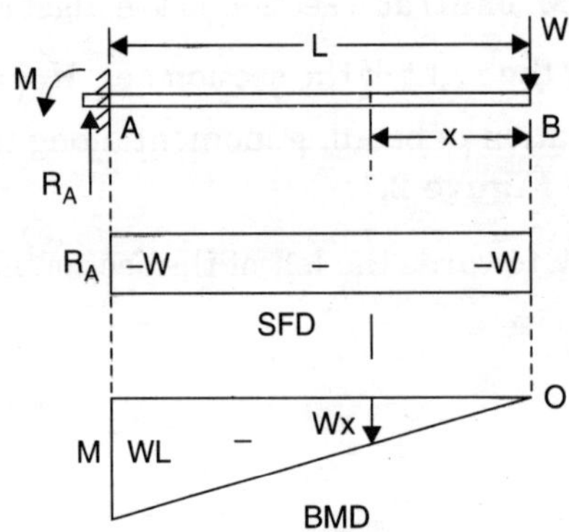

Figure 2.3.

Start from the free end *B*

(*i*) SFD : $F_B = W, \quad F_x = W, \quad F_A = W \downarrow \quad R_A = W \uparrow$

(*ii*) BMD : $M_B = 0, M_x = W \cdot x$ ↻ (– ve), $\quad M_A = WL$

Note. When S.F. = constant, B.M. = a straight line according to relation (*i*).

2. Cantilever with uniformly distributed load (udl) of w per m, see Figure 2.4.

Start from the free end *B*.

(*i*) SFD : $F_B = 0, \quad F_x = w \cdot x, \quad F_A = wL \downarrow \quad R_A = wL \uparrow$

(*ii*) BMD : $M_B = 0, M_x = wx \cdot \dfrac{x}{2} = \dfrac{wx^2}{2}$ ↻ (– ve)

$$M_A = wL \cdot \frac{L}{2} = \frac{wL^2}{2}$$

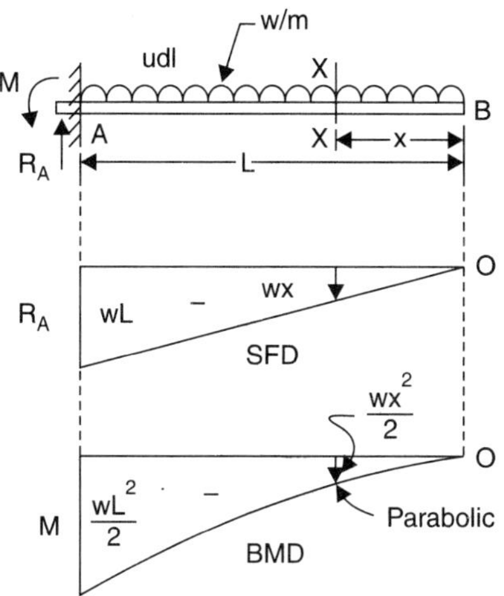

Figure 2.4.

Note. When S.F. = a straight line, B.M. = a parabola, according to relation (*i*).

3. Simply supported beam with concentrated load W, at mid-span, see Figure 2.5.

(*i*) Reactions : $\sum Y = 0: \quad R_A + R_B = W$

$$\text{Due to symmetry, } R_A = R_B = \frac{W}{2} \uparrow$$

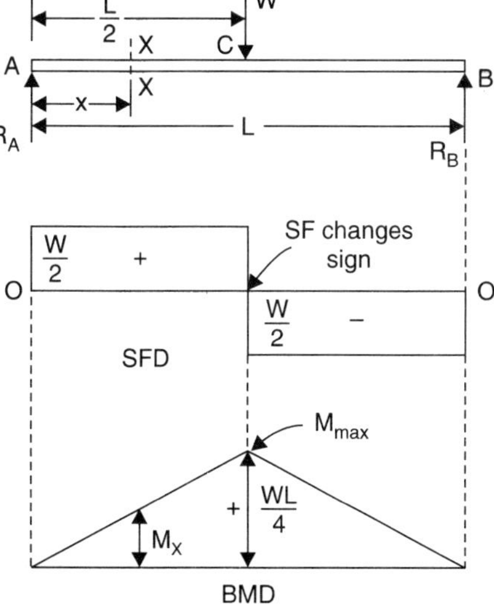

Figure 2.5.

(*ii*) SFD : Start from the left end *A*

$$F_A = R_A = \frac{W}{2} = F \text{ upto } C$$

$$F_C = \frac{W}{2} - W = -\frac{W}{2} = F \text{ upto } B$$

$$F_B = -\frac{W}{2} + \frac{W}{2} = 0 \text{ (check)}$$

***(iii)* BMD:** $M_A = 0$, $M_x = R_A \cdot x = \frac{W}{2} x$ (+ ve) (linear)

$$M_C = \frac{W}{2} \cdot \frac{L}{2} = \frac{\mathbf{WL}}{\mathbf{4}} = \mathbf{M_{max}}$$

$$M_B = \frac{W}{2} \times L - W \cdot \frac{\mathbf{L}}{\mathbf{2}} = 0 \text{ (check)}$$

Note. When S.F. = constant, B.M. = a straight line.

4. Simply supported beam with udl, see Figure 2.6.

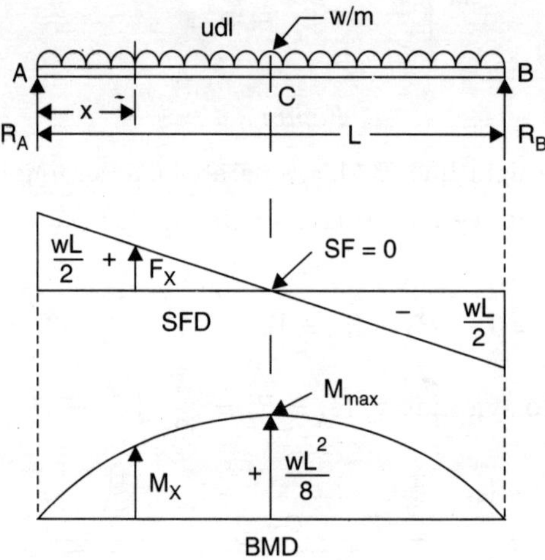

Figure 2.6.

***(i)* Reactions :** $Y = 0$: $R_A + R_B = wL$

Due to symmetry, $R_A = R_B = \frac{wL}{2} \uparrow$

***(ii)* SFD :** Start from the left end A

$$F_A = 0, F_x = R_A - wx = \frac{wL}{2} - wx \text{ (linear)}$$

$$F_C = \frac{wL}{2} - \frac{wL}{2} = 0$$

$$F \text{ upto } B = \frac{wL}{2} - wL = -\frac{wL}{2}$$

$$F_B = -\frac{wL}{2} + \frac{wL}{2} = 0 \text{ (check)}$$

***(iii)* BMD :** $M_A = 0, M_x = R_A \cdot x - wx \cdot \frac{x}{2}$

$$M_x = \frac{wL}{2} \cdot x - \frac{wx^2}{2} \text{ (parabolic)}$$

$$M_C = \frac{wL}{2}\cdot\frac{L}{2} - \left(w\cdot\frac{L}{2}\right)\frac{L}{4} = \frac{\mathbf{wL^2}}{\mathbf{8}} = \mathbf{M_{max}}$$

Note : (*i*) When S.F. = a straight line, B.M. = a parabola.

(*ii*) When S.F. = 0, B.M. = maximum.

5. Simply supported beam with concentrated load W, at 'a' from A and 'b' from 'B' see Figure 2.7.

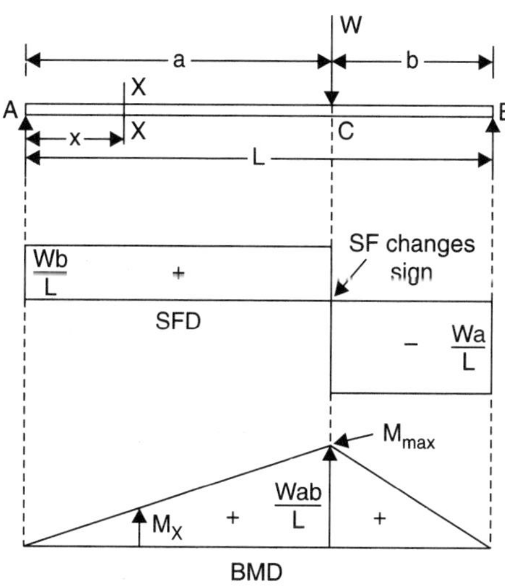

Figure 2.7.

$$L = a + b$$

(*i*) Reactions:

$$\sum Y = 0 : \quad R_A + R_B = W$$

$$\sum M_A = 0 : \quad R_B \cdot L - W \cdot a = 0$$

$$R_B = \frac{Wa}{L}\uparrow, \quad R_A = W - \frac{Wa}{L} = \frac{Wb}{L}\uparrow$$

(*ii*) SFD: Start from the left end.

$$F_A = R_A = \frac{Wb}{L}, \quad F_X = R_A = \frac{Wb}{L} = \text{constant}$$

$$F_C = \frac{Wb}{L} - W = -\frac{Wa}{L}$$

$$F_B = -\frac{Wa}{L} + \frac{Wa}{L} = 0 \text{ (check)}$$

(*iii*) BMD: $M_A = 0 : \quad M_x = R_A \cdot x = \frac{Wb}{L}\cdot x$ (linear)

$$M_C = \frac{Wb}{L} \cdot a = \mathbf{W}\frac{\mathbf{ab}}{\mathbf{L}} = \mathbf{M_{max}}$$

$$M_B = \frac{Wb}{L} \cdot L - W \cdot b = 0 \text{ (check)}$$

6. Simply supported beam with udl with equal overhang at each end, see Figure 2.8.

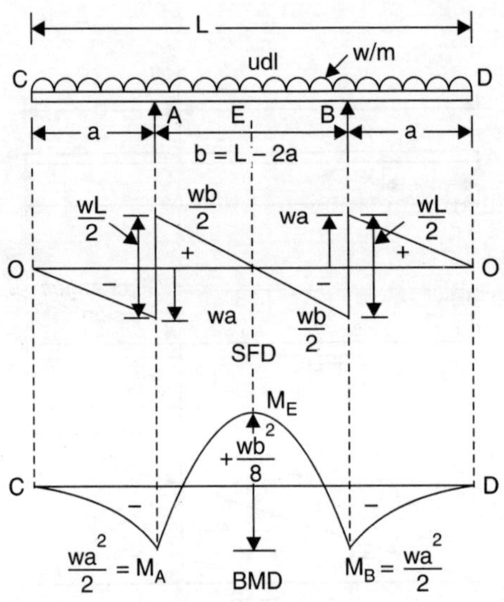

Figure 2.8.

$$b = L - 2a$$

(*i*) Reactions : $\sum Y = 0 : \quad R_A + R_B = wL$

Due to symmetry, $R_A = R_B = \frac{wL}{2} \uparrow$

(*ii*) SFD : $\quad F_A = -wa + \frac{wL}{2}$

$$F_E = -\frac{wL}{2} + \frac{wL}{2} = 0$$

The other half is symmetrical.

(*iii*) BMD : $M_A = -wa \cdot \frac{a}{2} = -\frac{\mathbf{wa^2}}{\mathbf{2}}$ (parabolic)

$$M_E = -w\frac{L}{2} \cdot \frac{L}{4} + \frac{wL}{2}\left(\frac{L}{2} - a\right)$$

$$= \frac{wL^2}{8} - \frac{wLa}{2} = \frac{wb^2}{8} - \frac{wa^2}{2}$$

The other half is symmetrical.

$$M_E = 0, \text{ when } a = \frac{b}{2}; \quad \text{then } M_{A/B} = -\frac{wb^2}{8},$$

when the point M_E would just touch the line CD.

The bending moment on the beam is the least when the numerical values at the supports and the centre are equal i.e.,

$$\frac{wa^2}{2} = \frac{wb^2}{8} - \frac{wa^2}{2}, \quad \text{or} \quad a = \frac{b}{2\sqrt{2}} \simeq \mathbf{0.354b}$$

The railway sleepers are subjected to uniform pressure on the ground (ballast) due to two equal point loads i.e., rail loads at Gauge G (= b) apart.

For broad gauge, $G = 1.676 \text{ m} = b$

$$a = \frac{b}{2\sqrt{2}} = \frac{1.676}{2\sqrt{2}} = 0.6 \text{ m}$$

Then, the most economical length of the sleeper

$$L = a + b + a = 0.6 + 1.676 + 0.6 = \mathbf{2.876 \text{ m}}$$

Similarly, for metre gauge, $G = 1\ m = b$

$$a = \frac{1}{2\sqrt{2}} = 0.354 \text{ m}, \quad \text{and } L = 1 + 0.354 \times 2 \simeq \mathbf{1.71 \text{ m}}$$

7. Simply supported beam with equal overhang and concentrated load at each end, see Figure 2.9.

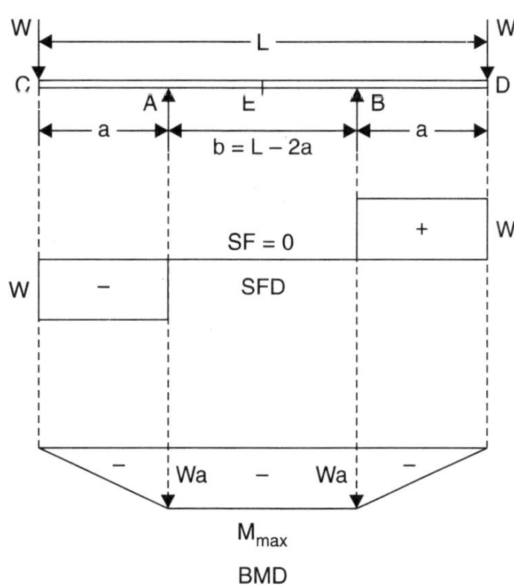

Figure 2.9.

$$b = L - 2a$$

(*i*) Reactions : $Y = 0 : R_A + R_B = 2W$

Due to symmetry, $R_A = R_B = W$

(*ii*) SFD : $F_{CA} = -W, \quad F_{AB} = 0, \quad F_{BD} = +W$

***(iii)* BMD :** $M_A = -Wa, \quad M_B = -W\mathring{a}$

$$M_E = -W\frac{L}{2} + W\left(\frac{L}{2} - a\right) = -Wa$$

$$M_{AB} = -Wa = \text{constant.}$$

Example 2.1. *A simply supported beam with overhang on the right is loaded as shown in Figure 2.10. Draw the S.F. and B.M. diagrams. Determine the maximum positive and negative bending moments and the point of contraflexure.*

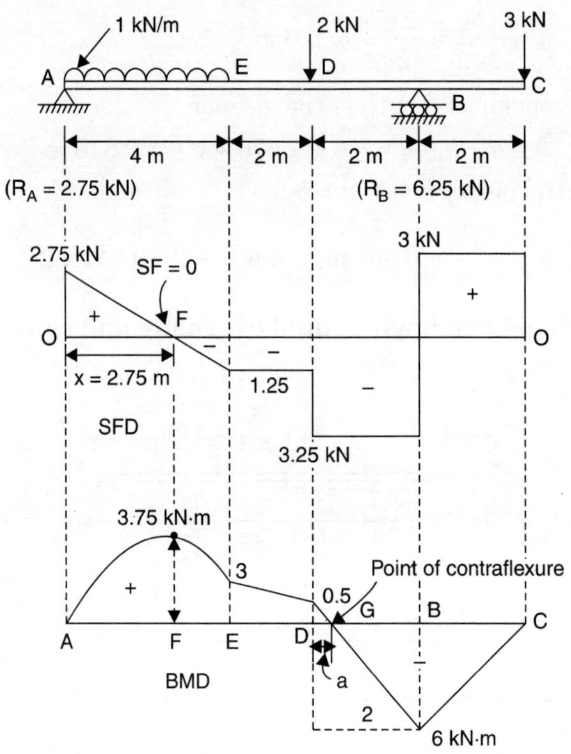

Figure 2.10.

Solution. A : Simply supported, B : on rollers, BC = overhang

***(i)* Reactions :** $\sum Y = 0: \quad R_A + R_B = 1 \times 4 + 2 + 3 = 9 \text{ kN}$

$$+\circlearrowright \sum M_A = 0: \quad (1 \times 4)\frac{4}{2} + 2 \times 6 + 3 \times 10 - R_B \times 8 = 0$$

$$R_B = \mathbf{6.25\ kN} \uparrow, \quad R_A = 9 - 6.25 = \mathbf{2.75\ kN} \uparrow$$

***(ii)* SFD :** Start from left end A.

$$F_A = R_A = 2.75 \text{ kN (+ ve)}, F_E = 2.75 - 1 \times 4 = -1.25 \text{ kN},$$

$$F_D = -1.25 - 2 = -3.25 \text{ kN}, F_B = -3.25 + 6.25 = 3 \text{ kN}$$

$$F_C = 3 - 3 = 0 \text{ (check).}$$

(*iii*) BMD: Start from the left end *A* ; $M \circlearrowright$ + ve

$$M_A = 0$$

M_{max} occurs when S.F. = 0. In the portion *AE* of the beam,

$$F_x = 2.75x - 1 \times x = 0, \quad x = 2.75 \text{ m}$$

$$M_x = 2.75 \times 2.75 - (1 \times 2.75)\frac{2.75}{2} = \mathbf{3.78 \text{ kN·m} = M_{max} \text{ (+ ve)}}$$

$$M_E = 2.75 \times 4 - (1 \times 4)\frac{4}{2} = 3 \text{ kN·m.}$$

$$M_D = 2.75 \times 6 - \left(1 \times 4\left(\frac{4}{2} + 2\right)\right) = 0.5 \text{ kN·m}$$

$$M_B = 2.75 \times 8 - (1 \times 4)6 - 2 \times 2 = \mathbf{- 6 \text{ kN·m} = M_{max} \text{ (– ve)}}$$

Alternatively, working from the right end *C*,

$$M_B = 3 \times 2 = \mathbf{6 \text{ kN·m}} \circlearrowright \text{ (hogging B.M., – ve)}$$

which is the maximum negative B.M.

$$M_C = 0$$

The point of contraflexure *G* occurs when the B.M. changes sign i.e., B.M. = 0. In the portion *DB* of the beam, from the similar triangles in BMD.

$$\frac{a}{0.5} = \frac{2}{6 + 0.5}, \quad \mathbf{a = 0.154 \text{ m}} \text{ from } D.$$

Example 2.2. *Draw the B.M. and S.F. diagrams for the loaded beam shown in Figure 2.11. Determine the maximum B.M. and the point of contraflexure.*

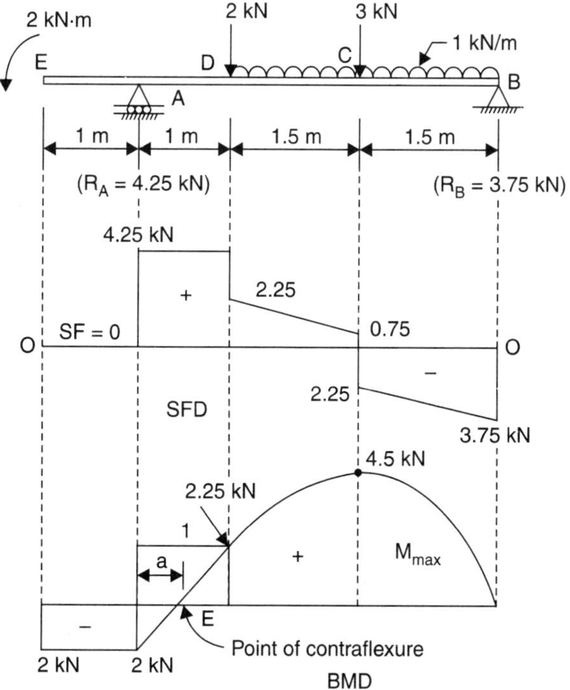

Figure 2.11.

Solution. (*i*) Reactions:

$$Y = 0: \quad R_A + R_B = 2 + 3 + 1 \times 3 = 8 \text{ kN}$$

$$M_B = 0: \quad -2 + R_A \times 4 - 2 \times 3 - 3 \times 1.5 - (1 \times 3) \times 1.5 = 0$$

$$R_A = \mathbf{4.25\ kN} \uparrow, \quad R_B = 8 - 4.25 = \mathbf{3.75\ kN} \uparrow$$

(*ii*) SFD : Start from *A*.

$$F_{EA} = 0, F_A = R_A = 4.25 \text{ kN}, F_D = 4.25 - 2 = 2.25 \text{ kN}$$

$$F \text{ upto } C = 2.25 - 1 \times 1.5 = 0.75 \text{ kN}, F_C = 0.75 - 3 = -2.25 \text{ kN}$$

$$F \text{ upto } B = -2.25 - 1 \times 1.5 = -3.75 \text{ kN}$$

$$F_B = -3.75 + 3.75 = 0 \text{ (check)}$$

(*iii*) BMD : Start from *A* ; $M\circlearrowright$ + ve.

$$M_{EA} = -2 \text{ kN·M (hogging B.M.)}$$

$$M_D = -2 + 4.25 \times 1 = 2.25 \text{ kN·m}$$

$$M_C = -2 + 4.25 \times 2.5 - 2 \times 1.5 - (1 \times 1.5) \times \frac{1.5}{2} = 4.5 \text{ kN·m}$$

$$M_B = -2 + 4.25 \times 4 - 2 \times 3 - 3 \times 1.5 - (1 \times 3) \times 1.5 = 0 \text{ (check)}$$

The BMD can be completed by seeing that when S.F. = 0, B.M. = constant;

When S.F. is constant, B.M. is a straight line; and when SFD is a straight line, BMD is parabolic.

[**Note.** When S.F.D. is parabolic, B.M.D. is cubic.]

To find the point of contraflexure *E* (where B.M. = 0), from similar triangles,

$$\frac{a}{2} = \frac{1}{2 + 2.25} \quad \Rightarrow \quad a = \mathbf{0.47\ m} \text{ from the support } A.$$

Example 2.3. *Draw the B.M. and S.F. diagrams for the beam loaded as shown in Figure 2.12. Determine the point of contraflexure.*

Solution.

(*i*) Reactions : $Y = 0: \quad R_A + R_B = \frac{20}{2} \times 2 + 15 + 15 + 10 \times 4 = 90 \text{ kN}$

$$M_A = 0: \quad (10 \times 4)2 + 15 \times 6 + 15 \times 8 + \left(\frac{20}{2} \times 2\right)\left(10 + \frac{2}{3}\right)\circlearrowright = R_B \times 10 + 50\circlearrowleft$$

$$R_B = \mathbf{45.4\ kN} \uparrow, \quad R_A = 90 - 45.4 = \mathbf{44.6\ kN} \uparrow$$

(*ii*) SFD : Starting from the left end *A* as shown in *Figure 2.12 (b).*

(*iii*) BMD : Start from *A* ; $M\circlearrowright$ + ve.

$$M_A = 0, M \text{ upto } F = 44.6 \times 4 - (10 \times 4) \times 2 = 98.4 \text{ kN·m}$$

$$M_F = 98.4 - 50 = 48.4 \text{ kN·m}$$

$$M_C = 44.6 \times 4 - (10 \times 4)(2 + 2) - 50 = 57.6 \text{ kN·m}$$

$$M_D = 44.6 \times 8 - (10 \times 4)(2 + 4) - 50 - 15 \times 2 = 36.8 \text{ kN·m}$$

$$M_B = \left(\frac{20}{2} \times 2\right)\left(\frac{1}{3} \times 2\right) = -13.3 \text{ kN·m (hogging)}$$

$$M_E = 0.$$

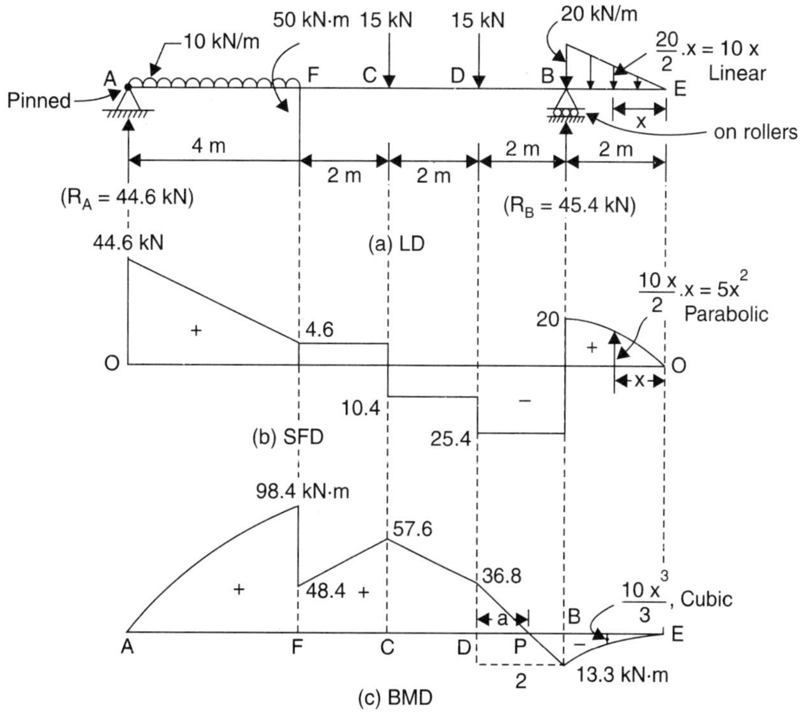

Figure 2.12.

Point of contraflexure (*P*)

$$\frac{a}{36.8} = \frac{2}{(13.3 + 36.8)}, \quad a = \mathbf{1.47\ m\ from\ D.}$$

Note. When S.F. changes sign, B.M. is maximum at *C* of 57.6 kN·m which is < 98.4 kN·m at *F*, and so at *B*.

Example 2.4. *For th^ cantilever loaded as shown in Figure 2.13, draw the S.F. and B.M. diagrams.*

Solution.

***(i)* SFD :** Start from the free end *B* and complete as shown in Figure 2.13 (*b*).

***(ii)* BMD :** Start from the free end *B*.

$$M_C = 2 \times 2 = 4 \text{ kN·m}, M_D = 2 \times 6 + (2 \times 4) \times 2 = 28 \text{ kN·m}.$$

$$M_E = 2 \times 8 + (2 \times 4)(2 + 2) + 2 \times 2 = 52 \text{ kN·m}.$$

$$M_A = 2 \times 10 + (2 \times 4)(2 + 4) + 2 \times 4 + 2 \times 2 = 80 \text{ kN·m}.$$

Complete the BMD by knowing the relation between B.M. and S.F. as shown in Figure 2.13 (*c*).

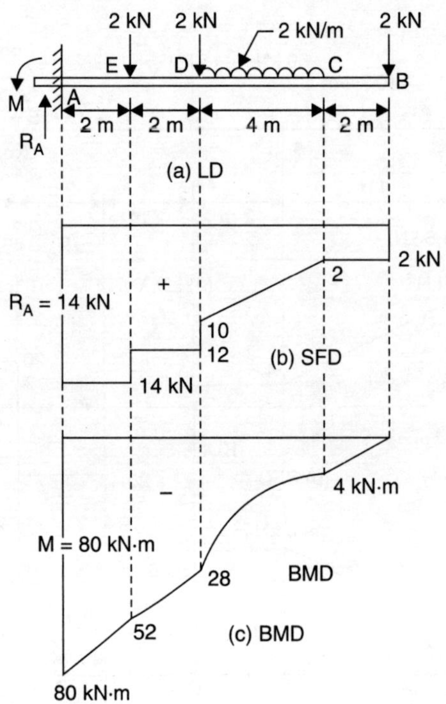

Figure 2.13.

Example 2.5. *For the loaded beam shown in Figure 2.14, draw the S.F. and B.M. diagrams. What is the maximum B.M. anywhere in the beam?*

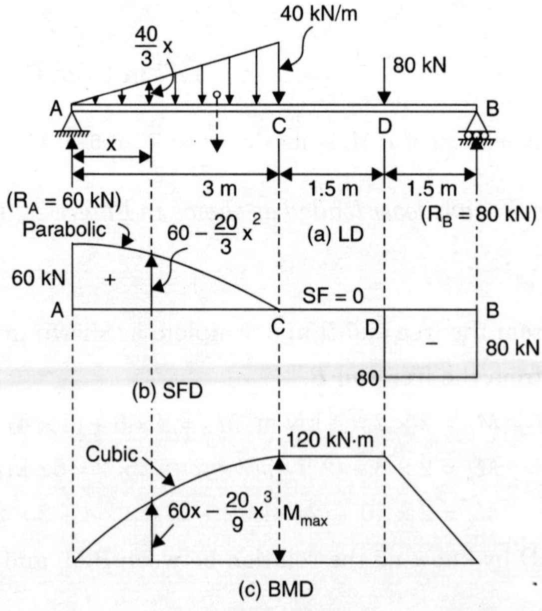

Figure 2.14.

Solution.

(*i*) Reactions : $Y = 0 : R_A + R_B = \frac{40}{2} \times 3 + 80 = 60 + 80 = 140 \text{ kN}$

$$M_A = 0 : R_B \times 6 = 80 \times 4.5 + 60\left(\frac{2}{3} \times 3\right) = 480$$

$$R_B = \mathbf{80\ kN} \uparrow, \quad R_A = 140 - 80 = \mathbf{60\ kN} \uparrow$$

(*ii*) SFD : Start from the left support *A*.

$$F_A = 60 \text{ kN}, F_C = 60 - \frac{40}{2} \times 3 = 0$$

$$F_{CD} = 0, \quad F_D = 0 - 80 = -80 \text{ kN}$$

$$F_B = -80 + 80 = 0 \text{ (check)}$$

Between *AC*, $W_x = \frac{40}{3} x$

$$F_x = 60 - \frac{40x}{3 \times 2} \times x = 60 - \frac{20}{3} x^2 \text{ (parabolic)}$$

(*iii*) BMD : Start from *A*, $M \circlearrowright$ + ve

$$M_A = 0, M_C = 60 \times 3 - \left(\frac{40}{3} \times 3\right)\frac{1}{3} \times 3 = 120 \text{ kN·m}$$

$$M_{CD} = \mathbf{120\ kN{\cdot}m = M_{max}} \text{ (where S.F. = 0)}$$

$$M_D = 80 \times 1.5 = 120 \text{ kN· m (check)}$$

Between *AC*, $W_x = \frac{40}{3} x$

$$M_x = 60x - \left(\frac{40x}{3 \times 2} \times x\right) \times \frac{1}{3} x = 60\,x - \frac{20x^3}{9} \text{ (cubic)}$$

Note. When SFD is parabolic, BMD is cubic.

Example 2.6. *For the cantilever loaded as shown in Figure 2.15, draw the SFD and BMD.*

Solution. (*i*) Support reactions and fixing moment

$$R_x = 2 \text{ kN}$$

$$R_y = 1 + \frac{1 + 0.5}{2} \times 5 = \mathbf{4.75\ kN} \uparrow$$

$$M_A = 1 \times 10 + (0.5 \times 5)\left(\frac{5}{2} + 2.5\right) + \left(\frac{0.5}{2} \times 5\right)\left(\frac{2}{3} \times 5 + 2.5\right) + 5$$

$$= \mathbf{34.8\ kN{\cdot}m}$$

(*ii*) SFD : Start from the free end *B*.

$$F_B = 1 \text{ kN} = F_{BD}$$

$$F_E = 1 + \frac{1 + 0.5}{2} \times 5 = 4.75 \text{ kN} = F_A$$

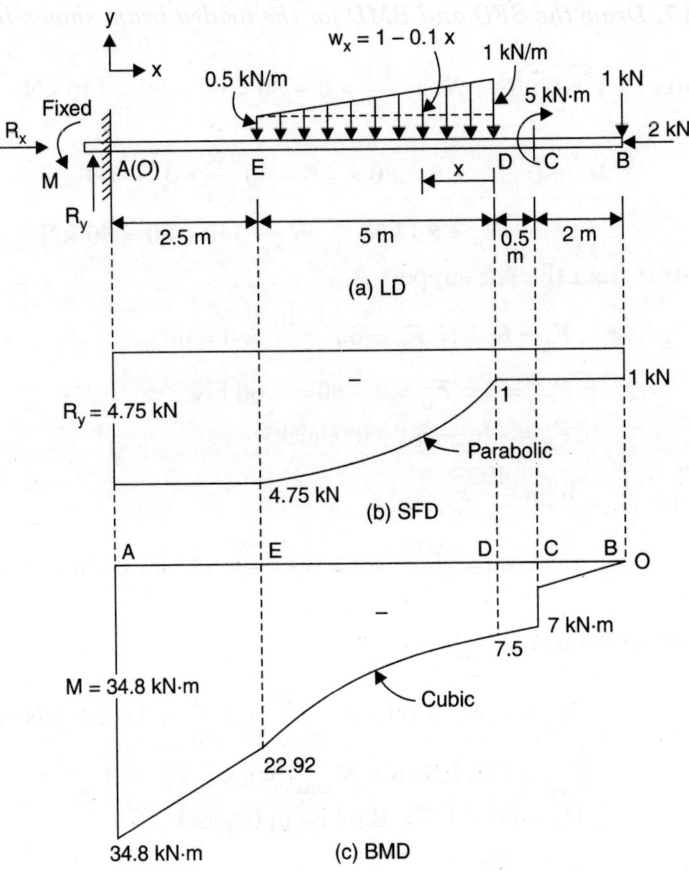

Figure 2.15.

In the portion *DE,* at any distance *x* from *D*,

$$w_x = 1 - \frac{0.5}{5} \times x = 1 - 0.1x$$

$$F_x = 1 + \frac{1 - 0.1x + 1}{2}\,x = 1 + x - 0.05\,x^2 \text{ (parabolic)}$$

***(iii)* BMD :** Start from *B* ; *M* ↻ –ve, hogging (i.e. tension on top)

$$M_B = 0, M \text{ upto } C = 1 \times 2 = 2 \text{ kN·m.}$$

$$M_C = 2 + 5 = 7 \text{ kN·m}$$

$$M_D = 1 \times 2.5 + 5 = 7.5 \text{ kN·m}$$

$$M_E = 1 \times 7.5 + 5 + (0.5 \times 5)\,\frac{5}{2} + \left(\frac{0.5}{2} \times 5\right)\left(\frac{2}{3} \times 5\right) = 22.92 \text{ kN·m}$$

$$M_A = 34.8 \text{ kN·m}$$

Between the portion *DE* of the beam,

$$w_x = 1 - 0.1x$$

$$M_x = 1 \times (2.5 + x) + 5 + (1 - 0.1x) \times \frac{x}{2} + \left(\frac{0.1\,x}{2}\,x\right)\frac{2}{3}\,x$$

$$M_x = 7.5 + x + 0.5\,x^2 - 0.017\,x^3 \quad \text{(cubic)}$$

Example 2.7. *Draw the SFD and BMD for the loaded beam shown in Figure 2.16.*

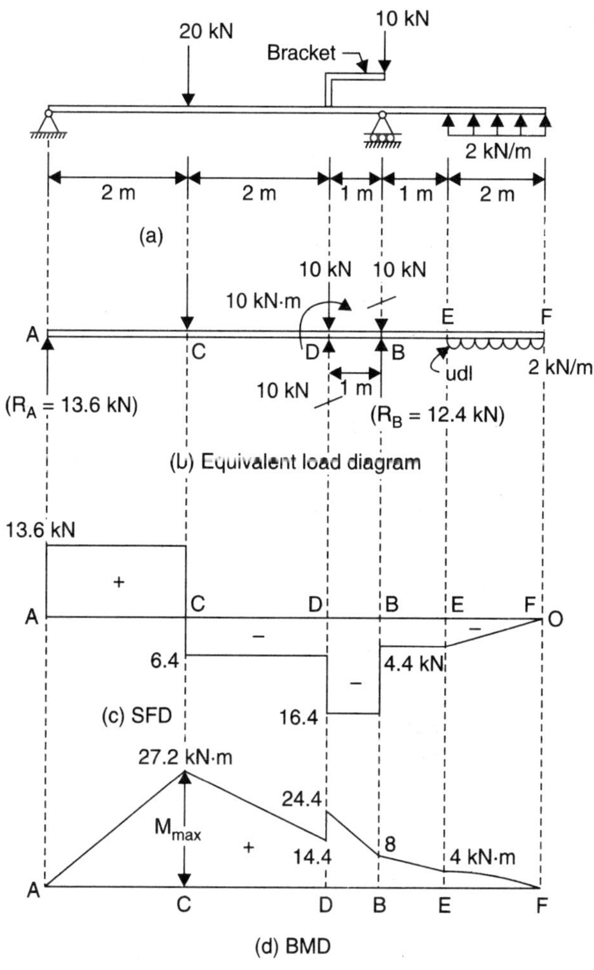

Figure. 2.16.

Solution. The effect of the bracket is to apply a load of 10 kN and a *B.M.* of 10 kN·m at a point 4 m from the left support, i.e. a force ≡ another force (moved) + a couple

(*i*) Reactions : $\sum Y = 0: \quad R_A + R_B = 20 + 10 - 2 \times 2 = 26$

$$\sum M_A = 0: \quad 20 \times 2 + 10 \times 4 + 10 - R_B \times 5 - (2 \times 2) \times 7 = 0$$

$$R_B = \mathbf{12.4\ kN} \uparrow, \quad R_A = 26 - 12.4 = \mathbf{13.6\ kN} \uparrow$$

(*ii*) SFD : Start from *A* and complete as shown in *Figure 2.16 (c).*

(*iii*) BMD : Start from *A* ; $M\circlearrowright$ +ve.

$M_A = 0, M_C = 13.6 \times 2 = \mathbf{27.2\ kN{\cdot}m = M_{max}}$

M upto $D = 13.6 \times 4 - 20 \times 2 = 14.4$ kN·m

$M_D = 14.4 + 10 = 24.4$ kN·m

$M_B = (2 \times 2) \times 2 = 8$ kN·m (working from the R.S.)

$M_E = (2 \times 2) \times 1 = 4$ kN·m

The BMD is completed by seeing the relation between B.M. and S.F.

Example 2.8. *The SFD of a simply supported beam is shown in Figure 2.17. Draw the load diagram and BMD. Determine the value of maximum B.M. and the point of contraflexure.*

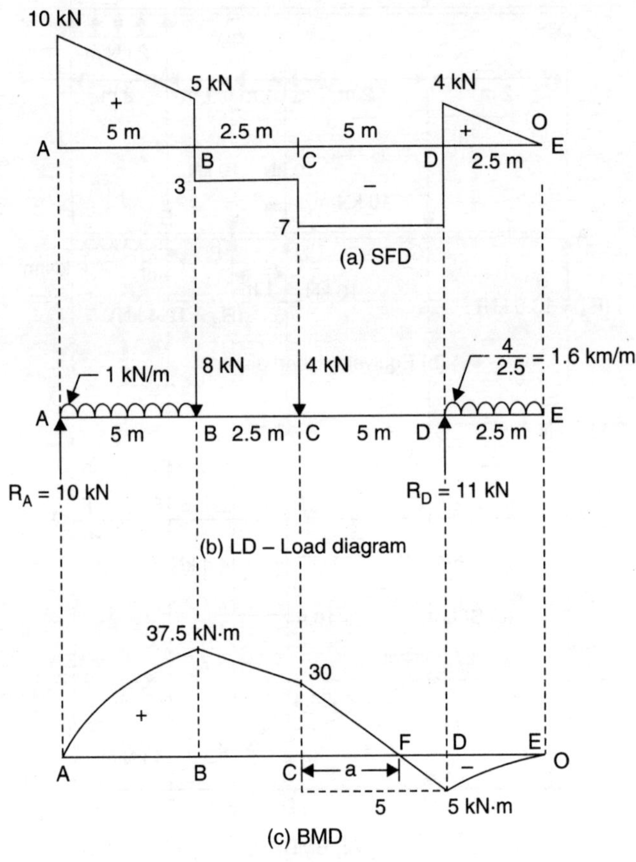

Figure 2.17.

Solution. (*i*) The vertical upward reactions and the downward loads are determined by inspection of the given SFD. For instance the *udl* on the right end overhang

$$w = \frac{4}{2.5} = 1.6 \text{ kN/m}$$

(*ii*) The BMD is drawn from the load diagram and seeing the relation between B.M. and S.F.

The max. + ve B.M. = 37.5 kN·m at *B*.

The max. – ve B.M. = 5 kN·m at *D*

The point of contraflexure *F*, in the portion *CD* of the beam is determined from the similar triangles as

$$\frac{a}{30} = \frac{5}{30+5} \quad \Rightarrow \quad a = \mathbf{4.3 \text{ m from C.}}$$

PROBLEMS

2.1. Draw the S.F. and B.M. diagrams for the loaded beams given below (Figure P.2.1) :

(a)

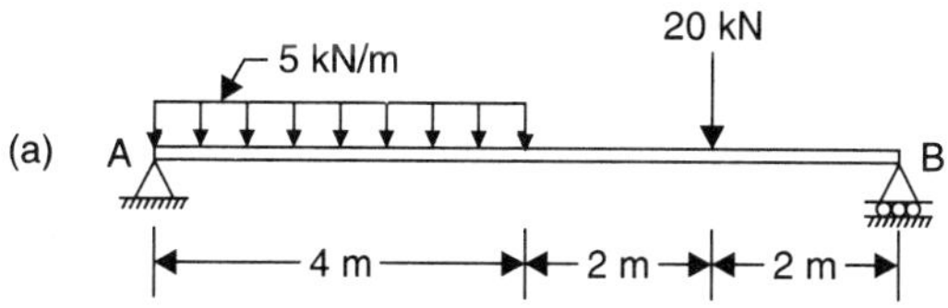

(b)

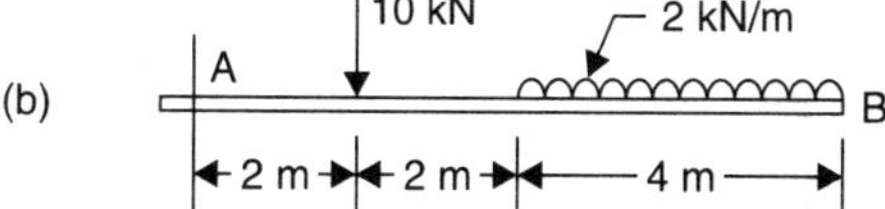

(c)

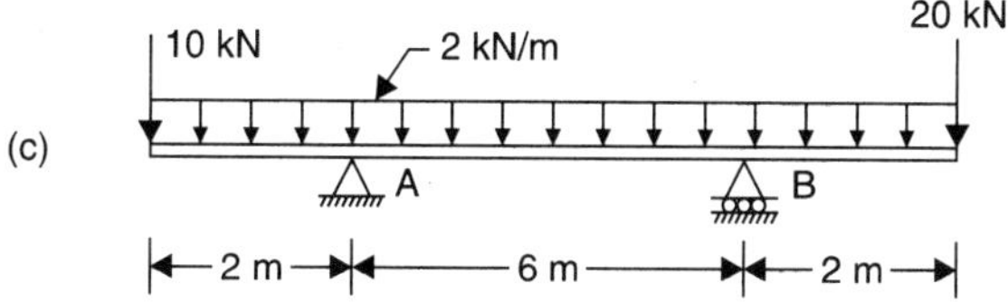

(d)

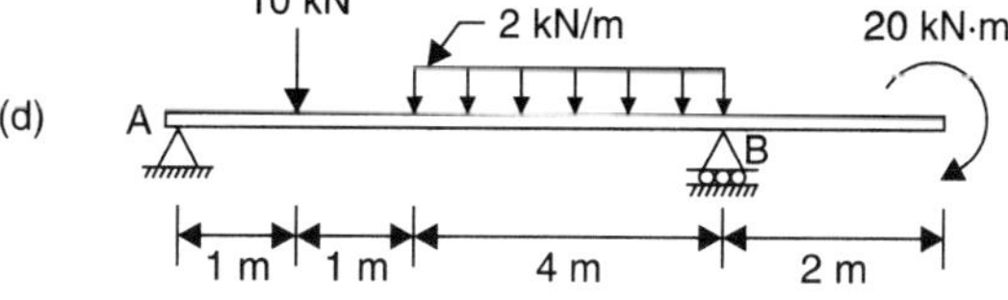

(e)

(f)

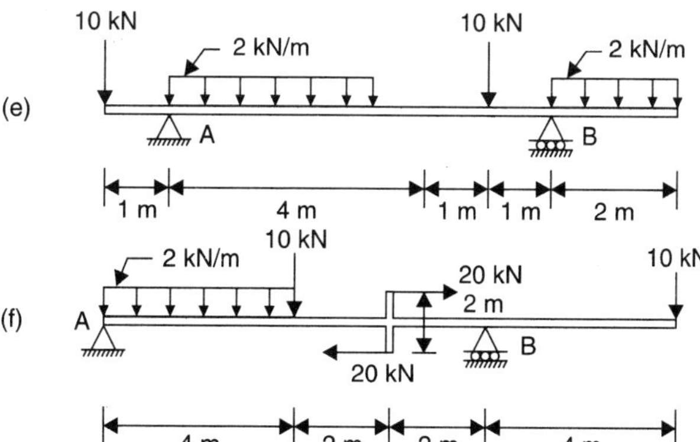

(g)

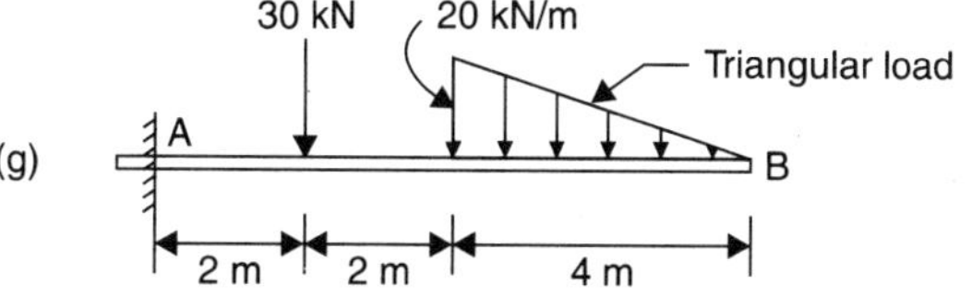

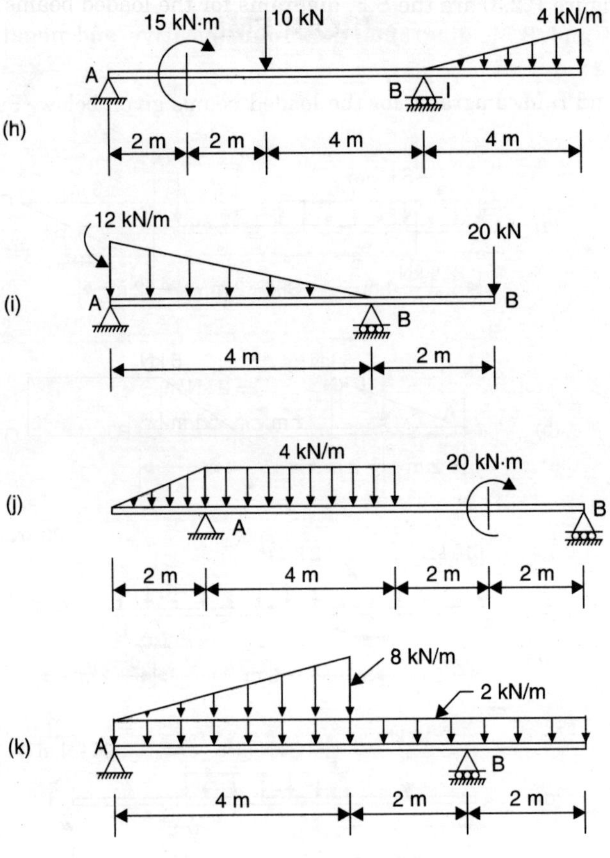

Figure P.2.1

2.2. A flat plate 4 × 1 m hinged at top retains oil (sp. gr. 0.9) to a depth of 3 m as shown in Figure P.2.2. What is the reaction (*F*) of the floor at the bottom ? Draw the S.F. and B.M. diagrams for the plate.

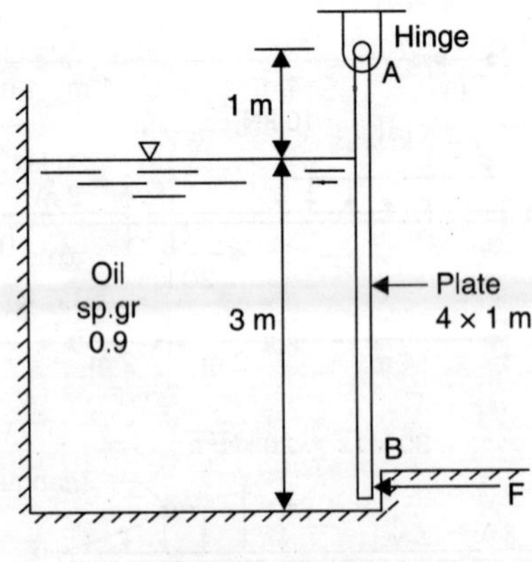

Figure P.2.2

2.3. Given below (Figure P.2.3) are the S.F. diagrams for the loaded beams supported at two points. Draw the load and B.M. diagrams, maximum positive and negative B.M. and points of contraflexure.

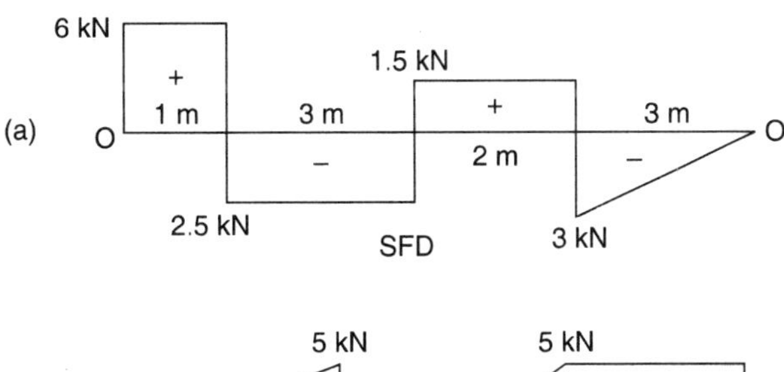

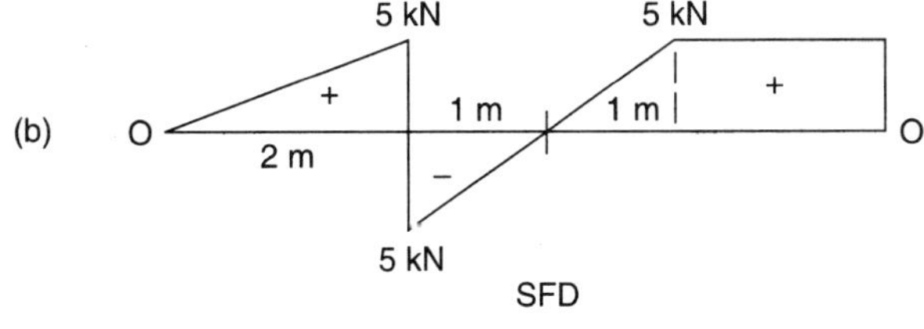

Figure P.2.3

2.4. The B.M. diagram for a loaded beam is shown in Figure P.2.4. Draw the corresponding S.F. and load diagrams.

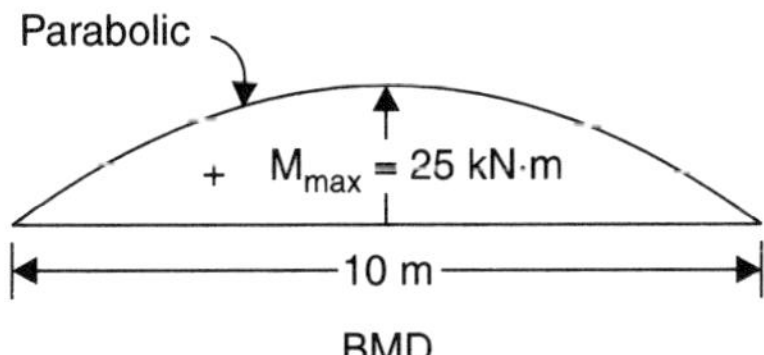

Figure P.2.4

3 Stresses in Beams

A. Bending or Flexure Stresses—In Figure 3.1, a beam initially straight subjected to pure bending (i.e., constant B.M. = M, and S.F. = 0), has bent with top concave edge in compression and the bottom convex edge in tension. There is an intermediate surface which is **not stressed**, and is called neutral surface (*n.s.*) ; the neutral surface cuts the cross-section of the beam (*c/s*) along a line called the neutral axis (N.A.).

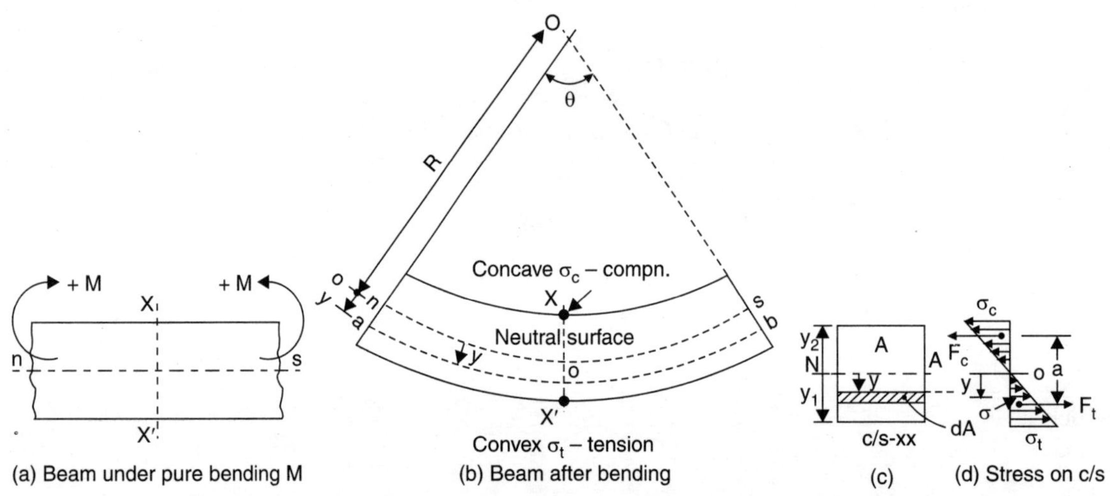

Figure 3.1. Bending stresses in beam

Let O = centre of curvature,

R = radius of curvature of neutral surface (*n.s.*),

θ = angle subtended by the length of the beam at centre O,

M = B.M. on the length of the beam,

σ = stress due to bending, normal to the cross-section,

dA = elemental cross-sectional (*c/sl*) area at a distance y from the N.A.,

XX = cross-sectional plane before bending,

XX' = cross-sectional plane after bending,

σ_c = maximum compressive stress on top fibre,

σ_t = maximum tensile stress on bottom fibre.

$$\text{Strain in fibre } ab = \frac{\sigma}{E} = \frac{ab - ns}{ab} = \frac{(R+y)\theta - R\theta}{R\theta} = \frac{y}{R}$$

$$\frac{\sigma}{E} = \frac{y}{R} \qquad \text{...}(i)$$

For pure bending, the net normal force on the *c/s* must be zero i.e.,

$$\int_A \sigma \cdot dA = 0$$

Substituting for 'σ' from (*i*), $\dfrac{E}{R}\displaystyle\int_A y \cdot dA = 0$

$$\int_A y \cdot dA = \text{first moment of area about the N.A.} = 0$$

which implies the condition the N.A. passes through the centroid G, of the *c/sl* area.

The B.M. (M) is resisted by the moment of the normal forces (F_C and F_T form the resisting couple) about the N.A., i.e.,

$$M = \int_A (\sigma \cdot dA) \cdot y$$

From (*i*), $$M = \frac{E}{R}\int_A y^2 \cdot dA$$

$\displaystyle\int_A y^2 \cdot dA$ = second moment of the *c/sl* area or moment of inertia about the N.A. = I

$$\therefore \qquad M = \frac{E}{R} \cdot I \qquad \text{...}(ii)$$

From (*i*) and (*ii*), $$\boxed{\frac{\mathbf{M}}{\mathbf{I}} = \frac{\sigma}{\mathbf{y}} = \frac{\mathbf{E}}{\mathbf{R}}} \qquad \text{...(3.1)}$$

which is called the **flexure equation**.

From Eqn. (3.1), $$M = \sigma \cdot \frac{I}{y} = \sigma \cdot z \qquad \text{...(3.2)}$$

where $z = \dfrac{I}{y}$ = section modulus.

For a limiting maximum stress 'σ' (i.e., allowable or safe permissible stress), the *c/sl* dimensions which give the maximum value of z is the strongest section to resist a maximum B.M. see Example 3.1.

Assumptions in the Simple Bending Theory

1. The material is homogeneous, isotropic and E has the same value in tension and compression.
2. A transverse section of the beam which is a plane before bending, will remain a plane after bending.
3. There are no lateral stresses and any lateral strain is neglected.

4. Radius of curvature is large compared with the c/sl dimensions of the beam.
5. Loads are applied in the plane of bending.

Example 3.1. *Find the c/sl dimensions b × d of a rectangular beam to resist a maximum B.M. (M), that can be cut from a log of wood of diameter D.*

Solution. See Figure 3.2.

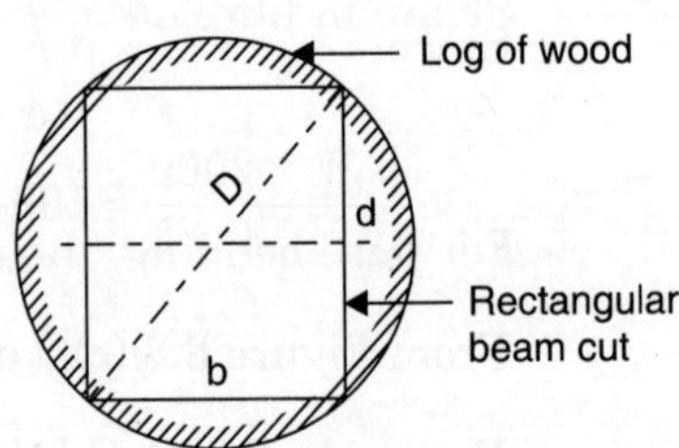

Figure 3.2.

$M = \sigma \cdot z$; for a limiting value of 'σ', M is maximum when z is maximum.

$$z = \frac{I}{y} = \frac{bd^3/12}{d/2} = \frac{bd^2}{6}, \quad ...(i)$$

and $D^2 = b^2 + d^2$ (from Fig. 3.2) ...(*ii*)

From (*i*) and (*ii*), $z = \frac{1}{6} \cdot b \cdot (D^2 - b^2) = \frac{1}{6}(bD^2 - b^3)$

For z_{max}, $\frac{dz}{db} = 0 : \; D^2 - 3b^2 = 0$

$$\therefore \quad b = \sqrt{\frac{D^2}{3}} = \mathbf{0.577\ D} \quad ...(3.3)$$

$$d^2 = D^2 - b^2 = 3b^2 - b^2 = 2b^2 \quad [\because \; D^2 = 3b^2]$$

$$\therefore \quad d = \sqrt{2}\,b = \mathbf{1.414\ b = 0.816\ D} \quad ...(3.4)$$

For example, if D = 180 mm, b = 0.577 × 180 = **104 mm** and d = 0.816 × 180 = **147 mm.**

Example 3.2. *A T-beam of c/s 100 × 150 × 10 mm is simply supported over a span of 2 m, the flange of 100 mm being horizontal. What concentrated load can be applied at mid-span if the maximum tensile stress is not to exceed 100 N/mm². What is then the greatest bending stress in the flange ?*

Solution. See Figure 3.3.

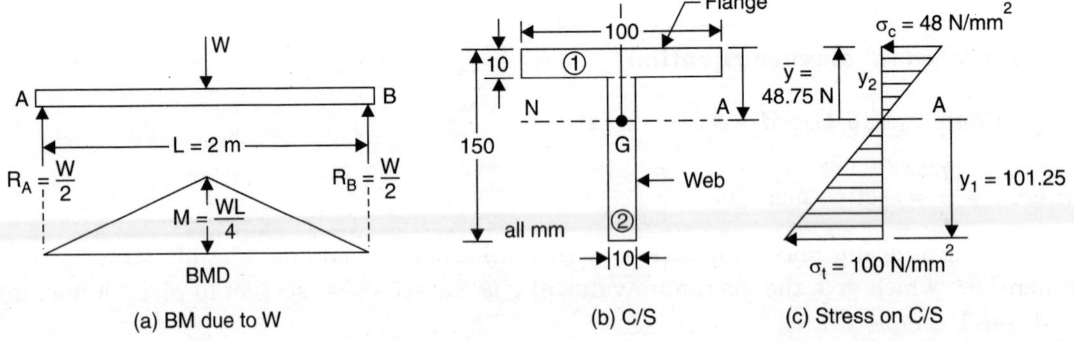

Figure 3.3.

(*i*) Position of N.A. (*G*) : $A\bar{y} = a_1 y_1 + a_2 y_2$

$$\bar{y} = \frac{(100 \times 10) \times 5 + (140 \times 10) \times 80}{(100 \times 10) + (140 \times 10)} = 48.75 \text{ mm}$$

(ii) $I_{N.A.} = \left\{\frac{100 \times 10^3}{12} + (100 \times 10) \times 43.75^2\right\} + \left\{\frac{10 \times 140^3}{12} + (140 \times 10) \times 31.25^2\right\}$

$= 5.576 \times 10^6 \text{ mm}^4$

Since, $M = \sigma \frac{I}{y}$, But $M = \frac{WL}{4}$, $\sigma = \sigma_t$ (max. stress) at $y = y_1$

$\frac{W \times 2000}{4} = 100 \times \frac{5.576 \times 10^6}{101.25}$, $W = 11014.3$ N or **11 kN**

From Figure 3.3 (c), $\sigma_c = \sigma_t \times \frac{y_2}{y_1} = 100 \times \frac{48.75}{101.25}$ = **48 N/mm²**

Example 3.3. *A.C.I. beam of I-section shown in Figure 3.4, is simply supported over a span of 6 m. If the allowable stresses are 30 MN/m² in tension and 150 MN/m² in compression, find the **udl** the beam can carry. What are the actual extreme fibre stresses?*

Solution. See Figure 3.4.

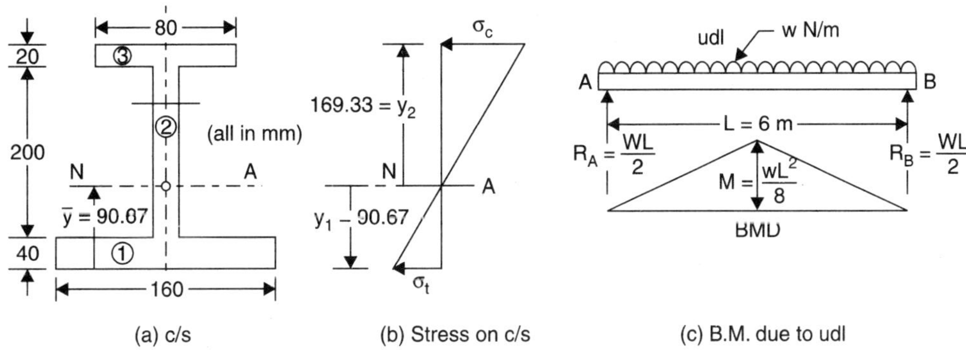

Figure 3.4.

(i) To find N.A. : $A\bar{y} = a_1y_1 + a_2y_2 + a_3y_3$

$$\bar{y} = \frac{(160 \times 40)^3 \times 20 + (200 \times 20) \times 140 + (80 \times 20) \times 250}{(160 \times 40) + (200 \times 20) + (80 \times 20)} = 90.67 \text{ mm}$$

(ii) To find $I_{N.A.} = \left\{\frac{160 \times 40^3}{12} + (160 \times 40) \times 70.67^2\right\} + \left\{\frac{20 \times 200^3}{12} + (200 \times 20) \times 49.33^2\right\}$

$$+ \left\{\frac{80 \times 20^3}{12} + (80 \times 20) \times 159.33\right\} = 96.56 \times 10^6 \text{ mm}^4$$

(iii) Working stresses : $\frac{\sigma_t}{y_1} = \frac{\sigma_c}{y_2}$

For, $\sigma_t = 30$ N/mm², $\sigma_c = \sigma_t \times \frac{y_2}{y_1} = 30 \times \frac{169.33}{90.67} = 56.02$ N/mm² < 150 N/mm² (σ_c allow)

For, $\sigma_c = 150$ N/mm², $\sigma_t = \sigma_c \times \frac{y_1}{y_2} = 150 \times \frac{90.67}{169.33} = 80.32$ N/mm² > 30 N/mm² (σ_t allow)

Hence, the tensile stress is the limiting one.

$\therefore$ Working stresses are : σ_t = **30 N/mm², σ_c = 56.02 N/mm².**

(*iv*) UDL the beam can carry, see Figure 3.4 (*c*),

From Eqn. (3.1), $\dfrac{M}{I} = \dfrac{\sigma}{y} = \dfrac{\sigma_t}{y_1}, M = \dfrac{wL^2}{8}$

$$\frac{w \times 6^2 \times 10^3}{8 \times (96.56 \times 10^6)} = \frac{30}{90.67}, \quad w = 7.1 \times 10^3 \text{ N/m or } \mathbf{7.1\ kN/m}$$

Example 3.4. *A horizontal cantilever 3 m long is of rectangular section 60 mm wide throughout its length, the depth varying uniformly from 6 mm at the free end to 180 mm at the fixed end. A load of 4 kN is applied at the free end. Find the maximum bending stress induced anywhere in the cantilever. Neglect the weight of the cantilever itself.*

Solution. See Figure 3.5.

The depth d at a distance x (in metre) from the free end

$$d = 60 + (180 - 60) \times \frac{x}{3} = 60 + 40x$$

c/sl dimensions = 60 × (60 + 40*x*) mm

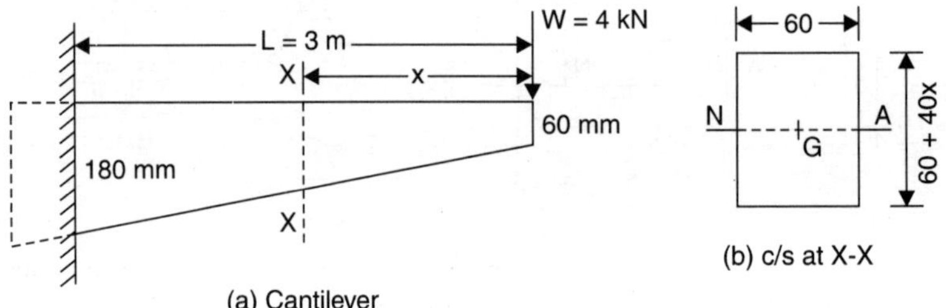

Figure 3.5.

$$\sigma = M \times \frac{y}{I} = (4 \times 10^3 \times (x \times 10^3)) \times \frac{(60 + 40x)/2}{60 \times (60 + 40x)^3/12} \qquad ...(i)$$

For σ_{max}, $\dfrac{d\sigma}{dx} = 0 : \quad \dfrac{d}{dx}\left\{\dfrac{x}{(60+40x)^2}\right\} = 0$

$$(60 + 40x)^2 - 2x\,(60 + 40x)\,40 = 0$$

$$(60 + 40x)\,(60 + 40x - 80x) = 0, \quad x = 1.5 \text{ m}$$

From (*i*), $\sigma_{max} = \dfrac{(4 \times 10^3)(1.5 \times 10^3) \times 12}{60(60 + 40 \times 1.5)^2 \times 2}$ = **41.67 N/mm²**

at 1.5 m from the free end of the cantilever.

Moment of Inertia of Symmetrical Sections:

For the three sections shown in Fig. 3.6 , $I_{XX} = \dfrac{1}{12}\,(BD^3 - bd^3)$ ...(3.5)

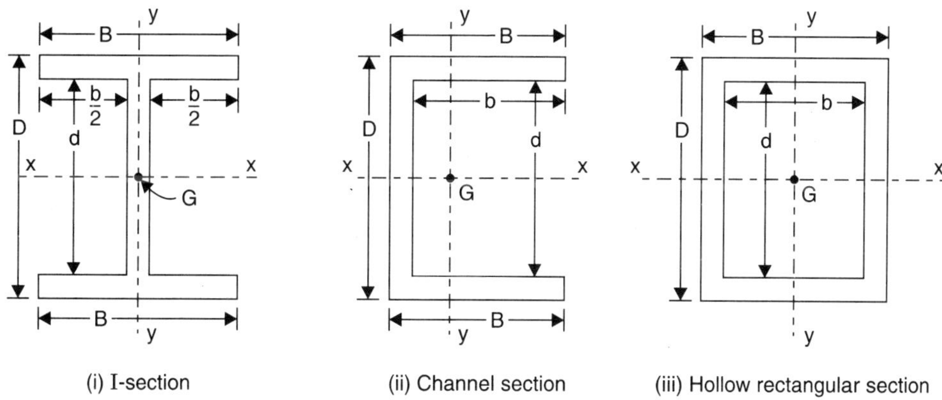

Figure 3.6. M.I. of symmetrical sections

B. Shear Stress in Beams—Due to S.F. at any section of a beam, there is vertical shear stress τ, and this is accompanied at any point in the section by a complimentary shear stress on a horizontal section, see Figure 3.7. The intensity of shear stress varies across the depth of the beam, but for all practical purposes assumed constant across the width.

To get an expression for the shear stress at any depth, consider two sections of a beam δx apart subjected to bending moment and shear, see Figure 3.7.

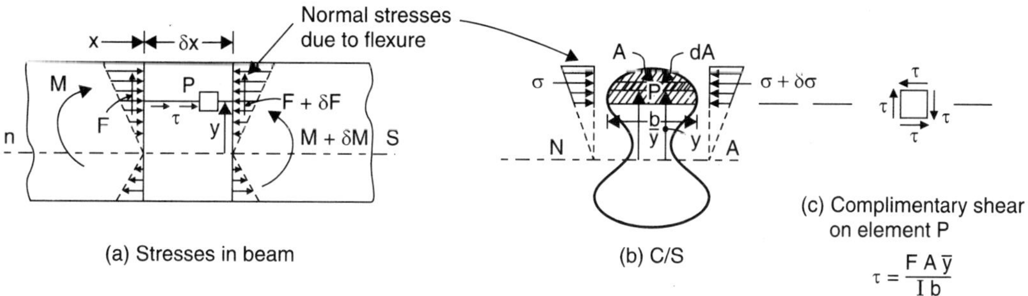

Figure 3.7. Shear stress in beams

Let τ be the shear stress at a distance *y* from the N.A. where the width of cross-section is *b*, and *A* is the area of cross-section above the horizontal line *b*. Then, for equilibrium.

S.F. on horizontal *c/s* = difference of normal forces

$$\tau\,(b\,.\,dx) = \int_A \delta\sigma \cdot dA$$

$$\sigma = \frac{M \times y}{I}, \quad \sigma + \delta\sigma = (M + \delta M)\,\frac{y}{I}, \quad \delta\sigma = \delta M \times \frac{y}{I}$$

$$\tau\,(b \cdot dx) = \frac{\delta M}{I}\int_A y \cdot dA, \quad \int_A y \cdot dA = A\bar{y}$$

$$\tau = \frac{dM}{dx} \times \frac{A\bar{y}}{Ib}, \quad \frac{dM}{dx} = F \text{ at the section}$$

$$\therefore \quad \boxed{\tau = \frac{\mathbf{FA\bar{y}}}{\mathbf{Ib}}} \qquad ...(3.6)$$

where $\bar{y}$ = distance to the centroid of the area A,

I = moment of inertia of the whole section of the beam about the N.A.,

In many cases, it will be convenient to determine $A\bar{y}$ as

$$A\bar{y} = a_1y_1 + a_2y_2$$

***(a)* Rectangular Section,** See Figure 3.8.

If the S.F. at the section is F, $\tau_{av} = \frac{F}{A_{c/s}} = \frac{F}{bd}$ ...(*i*)

At a distance y from N.A., area above the horizontal line at y,

$$A = b\left(\frac{d}{2} - y\right);$$

Figure 3.8.

C.G., of the shaded area A,

$$\bar{y} = y + \frac{1}{2}\left(\frac{d}{2} - y\right) = \frac{1}{2}\left(\frac{d}{2} + y\right)$$

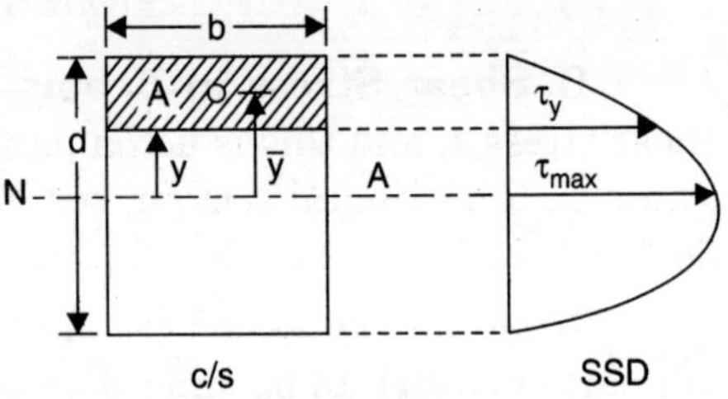

$$\tau_y = \frac{FA\bar{y}}{Ib} = \frac{F \cdot b\left(\frac{d}{2} - y\right)\left(\frac{d}{2} + y\right)}{(bd^3/12) \times b \times 2}$$

$$\tau_y = \frac{6F}{bd^3}\left(\frac{d^2}{4} - y^2\right)$$, which is parabolic.

τ_y is maximum at $y = 0$, i.e., at the N.A.

$$\therefore \quad \tau_{max} = \frac{6F}{bd^3} \times \frac{d^2}{4} = \frac{3}{2} \times \frac{F}{bd} = \frac{\mathbf{3}}{\mathbf{2}}\,\tau_{\mathbf{av}} \qquad ...(3.7)$$

Thus, the maximum shear stress is 1.5 times the average shear stress over the section.

***(b)* Circular Section.** See Figure 3.9.

$$\tau = \frac{FA\bar{y}}{Ib} = \frac{F\int (2x \cdot dy)\, y}{(\pi R^4/4)\, 2R\cos\theta} = \frac{4F\int \sqrt{R^2 - y^2}\; y\, dy}{\pi R^5 \cos\theta}$$

Put $R^2 - y^2 = u$, $\quad -2y\, dy = du$, $\quad ydy = -\frac{1}{2}\, du$

$$\tau = \frac{4F}{\pi R^5 \cos\theta}\int u^{1/2}\left(-\tfrac{1}{2}\right) du = \frac{4F}{\pi R^5 \cos\theta} \times \frac{1}{2} \times \frac{2}{3}\left[(R^2 - y^2)^{3/2}\right]_R^{R\sin\theta}$$

$$= \frac{4F}{\pi R^5 \cos\theta} \times \frac{1}{3}\,(R^2 - R^2\sin^2\theta)^{3/2}$$

$$\tau = \frac{4F}{3\pi R^2}\cos^2\theta, \quad \text{which is parabolic}$$

At N.A., $\tau_{max} = \dfrac{4F}{3\pi R^2} = \dfrac{\mathbf{4}}{\mathbf{3}}\tau_{av}$ ...(3.8)

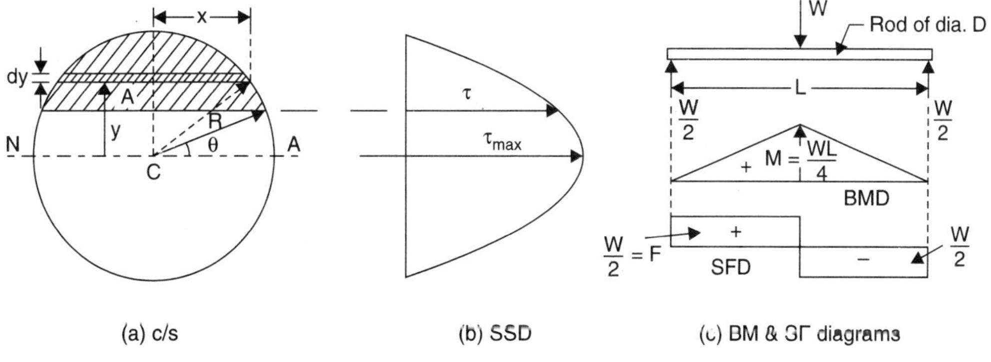

(a) c/s (b) SSD (c) BM & SF diagrams

Figure 3.9.

For a rod of circular section used as a beam simply supported at ends and with a concentrated load W at mid span (see Figure 3.9 (*c*)), the maximum bending stress due to $M = \dfrac{WL}{4}$ is

$$\sigma_{max} = M \times \frac{y}{I} = \frac{WL}{4} \times \frac{R}{(\pi R^4/4)} = \frac{WL}{\pi R^3}$$

and the ratio, $\dfrac{\sigma_{max}}{\tau_{max}} = \dfrac{WL/\pi R^3}{4F/3\pi R^2}, \quad F = \dfrac{W}{2}$

$$= \frac{3}{2} \times \frac{L}{R} \text{ or } 3 \times \frac{L}{D} \quad \text{...(3.9)}$$

i.e., three times the ratio of span to rod diameter.

(*c*) I-section, see Figure 3.10

In flange, $\tau = \dfrac{F}{IB} \cdot B\left(\dfrac{D}{2} - y\right) \cdot \dfrac{1}{2}\left(\dfrac{D}{2} + y\right)$

$$= \frac{F}{2I}\left(\frac{D^2}{4} - y^2\right), \text{ which is parabolic.}$$

At, $y = \dfrac{D}{2}, \quad \tau_{1-1} = 0$

At, $y = \dfrac{d}{2}, \quad \tau_{2-2} = \dfrac{F}{81}(D^2 - d^2)$

In web, $\tau = \dfrac{F}{Ib}\left\{\left(\dfrac{d}{2} - y\right) b \times \dfrac{1}{2}\left(\dfrac{d}{2} + y\right) + B\left(\dfrac{D-d}{2}\right) \times \dfrac{1}{2}\left(\dfrac{d+d}{2}\right)\right\}$

$$= \frac{F}{Ib}\left\{\frac{b}{2}\left(\frac{d^2}{4} - y^2\right) + \frac{B}{8}(D^2 - d^2)\right\}$$

$$\tau = \frac{B}{b} \times \frac{F}{8I}(D^2 - d^2) + \frac{F}{2I}\left(\frac{d^2}{4} - y^2\right), \quad \text{which is parabolic.}$$

$$\text{At } y = \frac{d}{2}, \quad \tau_{3\text{-}3} \; \frac{B}{b} \cdot \frac{F}{8I}(D^2 - d^2)$$

$$\text{At N.A., } y = 0, \quad \tau_{max} = \frac{B}{b} \cdot \frac{F}{8I}(D^2 - d^2) + \frac{Fd^2}{8I}$$

Average shear stress in web

$$= \tau_{3\text{-}3} + \frac{2}{3}(\tau_{max} - \tau_{3\text{-}3})$$

$$\tau_{av} = \frac{B}{b} \cdot \frac{F}{8I}(D^2 - d^2) + \frac{2}{3} \cdot \frac{Fd^2}{8I}$$

$$[\tau_{max} - \tau_{mean}]_{web} = \frac{Fd^2}{24I} \qquad \text{...(3.10)}$$

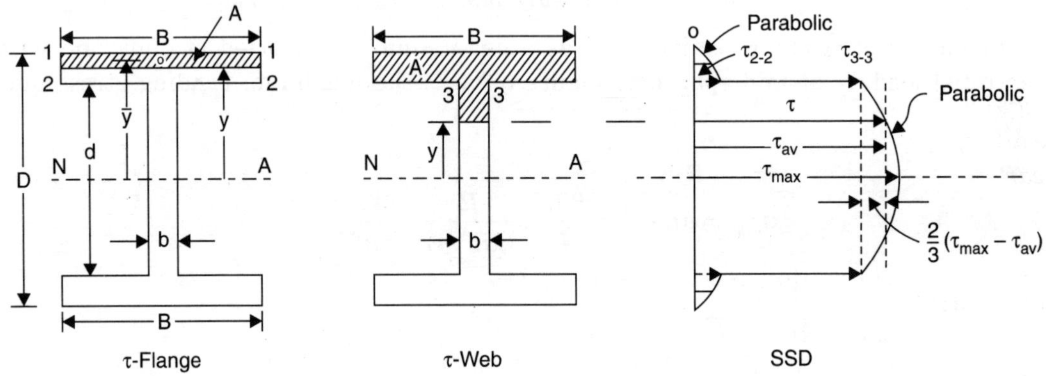

Figure 3.10. SSD in I-section

Nearly, 95% of the S.F. at the section is carried by the web, and the shear stress in the flanges is negligible. Since, the variation of shear stress is small over the web, it is assumed, in practice, that all the S.F. is carried by the web and the shear stress is uniformly distributed over it, and the B.M. at the section is carried wholly by the flanges. The practical utility of the *I*-section is thus evident.

***(d)* Beam of square cross-section with one of its diagonals horizontal,** see Figure 3.11.

$$I = 2\left\{\frac{1}{12} \times \sqrt{2}\, a \left(\frac{a}{\sqrt{2}}\right)^3\right\} = \frac{a^4}{12} \qquad \text{...(3.11)}$$

$$y = \frac{FA\bar{y}}{Ib} = \frac{F\left(\frac{1}{2} \times 2h \times h\right)\left(\frac{a}{\sqrt{2}} - \frac{2}{3}h\right)}{(a^4/12) \times 2h}$$

$$= \frac{Fh}{a^4}(3\sqrt{2}\, a - 4h), \quad \text{which is parabolic}$$

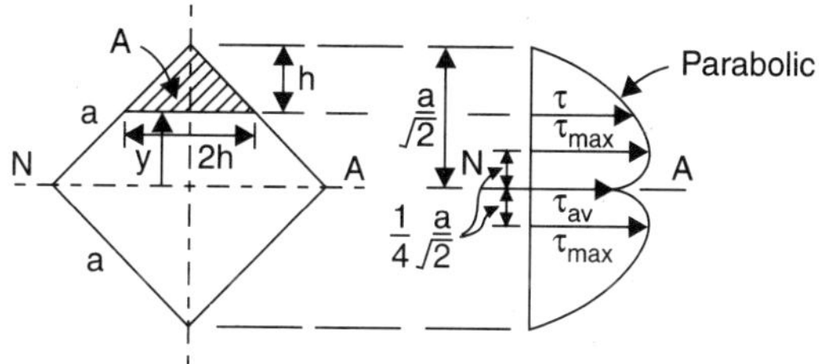

Figure 3.11.

At apex, $h = 0, \tau = 0$

At N.A., $h = \dfrac{a}{\sqrt{2}}$, $\tau = \dfrac{F}{a^2} = \tau_{av}$

For, τ_{max}, $\dfrac{\partial \tau}{\partial h} = 0$, $\dfrac{3\sqrt{2}\,F}{a^3} - \dfrac{8Fh}{a^4} = 0$, or $h = \dfrac{3}{4} \cdot \dfrac{a}{\sqrt{2}}$

$$\tau_{max} = \frac{F}{a^4}\left[\frac{3 \cdot a}{4\sqrt{2}}\left(3\sqrt{2}\,a - 4 \cdot \frac{3}{4} \cdot \frac{a}{\sqrt{2}}\right)\right] = \frac{9}{8} \cdot \frac{F}{a^2} = \mathbf{\frac{9}{8}}\,\tau_{av} \quad \text{...(3.12)}$$

Example 3.5. *A vertical flag mast 10 m high is of square section 150 mm × 150 mm at ground level tapering to 50 mm × 50 mm at the top. A horizontal pull of 300 N acts at the top in the direction of a diagonal. Determine the ratio of the maximum bending stress to the maximum shear stress.*

Solution. The side of the square section at a depth z (in metre) from the top

$$a_z = 50 + (150 - 50) \times \frac{z}{10} = 50 + 10z, \quad M_z = 300\,(z \times 10^3) \text{ N·mm}$$

$$\frac{M}{I} = \frac{\sigma}{Y}, \quad \sigma_b = M\frac{Y}{I} = \frac{(3 \times 10^5 z)(50 + 10z)/\sqrt{2}}{(50 + 10z)^4/12} \quad \text{...}(i)$$

For σ_b to be maximum, $\dfrac{d\sigma_b}{dz} = 0$, i.e., $\dfrac{d}{dz}\left\{\dfrac{z}{(50 + 10z)^3}\right\} = 0$

$$(50 + 10z)^3 - 3z\,(50 + 10z)^2 \times 10 = 0$$

or

$$(50 + 10z)^2\,(50 + 10z - 30z) = 0, \quad z = \frac{50}{20} = 2.5 \text{ m}$$

From (*i*), $\sigma_{b\,max} = \dfrac{(3 \times 10^5 \times 2.5) \times 12}{(50 + 10 \times 2.5)^3 \times \sqrt{2}} = 15.08 \text{ N/mm}^2$

At $z = 2.5$ m, $\tau_{max} = \dfrac{9}{8} \times \dfrac{F}{a_z^2} = \dfrac{9}{8} \times \dfrac{300}{75^2} = 0.06 \text{ N/mm}^2$

$$\frac{\sigma_{b\,max}}{\tau_{max}} = \frac{15.08}{0.06} = \mathbf{251.3}$$

Note. τ_{max} at top $= \dfrac{9}{8} \times \dfrac{300}{50^2} = 0.135 \text{ N/mm}^2$.

Example 3.6. *Determine the shear stress distribution across the cross-section of an R.S.J. of I-section 200 × 100 × 10 mm. S.F. at the section = 100 kN. Determine the percentage of the S.F. carried by the web.*

Solution. For the symmetrical *I*-section, See Figure 3.12.

$$I = \frac{1}{12}(BD^3 - bd^3) = \frac{1}{12}(100 \times 200^3 - 90 \times 180^3) = 22.93 \times 10^6 \text{ mm}^4$$

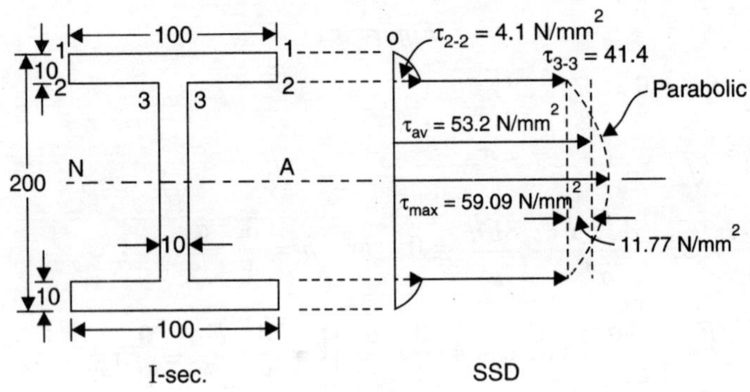

Figure 3.12.

$$\tau_{1-1} = 0$$

$$\tau_{2-2} = \frac{FA\bar{y}}{Ib} = \frac{(100 \times 10^3)(100 \times 10) \times 95}{(22.93 \times 10^6) \times 100} = 4.143 \text{ N/mm}^2$$

$$\tau_{3-3} = \frac{FA\bar{y}}{Ib} = 4.143 \times \frac{100}{10} = 41.43 \text{ N/mm}^2$$

$$\tau_{N.A.} = \frac{FA\bar{y}}{Ib} = \frac{(100 \times 10^3)\left(90 \times 10 \times \frac{90}{2} + 100 \times 10 \times 95\right)}{(22.93 \times 10^6) \times 10} = 59.09 \text{ N/mm}^2$$

τ_{av} in web = $41.43 + \frac{2}{3}(59.09 - 41.43)$ = **53.205 N/mm²**

% of S.F. carried by web = $\frac{53.205\,(180 \times 10)}{100 \times 10^3} \times 100$ = **95.77%**

The shear stress distribution (SSD) is symmetrical about the N.A. and is shown in Figure 3.12.

Example 3.7. *A T-Joist of 80 × 150 × 10 mm, with 80 mm side horizontal, is simply supported over a length of 2 m and carries a udl of 4 kN/m. Sketch the distribution of maximum shear stress intensities.*

Solution. See Figure 3.13.

$\bar{y}$ to N.A. from *AB* : $A\bar{y} = a_1y_1 + a_2y_2$

$$\bar{y} = \frac{(80 \times 10) \times 5 + (140 \times 10) \times 80}{80 \times 10 + 140 \times 10} = 52.73 \text{ mm}$$

$$I_{N.A.} = \left\{\frac{80 \times 10^3}{12} + (80 \times 10) \times 47.73^2\right\} + \left\{\frac{10 \times 140^3}{12} + (140 \times 10) \times 27.27^2\right\}$$

$= 5.16 \times 10^6$ mm^4

Max. S.F. occurs at supports, see Figure 3.13 (*c*),

$$F = \frac{WL}{2} = \frac{4000 \times 2}{2} = 4000 \text{ N}$$

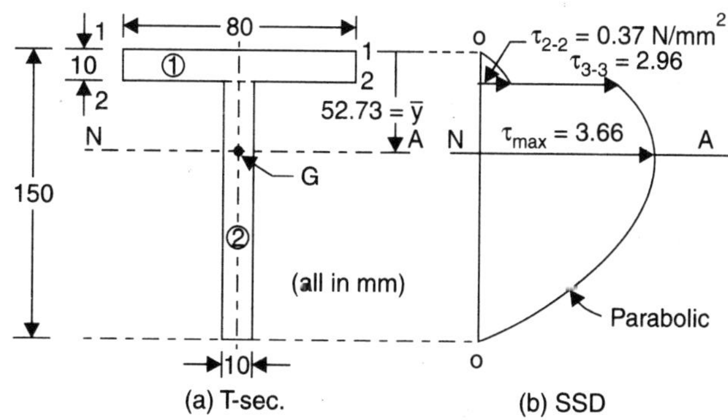

Figure 3.13.

$$\tau_{AB} = 0$$

$$\tau_{2\text{-}2} = \frac{FA\bar{y}}{Ib} = \frac{4000 \times (80 \times 10) \times 47.73}{(5.16 \times 10^6) \times 80} = 0.37 \text{ N/mm}^2$$

$$\tau_{3\text{-}3} = \frac{FA\bar{y}}{Ib} = 0.37 \times \frac{80}{10} = 2.96 \text{ N/mm}^2$$

$$\tau_{N.A.} = \frac{FA\bar{y}}{Ib} = \frac{4000\left\{(80 \times 10) \times 47.73 + (42.73 \times 10) \times \frac{42.73}{2}\right\}}{(5.16 \times 10^6) \times 10} = 3.66 \text{ N/mm}^2$$

The shear stress distribution is shown in Figure 3.13 (*b*).

PROBLEMS

3.1. A R.S.J. 450 × 300 × 20 mm (overall depth = 450 mm) is simply supported over a span 4 m. The flanges are strengthened by two 300 × 20 mm plates one riveted to each flange 300 mm wide. Determine :

(*i*) the greatest central load the beam can carry if the stress in the plate is not to exceed 120 MN/m^2.

(*ii*) the *udl* inclusive of self weight, the beam can carry for the same limitation of stress in plates.

3.2. A symmetrical *I*-section has a moment of inertia of 10 × 10^6 mm^4 and an overall depth of 150 mm. It is used as a beam simply supported at its ends and carries a *udl* of 4 kN/m. If the allowable maximum stress is 60 N/mm^2, determine the span.

What additional central load can be placed if the limiting stress is 90 N/mm^2.

3.3. A *T*-beam of cross-section 100 × 170 × 20 mm (overall depth = 170 mm) is subjected to a BM = 1 kN·m and S.F. = 50 kN. Determine the variation of flexure and shear stresses along the depth.

3.4. A beam of rectangular section 100 mm wide × 200 mm deep is loaded as shown in Figure P.3.4. The total *udl* of $4w = P$. Determine :

(*i*) the value of P, if the maximum bending stress should not exceed 120 N/mm^2 ;

(*ii*) the maximum shear stress in the section at the free end.

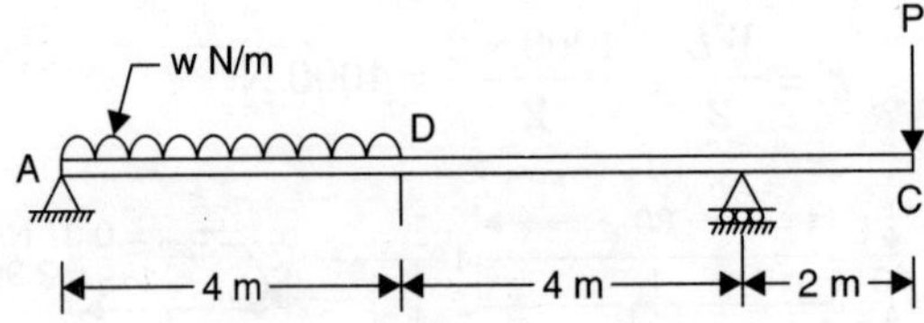

Figure P.3.4

3.5. A R.S.J. is shown in Figure P.3.5. Determine :

(*a*) the ratio of its moment of resistance to bending in the plane *Y-Y* to that for bending in the plane *X-X*, if the maximum stress due to bending is the same in both cases. [For a semi-circle of radius r, the centroid is at a distance of $\dfrac{4r}{3\pi}$ from the centre].

(*b*) the ratio of maximum to the mean shear stress if the section is subjected to a S.F. = 100 kN in the plane *Y–Y*. Sketch the variation of stress along the depth.

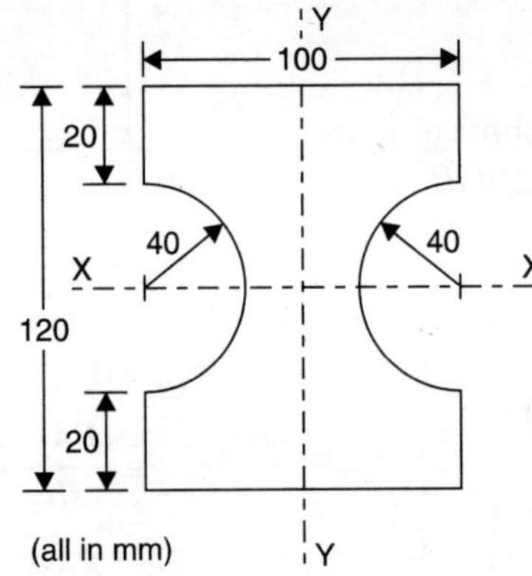

Figure P.3.5

3.6. A rolled bar of regular hexagonal cross-section of side 30 mm is used as a cantilever with one of its diagonals placed horizontal. If the S.F. at a section is 8 kN, sketch the shear stress distribution of the cross-section.

3.7. A vertical flagstaff 10 m high, is of square section tapering uniformly from 120 mm × 120 mm at bottom to 60 mm × 60 mm at top. A horizontal pull of 300 N is applied at the top along a diagonal of the section. Calculate the maximum stress due to bending.

[**Hint:** Moment of inertia of a square of side a about a diagonal = $\dfrac{a^4}{12}$.]

(**Ans.** 17.5 N/mm^2 at mid-height)

4 Torsion of Circular Shafts

Due to an applied torque T (twisting moment), applied at the end of the solid circular shaft of length L, see Figure 4.1, a line AB is twisted to AB', and the shear strain = $BAB' = \gamma$. If $\angle BOB' = \theta$, then

$$\text{arc } BB' = R\theta = \gamma L \qquad ...(i)$$

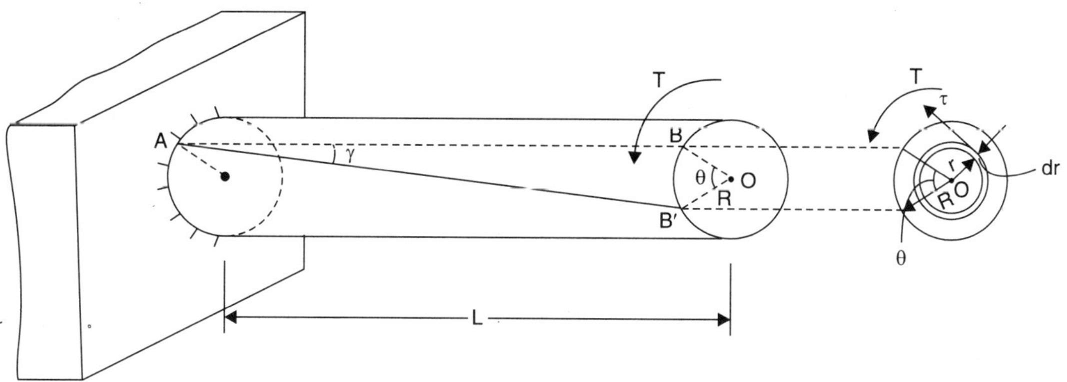

Figure 4.1. Twisting of shaft

$$\text{Rigidity modulus } (G) = \frac{\tau}{\gamma} \qquad ...(ii)$$

From (*i*) and (*ii*), $\tau = \gamma G = \dfrac{R\theta}{L} G = \tau_0$, see Fig. 4.2 (a).

For a given twist (θ), $\tau = Kr$, $K = \dfrac{G\theta}{L}$, indicating that the shear stress is proportional to the radial distance from the centre of the shaft and is maximum at the outer periphery, see Figure 4.2 (*a*).

$$\therefore \qquad \frac{\tau}{r} = \frac{G\theta}{L} \qquad ...(iii)$$

At a radial distance r, the shear stress τ, acts on circular elemental arc $(2\pi r \cdot dr)$ and the moment of the tangential force is $(2\pi r \cdot dr \cdot \tau) \cdot r$. Then, the applied torque

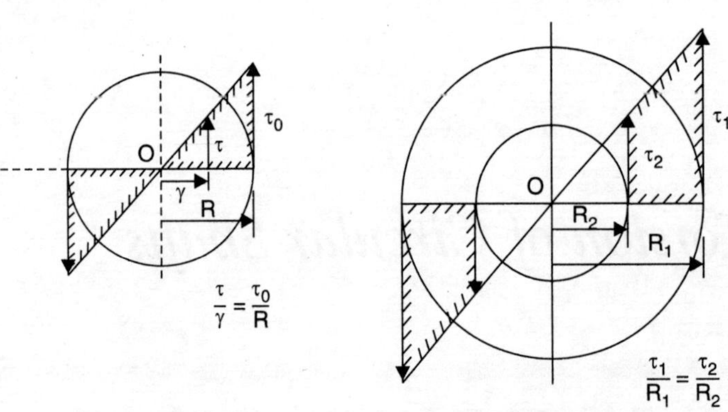

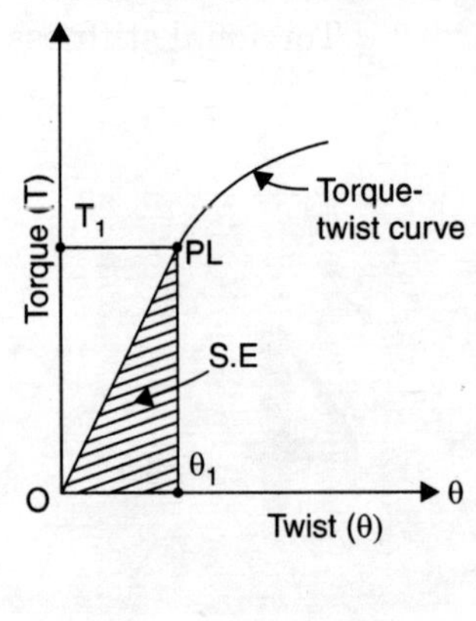

(a) τ-Solid circular shaft (b) τ-Hollow circular shaft (c) Torque-twist curve

Figure 4.2. Shear stress distribution

$$T = \int_0^R 2\pi r \cdot dr \cdot \tau \cdot r$$

Substituting for τ from (*iii*), i.e., $\tau = \frac{G\theta}{L} \cdot r$ in above equation,

$$T = \frac{G\theta}{L}\int_0^R (2\pi r \cdot dr) \times r^2$$

$$\int_0^R (2\pi\, r \cdot dr) \times r^2 = J \text{ (polar moment of inertia (about the axis) of the shaft)}$$

$$\therefore \qquad T = \frac{G\theta}{L} \cdot J \qquad \text{...}(iv)$$

From (*iii*) and (*iv*), the torsion equation is

$$\boxed{\frac{\mathbf{T}}{\mathbf{J}} = \frac{\mathbf{t}}{\mathbf{r}} = \frac{\mathbf{G\theta}}{\mathbf{L}}} \qquad \text{...(4.1)}$$

In the design of shafts, τ_0 should not exceed the maximum permissible shear stress given by

$$\frac{T}{J} = \frac{\tau_0}{R} = \frac{G\theta}{L} \qquad \text{...(4.1}a\text{)}$$

where, $R = \frac{D}{2}$, and the twist also is limited, see Example 4.1.

For hollow circular shafts, the τ-variation is shown in Figure 4.2(*b*).

For a solid shaft : $J = \frac{\pi D^4}{32}$, $\boxed{\tau_0 = \frac{\mathbf{16\,T}}{\pi \mathbf{D^3}}}$ ...(4.2)

For a hollow shaft : $J = \frac{\pi}{32}(D^4 - d^4)$,

$$\boxed{\tau_0 = \frac{\mathbf{16\,T}}{\pi\,\mathbf{D^3}\left[1-\left(\frac{\mathbf{d}}{\mathbf{D}}\right)^4\right]}} \qquad \text{..(4.3)}$$

Torsional stiffness k, is defined as torque per radian of twist i.e.,

$$k = \frac{T}{\theta} = \frac{GJ}{L} \quad \text{...(4.4)}$$

Strain energy (S.E.) in torsion is the work in twisting, see Figure 4.2 (*c*),

$$\text{S.E.} = \frac{1}{2}\, T\theta, \text{ and for a } \textbf{solid shaft}$$

From Eqn. (4.1*a* and 4.2), $= \dfrac{1}{2}\left(\dfrac{\pi D^3}{16}\cdot\tau_0\right)\left(\dfrac{2\,\tau_0\,L}{GD}\right) = \dfrac{\tau_0^{\;2}}{4G}\left(\dfrac{\pi D^2}{4}\cdot L\right)$

$$\therefore \qquad \text{S.E.} = \frac{\tau_0^{\;2}}{4G} \times \text{Volume} \quad \text{...(4.5)}$$

Resilience $U = \dfrac{\tau_0^{\;2}}{4G}$ ...(4.6)

which is equal to half that due to direct shear.

Example 4.1. (*a*) *A solid shaft has to transmit 500 kW at 200 rpm and the working conditions to be satisfied are (i) the shear stress must not exceed 70 kN/m², and (ii) the shaft must not twist more than 1° in a length of 15 diameters. Determine the diameter of the shaft and the actual working stress.*

(*b*) *If the solid shaft is replaced by a hollow shaft with diameter ratio 2 : 3 of the same material and working stress, determine the size of the shaft and the percentage saving in weight. Compare their stiffness.*

Take, $G = 80\ GN/m^2$.

Solution. Power, P (in kW) $= \dfrac{T\cdot w}{1000} = \dfrac{T}{1000}\cdot\dfrac{2\pi N}{60}$

$$P = \frac{2\pi NT}{60{,}000}, \text{ where } N = \text{rpm}$$

$$500 = \frac{2\pi\times 200\times T}{60{,}000}$$

$$\Rightarrow \qquad T = 2.387\times 10^4\ \text{N}\cdot\text{m or } 2.387\times 10^7\ \text{N}\cdot\text{mm}$$

(*a*) **For a solid shaft**

Condition (i): $T = \dfrac{\pi D^3}{16}\cdot\tau_0$, from Eqn. (4.2)

$$2.387\times 10^7 = \frac{\pi D^3}{16}\times 70 \quad , \quad D = 120.2 \text{ mm}$$

Condition (ii): $\dfrac{T}{J} = \dfrac{G\theta}{L}$, where θ in radians.

$$\frac{2.387\times 10^7}{\dfrac{\pi D^4}{32}} = \frac{(80\times 10^3)\dfrac{\pi}{180}}{15\,D} \quad , \quad D = 137.7 \text{ mm}$$

Adopt the larger diameter, D_1 = 137.7 mm, say **138 mm**

For the design diameter of 138 mm, the actual working stress is given by

$$T = \frac{\pi D^3}{16} \times \tau_0$$

$$2.387 \times 10^7 = \frac{\pi \times 138^3}{16} \cdot \tau_0 \, , \quad \tau_0 = \mathbf{46.26\ N/mm^2}$$

(*b*) **For a hollow shaft**

$$T = \frac{\pi D^3}{16} \cdot \tau_0 \left[1 - \left(\frac{d}{D}\right)^4\right], \quad \text{from Eqn. (4.3)}$$

$$2.387 \times 10^7 = \frac{\pi D^3}{16} \times 46.26 \left[1 - \left(\frac{2}{3}\right)^4\right]$$

$$\therefore \quad D = \mathbf{148.5\ mm}, \quad d = 148.5 \times \frac{2}{3} = \mathbf{99\ mm}$$

% saving in weight

$$= \frac{D_1^2 - (D^2 - d^2)}{D_1^2} \times 100 = \frac{137.7^2 - (148.5^2 - 99^2)}{137.7^2} \times 100 = \mathbf{35.39\%}$$

Stiffness $k = \dfrac{T}{\theta} = \dfrac{GJ}{L}$, i.e., $k \sim J$

$$\frac{k_{solid}}{k_{hollow}} = \frac{\frac{\pi}{32} D_1^4}{\frac{\pi}{32}(D^4 - d^4)} = \frac{137.7^4}{148.5^4 - 99^4} = \mathbf{0.92}$$

Thus, a hollow shaft is stronger and stiffer than the solid shaft of the same material, length and weight.

Example 4.2. *A stepped shaft is subjected to torque as shown in Figure 4.3. Determine the angle of twist at the free end. Also find the maximum shear stress in any step. Take, G = 80 GPa.*

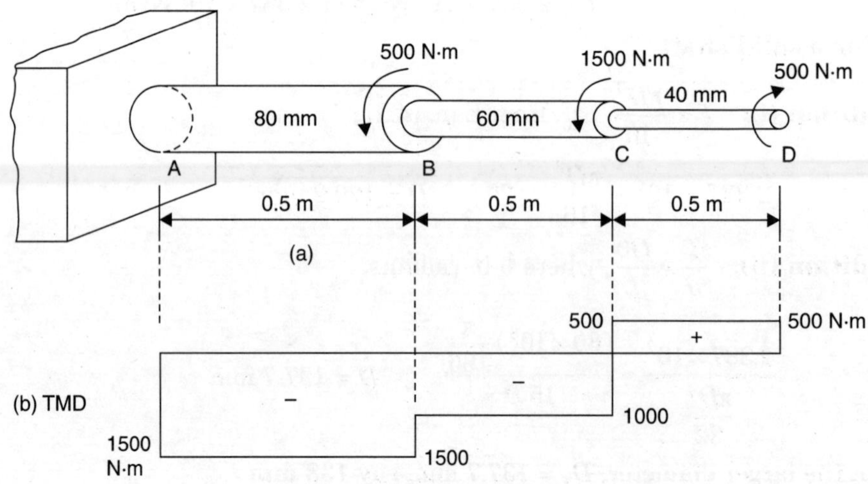

Figure 4.3. Stepped shaft

Solution. Twist at the free end $(\theta_D) = \frac{TL}{GJ}$, see Figure 4.3 (*b*)

$$= \frac{500}{(80 \times 10^3)\frac{\pi}{32}}\left[\frac{-1500 \times 10^3}{80^2} + \frac{-1000 \times 10^3}{60^4} + \frac{500 \times 10^3}{40^4}\right]$$

$$= -0.007324 - 0.015432 + 0.039062 = 0.0163 \text{ radian}$$

$$\text{or, } 0.0163 \times \frac{180^\circ}{\pi} \simeq \mathbf{1^\circ\ clockwise}$$

Further, $T = \frac{\pi D^3}{16}\tau \quad , \quad \tau = \frac{16T}{\pi D^3}$

$$\tau_{AB} = \frac{16(1500 \times 10^3)}{\pi \times 80^3} = 14.92 \text{ N/mm}^2$$

$$\tau_{BC} = \frac{16(1000 \times 10^3)}{\pi \times 60^3} = 23.58 \text{ N/mm}^2$$

$$\tau_{CD} = \frac{16(500 \times 10^3)}{\pi \times 40^3} = \mathbf{39.79\ N/mm^2}$$

∴ **Maximum shear stress** occurs in the **step CD.**

Example 4.3. *A compound shaft is attached to rigid supports at each end, see Figure 4.4. If the limiting shear stress in steel and bronze are 80 and 40 N/mm², respectively, what is the maximum torque T, that can be applied at the stepped end B. [Take, G = 80 GN/m² for steel and 40 GN/m² for bronze.]*

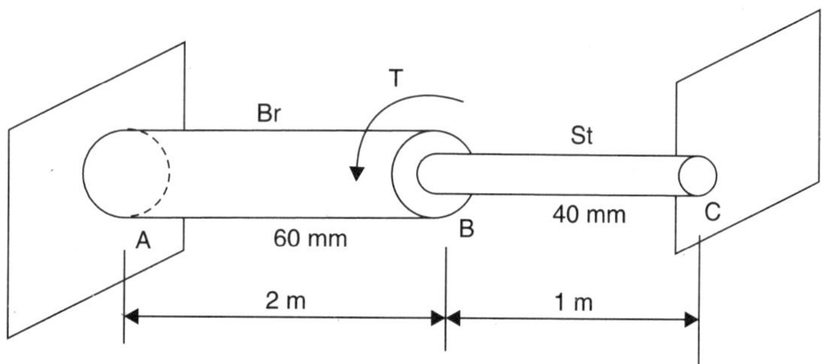

Figure 4.4. Compound shaft

Solution. The problem is statically indeterminate

$$T = T_{St} + T_{Br} \quad ...(i)$$

For compatibility, $\theta_{St} = \theta_{Br}$

i.e., $$\left(\frac{TL}{GJ}\right)_{St} = \left(\frac{TL}{GJ}\right)_{Br} \quad ...(ii)$$

From (*i*) and (*ii*), T can be obtained.

(*a*) Based on the permissible shear stress : $T = \frac{\pi D^3}{16} \cdot \tau_0$

$$T_{St} = \frac{\pi \times 40^3}{16} \times 80 = 1 \times 10^6 \text{ N· mm or 1 kN· m}$$

$$T_{Br} = \frac{\pi \times 60^3}{16} \times 40 = 1.7 \times 10^6 \text{ N· mm or 1.7 kN· m}$$

(*b*) Based on compatibility : $\theta_{St} = \theta_{Br}$

$$\frac{T_{St} \times 1000}{80\left(\frac{\pi}{32} \times 40^4\right)} = \frac{T_{Br} \times 2000}{40\left(\frac{\pi}{32} \times 60^4\right)}$$

$$\Rightarrow \qquad 1.265\, T_{St} = T_{Br}$$

Taking, $T_{St} = 1$ kN· m , $T_{Br} = 1.265 \times (1) = 1.265$ kN· m < 1.7 kN· m,

$$T = T_{St} + T_{Br} = 2.265\, T_{St}$$

$$= 2.265 \times (1) = \mathbf{2.265\ kN{\cdot}m}$$

Example 4.4. *A composite shaft consisting of a solid brass rod of 30 mm dia. enclosed in a steel tube of I.D. = 40 mm and O.D. = 50 mm. If the shaft is subjected to a pure torque of 1 kN· m, determine the maximum shear stress in steel and brass. Assume the angle of twist for the length of shaft is same and $G_{St} = 2G_{Br}$. Determine the angle of twist per metre length of the composite shaft. Determine the S.E. stored per metre length of shaft. [Take, G_{Br} = 40 GN/m².]*

Solution. For the composite shaft : $T = T_{St} + T_{Br}$ ...(*i*)

For compatibility : $\theta_{St} = \theta_{Br}$, i.e., $\left(\frac{TL}{GJ}\right)_{St} = \left(\frac{TL}{GJ}\right)_{Br}$ ...(*ii*)

$$\therefore \quad \frac{T_{St}\, L}{2G_{Br} \cdot \frac{\pi}{32}(50^4 - 40^4)} = \frac{T_{Br}\, L}{G_{Br}\left(\frac{\pi}{32} \cdot 30^4\right)}$$

$$\Rightarrow \qquad T_{St} = 9.11\, T_{Br} \qquad \text{...(iii)}$$

$$T_{St} + T_{Br} = T = 1 \times 10^6 \text{ N· mm (given)}$$

$$\therefore \qquad 9.11\, T_{Br} + T_{Br} = 1 \times 10^6 \qquad \text{...(iv)}$$

$$T_{Br} = \frac{1 \times 10^6}{10.11} = 0.099 \times 10^6 \text{ N· mm}$$

From Eqns. (*iii*) and (*iv*),

$$T_{St} = 9.11 \times (0.099 \times 10^6) = 0.902 \times 10^6 \text{ N· mm}$$

Since, $$T_{Br} = \frac{\pi d^3}{16} \cdot \tau_0$$

$$\therefore \qquad 0.009 \times 10^6 = \frac{\pi}{16} \cdot 30^3 \cdot \tau_{Br} \quad , \quad \tau_{Br} = \mathbf{18.67\ N/mm^2}$$

Also, $$T_{St} = \frac{\pi D^3}{16} \times \left[1 - \left(\frac{d}{D}\right)^4\right] \times \tau_{St}$$

$$\therefore \qquad 0.902 \times 10^6 = \frac{\pi \times 50^3}{16} \times \left[1 - \left(\frac{40}{50}\right)^4\right] \times \tau_{St}, \quad \tau_{St} = \mathbf{61.82\ N/mm^2}$$

$$\theta = \left(\frac{\tau}{R} \cdot \frac{L}{G}\right)_{Br} = \frac{18.67 \times 1000}{15 \times (40 \times 10^3)} = \mathbf{0.031 \ radian}$$

Also,
$$\theta = \left(\frac{\tau}{R} \cdot \frac{L}{G}\right)_{St} = \frac{61.82 \times 1000}{(50/2)(80 \times 10^3)} = \mathbf{0.031 \ radian}$$

Alternatively,
$$\theta = \left(\frac{TL}{GJ}\right)_{Br} = \frac{(0.099 \times 16^6) \times 1000}{40 \times 10^3 \times \left(\frac{\pi}{32} . 30^4\right)} = \mathbf{0.031 \ radian \ or \ 1.78°}$$

Note.
$$\theta = \left(\frac{TL}{GJ}\right)_{Br} = \left(\frac{TL}{GJ}\right)_{St} = \frac{TL}{(GJ)_{Br} + (GJ)_{St}}$$

$$= \frac{10^6 \times 10^3}{(40 \times 10^3)\left(\frac{\pi}{32} . 30^4\right) + (80 \times 10^3)\left(\frac{\pi}{32}(50^4 - 40^4)\right)} = \mathbf{0.031 \ radian}$$

$$\text{S.E.} = \frac{1}{2} T\theta = \frac{1}{2} \times 1000 \times (0.031) = \mathbf{15.2 \ N \cdot m \ or \ J}$$

Example 4.5. *A hollow circular shaft of O.D. = 80 mm, I.D. = 60 mm is subjected to a pure torque of 5 kN· m. Determine the maximum shear stress and tensile and compressive stresses, and indicate the stressed element on the shaft.*

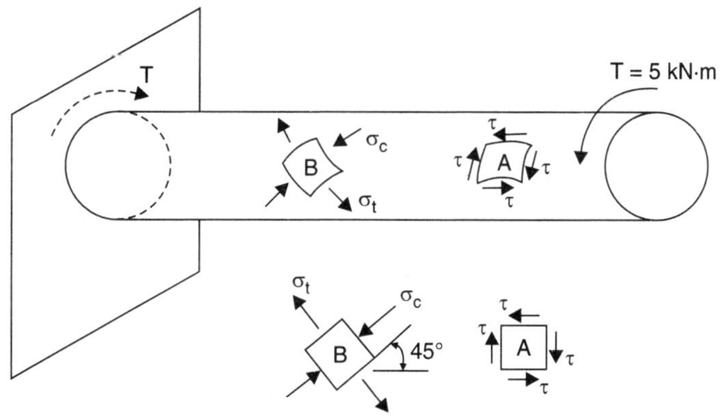

Figure 4.5. Torsion of hollow shaft

Solution. Since, $\frac{T}{J} = \frac{\tau_0}{R}$

$$\frac{5 \times 10^6}{\frac{\pi}{32}(80^4 - 60^4)} = \frac{\tau_0}{80/2}, \quad \tau_0 = \mathbf{72.76 \ N/mm^2}$$

$\sigma_t = \sigma_c =$ **72.76 N/mm²** on the diagonal planes, see Figure 4.5.

Combined Bending and Twisting. Shafts are generally subjected to B.M., torque, and shear force (S.F.); but S.F. is not so important since the maximum shear stress occurs at the neutral axis where the bending stress is zero. Sometimes, there is also an 'end thrust' as in the case of propeller shafts, and the stress due to this is assumed uniformly distributed over the cross-section.

For a solid shaft (see Figure 4.6),

Due to M: $\frac{M}{I} \times \frac{\sigma}{Y}$, $I = I_{XX} = \frac{\pi d^4}{64}$ and $Y = \frac{d}{2}$

$$\sigma = M \times \frac{Y}{I} = M \times \frac{d/2}{\pi d^4/64} = \frac{32\,M}{\pi d^3} \quad ...(i)$$

Due to T: $$\tau = \frac{16T}{\pi d^3} \quad ...(ii)$$

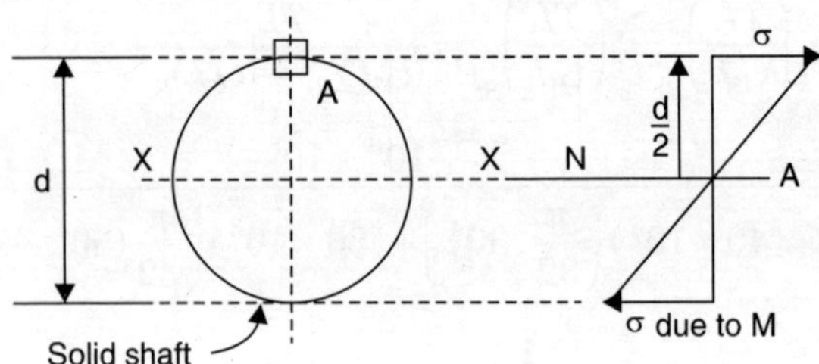

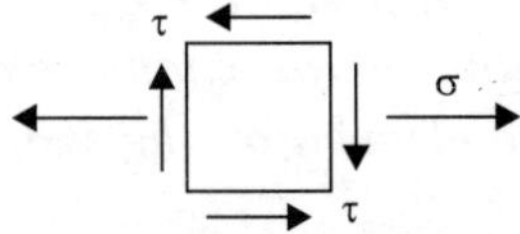

Element at A

Figure 4.6.

Maximum principal stress [See Eqn. (5.5)],

$$\sigma_1 = \frac{\sigma}{2} + \sqrt{\left(\frac{\sigma}{2}\right)^2 + \tau^2}$$

From (*i*) and (*ii*), $$\sigma_1 = \frac{16\,M}{\pi d^3} + \sqrt{\left(\frac{16M}{\pi d^3}\right)^2 + \left(\frac{16T}{\pi d^3}\right)^2}$$

$$\Rightarrow \quad \sigma_1 = \frac{16}{\pi d^3}[M + \sqrt{M^2 + T^2}] \quad ...(4.7)$$

Denoting:

Equivalent torque (T_e) = $\sqrt{M^2 + T^2}$, which would produce the same τ_{max} and

Equivalent moment (M_e) = $\frac{1}{2}$ ($M + T_e$), which would give the same max. stress σ_1

i.e., $$\tau_{max.} = \frac{16T_e}{\pi d^3}, \quad (\sigma_1)_{max.} = \frac{32M_e}{\pi d^3} \quad ...(4.8)$$

and the design diameter of the shaft is adopted as the larger of the two values of d obtained from the two criteria.

Example 4.6. *A hollow steel shaft 80 mm O.D. and 40 mm I.D. transmits 600 kW at 200 rpm, and is subjected to an end thrust of 10 kN. What bending moment may be safely applied to the shaft if the maximum principal stress does not exceed 80 N/mm². What is then the minor principal stress ?*

Solution. Since, $$P = \frac{2\pi NT}{60{,}000} \text{ kW}$$

$$\therefore \quad 600 = \frac{2\pi \times 200 \times T}{60{,}000}, \quad T = 2864 \text{ N}\cdot\text{m}$$

Also, $$T = \frac{\pi D^3}{16}\left[1 - \left(\frac{d}{D}\right)^4\right] \times \tau$$

$$2864 \times 10^3 = \frac{\pi \times 80^3}{16}\left[1 - \left(\frac{40}{80}\right)^4\right] \times \tau, \quad \tau = 30 \text{ N/mm}^2$$

The principal stresses

$$\sigma_{1,2} = \frac{\sigma}{2} \pm \sqrt{\left(\frac{\sigma}{2}\right)^2 + \tau^2}$$

$$\sigma_1 = 80 = \frac{\sigma}{2} + \sqrt{\left(\frac{\sigma}{2}\right)^2 + 30^2}$$

$$\left(\frac{\sigma}{2}\right)^2 + 30^2 = 80^2 - 80\sigma + \left(\frac{\sigma}{2}\right)^2$$

$$80\sigma = 80^2 - 30^2 = 5500, \quad \sigma = 68.75 \text{ N/mm}^2$$

Normal stress due to end thrust

$$\sigma_c = \frac{10 \times 10^3}{\frac{\pi}{4}(80^2 - 40^2)} = 2.65 \text{ N/mm}^2$$

The limiting case is obtained when both σ_c and σ_b are compressive, σ_1 also being compressive.

Stress due to bending, $\sigma_b = 68.75 - 2.65 = 66.10 \text{ N/mm}^2$

Since, $$\sigma_b = \frac{M \times Y}{I_{XX}},$$

$$\sigma_b = \frac{M \cdot D/2}{\frac{\pi}{64}(D^4 - d^4)} = \frac{32\,M}{\pi D^3\left[1 - \left(\frac{d}{D}\right)^4\right]}$$

$$66.10 = \frac{32M}{\pi \times 80^3\left[1 - \left(\frac{40}{80}\right)^4\right]}, \quad M = 3.12 \times 10^6 \text{ N}\cdot\text{m or } \mathbf{3.12\ kN\cdot m}$$

Minor principal stress

$$\sigma_2 = \frac{\sigma}{2} + \sqrt{\left(\frac{\sigma}{2}\right)^2 + \tau^2} = \frac{-68.75}{2} + \sqrt{\left(\frac{68.75}{2}\right)^2 + 30^2}$$

$= 11.25$ N/mm^2, i.e., **11.25 N/mm^2 (tensile).**

Example 4.7. *Design the diameter of a solid shaft subjected to a torque of 1.5 kN· m and a bending moment of 1 kN· m if the maximum shear stress is not to exceed 60 N/mm^2 and the maximum normal stress is not to exceed 100 N/mm^2.*

Solution. Equivalent torque

$$T_e = \sqrt{M^2 + T^2} = \sqrt{1^2 + 1.5^2} = 1.803 \text{ kN}\cdot\text{m}$$

Equivalent moment $(M_e) = \frac{1}{2}(M + T_e) = \frac{1}{2}(1 + 1.803) = 1.4025$ kN· m

Since, $\tau = \frac{16\,T_e}{\pi d^3}$

$$60 = \frac{16 \times (1.803 \times 10^6)}{d^3}, \quad d = 53.5 \text{ mm}$$

Also, $\sigma_b = \frac{32 M_e}{\pi d^3}$

$$\therefore \quad 100 = \frac{32 \times (1.4025 \times 10^6)}{\pi d^3}, \quad d = 52.3 \text{ mm}$$

Design diameter is the larger of the two values i.e., d = **53.5 mm**

PROBLEMS

4.1. A solid steel shaft of 30 mm dia. is used in the drilling rig of an oil well. The well is 5 km deep. If the drilling speed is 600 rpm and the maximum shear stress induced in the shaft is 50 N/mm^2, determine the power needed and the angle of twist at the drill end compared to the motor end. [Take, G = 80 GN/m^2.]

4.2. A solid shaft transmits 700 kW at 200 rpm. Determine the diameter of the shaft to satisfy the working conditions that

(*i*) the twist should not exceed 1° in a length of 15 diameters,

(*ii*) the shear stress must not exceed 80 N/mm^2.

[Take, G = 80 GN/m^2.]

4.3. A hollow shaft is subjected to a torque of 5 kN· m and bending moment 4 kN· m. The diameter ratio is 2/3. If the shear stress is not to exceed 80 N/mm^2 and the maximum normal stress 100 N/mm^2, what should be the design diameter of the shaft ?

4.4. A shaft of 50 mm dia. transmits 60 kW at 150 rpm. It is subjected to a bending moment of 10 kN· m and axial thrust of 20 kN. Find the principal stresses induced in the cross-section.

4.5. A hollow steel shaft of 80 mm O.D. and 40 mm I.D. transmits 500 kW at 250 rpm. What bending moment may safely be applied to the shaft so that the maximum principal stress does not exceed 90 N/mm^2. What is then the minimum principal stress ?

4.6. Design the diameter of a solid shaft subjected to a torque of 1.2 kN· m and a bending moment of 0.8 kN· m if the maximum shear stress is not to exceed 70 N/mm^2 and the maximum normal stress is not to exceed 110 N/mm^2. (**Ans.** 47.16 mm)

4.7. A torque of 2 kN· m is applied at the junction of the two shafts shown in Figure P. 4.7. What is the maximum shear stress developed in each shaft. [Take, G_{Al} = 26 GN/m^2, G_{St} = 80 GN/m^2.]

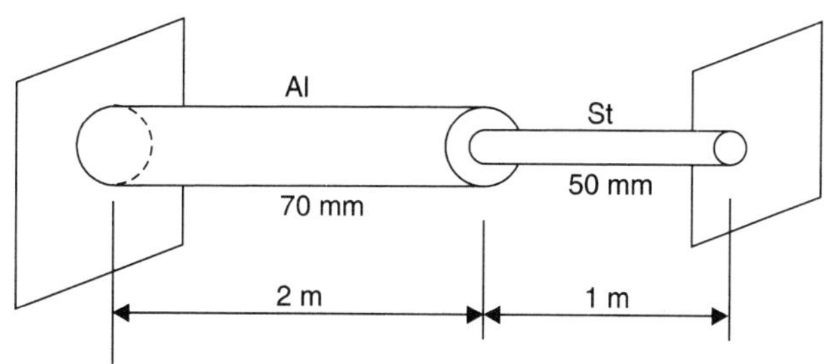

Figure P. 4.7

4.8. A compound shaft shown in Figure P. 4.8 is fixed to rigid supports. Determine the maximum shear stress developed in each segment. [Take, G_{Al} = 26 GPa, G_{Br} = 35 GPa and G_{St} = *80* GPa.]

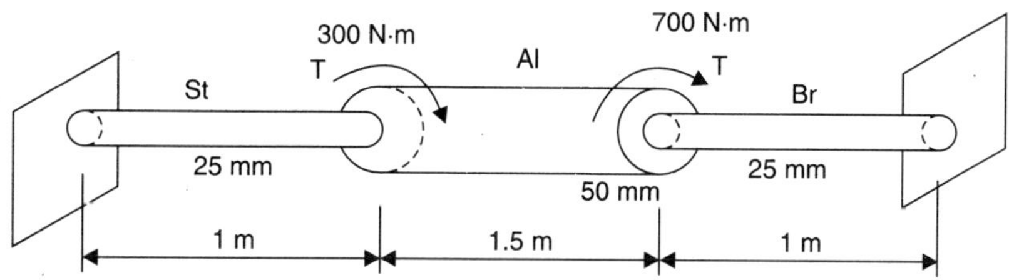

Figure P. 4.8

4.9. A solid aluminium shaft 1 m long and 50 mm outer diameter is to be replaced by a hollow circular steel shaft of the same length and O.D. such that the twist for the length of shaft is the same. What should be the I.D. of the steel shaft ? Take, G_{St} = 3 G_{Al}.

4.10. A solid shaft transmits 250 kW at 100 rpm. If the shear stress is not to exceed 75 N/mm^2, what should be the diameter of the shaft ?

If this shaft is to be replaced by a hollow one of the same length and material whose I.D. = 0.6 × O.D., determine the percentage saving in weight, the maximum shear stress being the same.

4.11. A 80 mm solid shaft is to be replaced by a hollow shaft of 100 mm O.D. Determine the thickness of metal for the new shaft (*i*) of equal strength, and (*ii*) of equal torsional stiffness. What is the percentage saving of material in each case ?

5 Compound Stresses and Strains

The general form of a two-dimensional stress system at a point in an elastic body is shown in Figure 5.1.

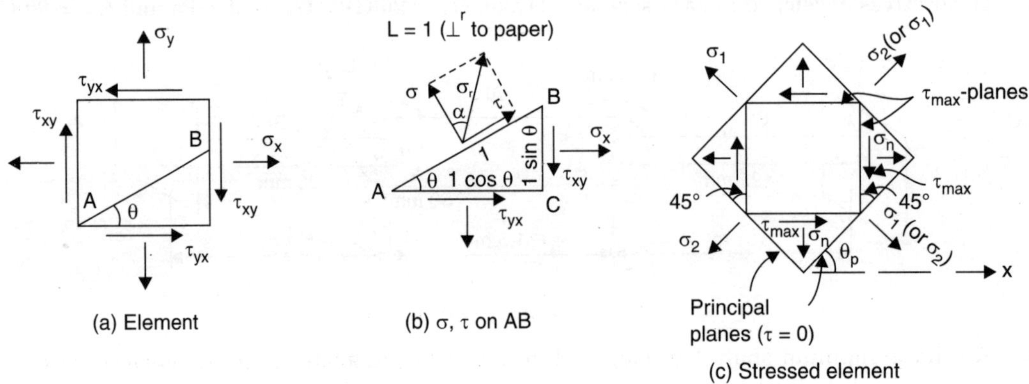

Figure 5.1. General two-dimensional stress system

The normal and shear stresses on a plane *AB* at an angle θ with the σ_x-direction can be found (by resolving the forces on the sides of the Δ*ABC* in Figure 5.1 (*b*), in σ and τ directions) as

$$\sigma = \frac{\sigma_x + \sigma_y}{2} + \frac{\sigma_y - \sigma_x}{2} \cos 2\theta + \tau_{xy} \sin 2\theta \qquad ...(5.1)$$

and

$$\tau = \frac{\sigma_y - \sigma_x}{2} \sin 2\theta - \tau_{xy} \cos 2\theta \qquad ...(5.2)$$

Their resultant is given by

$$\sigma_r = \sqrt{\sigma^2 + \tau^2} \qquad ...(5.3)$$

and

$$\tan \alpha = \frac{\tau}{\sigma} \qquad ...(5.4)$$

The planes (at $\theta = \theta_p$) on which there is no shear stress and there are only normal stresses are called principal planes; the plane on which the normal stress is maximum is called the major principal plane and of minimum normal stress, minor principal plane, and these planes are mutually at right angles to each other. The corresponding stresses are called the major and minor principal stresses, given by, see Figure 5.1 (*c*),

$$\sigma_{1,2} = \frac{\sigma_x + \sigma_y}{2} \pm \sqrt{\left(\frac{\sigma_x - \sigma_y}{2}\right)^2 + \tau_{xy}{}^2} \qquad \text{...(5.5)}$$

and their directions are given by

$$\tan 2\theta_p = -\frac{2\tau_{xy}}{\sigma_x - \sigma_y} \qquad \text{...(5.6)}$$

The planes of maximum shear stress τ_{max}, occur at 45° to the principal planes and their directions are given by

$$\tan 2\theta_s = \frac{\sigma_x - \sigma_y}{2\,\tau_{xy}} \qquad \text{...(5.7)}$$

and the normal stress on the τ_{max}-planes

$$\sigma_n = \frac{\sigma_x + \sigma_y}{2}$$

and

$$\tau_{max} = \sqrt{\frac{\sigma_x - \sigma_y}{2}^2 + \tau_{xy}{}^2} \qquad \text{...(5.8)}$$

Also,

$$\tau_{max} = \frac{\sigma_1 - \sigma_2}{2} \qquad \text{...(5.9)}$$

All these stresses can be graphically determined by constructing the classical Mohr's stress circle (after the German Engineer Otto Mohr), see Figure 5.2.

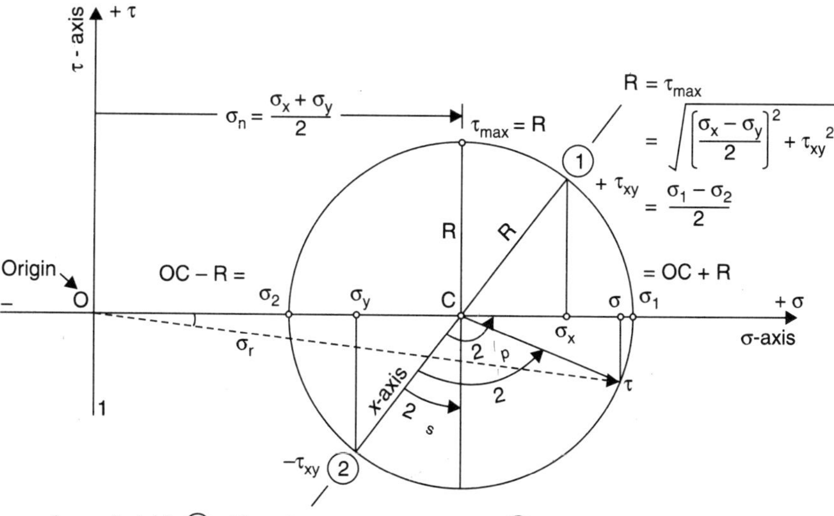

Origin-O; draw *x*-& *y*-axes; measure σ_x & σ_y and bisect to get centre C ; at σ_x draw vertical + τ_{xy} to get (1). With C-(1) as radius draw Mohr's circle (Note: σ_y-(2) = $-\tau_{xy}$)and obtain (2). C-(2) gives +ve *x*-axis. All angles are measured counter clockwise from C-(2) to get 2θ to different planes.

Figure 5.2. Mohr's stress circle [Element Fig. 5.1 (a)]

Example 5.1. (*a*) *A rectangular plate with a circular hole of 400 mm dia. is subjected to two mutually perpendicular direct stresses of 25 and 40 N/mm² along the x- and y- directions, respectively, and a shear stress of – 20 N/mm², see Figure 5.3 (a). The circular hole is deformed into the shape of an ellipse. If, E = 200 GN/m² and ν = 1/3, find the length of the major and minor axis.*

(*b*) *Also determine the normal and shear stresses on a plane inclined at 25° to the x-direction and their resultant.*

Solution. A. Analytical Solution.

(*a*) The principal stresses

$$\sigma_{1,2} = \frac{\sigma_x + \sigma_y}{2} \pm \sqrt{\left(\frac{\sigma_x - \sigma_y}{2}\right)^2 + \tau_{xy}^{\;2}}$$

$$= \frac{25 + 40}{2} \pm \sqrt{\left(\frac{25 - 40}{2}\right)^2 + (-20)^2} = 32.50 \pm 21.36$$

$$\sigma_1 = \mathbf{53.86\ N/mm^2},\ \sigma_2 = \mathbf{11.14\ N/mm^2}$$

$$\tau_{max} = \mathbf{21.36\ N/mm^2}$$

Note. $\tau_{max} = \frac{\sigma_1 - \sigma_2}{2} = \frac{53.86 - 11.14}{2} = \mathbf{21.36\ N/mm^2}$

and the directions of principal planes are given by

$$\tan 2\theta_p = -\frac{2\,\tau_{xy}}{\sigma_x - \sigma_y} = -\frac{2(-20)}{25 - 40} = -2.67$$

$\tan 69.5° = 2.67$, $2\theta_p = 180° - 69.44° = 110.56°$

$\tan (180 - 69.5°) = -2.67$, $\theta_p = \mathbf{55.28°}$, and $55.28° + 90° = \mathbf{145.28°}$

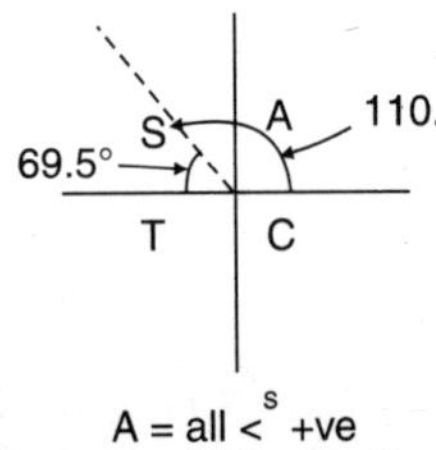

Sign convention : +ve < s : A, C, T, S

S A 110.5° 69.5° T C

A = all < s +ve
S = Sine, C = Cos, T = Tan +ve

$$\tan 2\theta_s = \frac{\sigma_x - \sigma_y}{2\,\tau_{xy}} = \frac{25 - 40}{2(-20)} = 0.375$$

$\Rightarrow \quad 2\theta_s = 20.55°, \quad \theta_s = 10.28°$, and $10.28° + 90° = 100.28°$

Note that θ_p and θ_s are at 45° to each other.

The normal stress on the maximum shear stress planes

$$\sigma_n = \frac{\sigma_x + \sigma_y}{2} = \frac{25 + 40}{2} = \mathbf{32.5\ N/mm^2}$$

To find, which of the two principal stresses act on these planes, for $\theta_p = 55.28°$, from Eqn. (5.1),

$$\sigma = \frac{\sigma_x + \sigma_y}{2} + \frac{\sigma_y - \sigma_x}{2} \cos 2\theta + \tau_{xy} \sin 2\theta$$

$$= \frac{25 + 40}{2} + \frac{40 - 25}{2} \cos 110.56° + (-20) \sin 110.56° = 11\ 14 = \sigma_2.$$

The stress system is shown in Figure 5.3 (*b*).

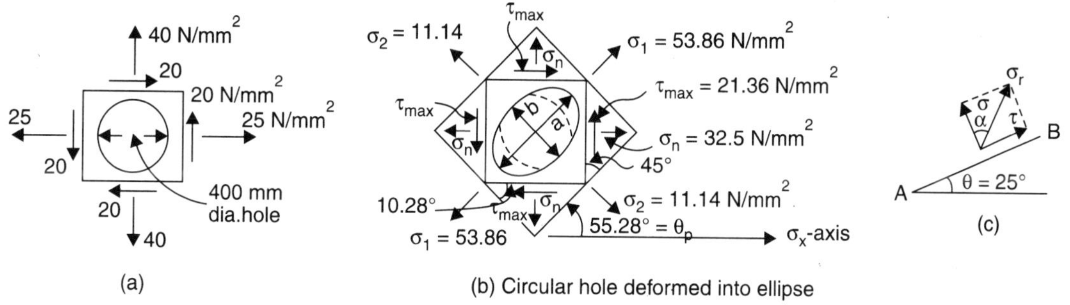

Figure 5.3. Stressed element in a plate

The circular hole of 400 mm dia is deformed into ellipse (see Figure 5.3 (*b*)), of dimensions $a \times b$. To find a and b :

$$\varepsilon_1 = \frac{\sigma_1}{E} - \nu\frac{\sigma_2}{E} - \frac{1}{E}\left(53.86 - \frac{1}{3} \times 11.14\right) - \frac{50.147}{200 \times 10^3} - 25.07 \times 10^{-5}$$

$$\because \qquad \varepsilon = \frac{\Delta L}{L}, \quad \therefore \quad a = 400 + 400 \times (25.07 \times 10^{-5}) = \mathbf{400.1\ mm}$$

$$\varepsilon_2 = \frac{\sigma_2}{E} - \nu\frac{\sigma_1}{E} = \frac{1}{E}\left(11.14 - \frac{1}{3} \times 53.86\right) = \frac{-6.81}{200 \times 10^3}$$

$$= -3.405 \times 10^{-5}$$

$$\therefore \qquad b = 400 + 400\,(-3.405 \times 10^{-5}) = \mathbf{399.9864\ mm}$$

(*b*) To find σ and τ on plane with θ = 25° :

From Eqns. (5.1) and (5.2),

$$\sigma = \frac{\sigma_x + \sigma_y}{2} + \frac{\sigma_y - \sigma_x}{2}\cos 2\theta + \tau_{xy}\sin 2\theta$$

$$= \frac{25 + 40}{2} + \frac{40 - 25}{2}\cos 50° + (-20)\sin(2 \times 25°)$$

$$= \mathbf{21.995\ N/mm^2}$$

and

$$\tau = \frac{\sigma_y - \sigma_x}{2}\sin 2\theta - \tau_{xy}\cos 2\theta$$

$$= \frac{40 - 25}{2}\sin 50° - (-20)\cos(2 \times 25°) = \mathbf{18.585\ N/mm^2}$$

The resultant stress $(\sigma_r) = \sqrt{\sigma^2 + \tau^2}$

$$= \sqrt{22^2 + 18.6^2} = 28.81\ \text{N/mm}^2$$

and its inclination $\tan\alpha = \frac{\tau}{\sigma} = \frac{18.6}{22} = 0.845$, $\alpha = 40.2°$

B. Graphical Solution. The Mohr's stress circle is constructed as shown in Figure 5.4, and all the planes and stresses are determined.

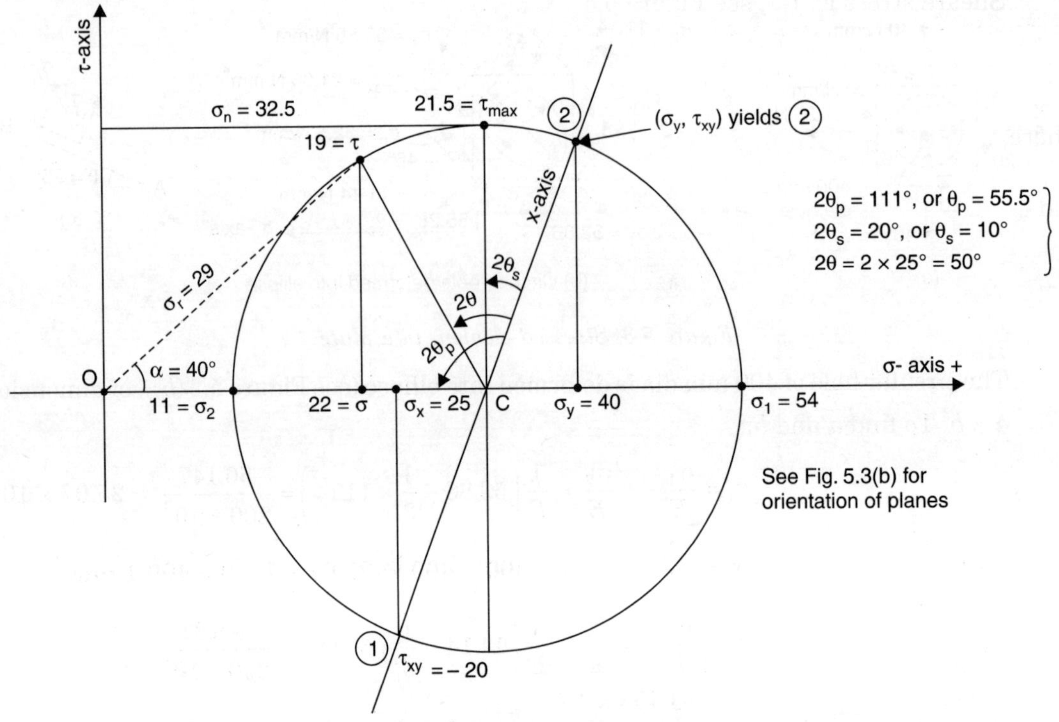

Figure 5.4. Mohr's stress circle

Example 5.2. *A simply supported beam of span 3 m carries a point load of 30 kN at a distance of 1 m from one support. The beam is of hollow square section with outer dimensions of 100 mm and wall thickness 25 mm. Determine the maximum bending stress and transverse shear stress in the beam at 25 mm from the neutral axis; hence determine the maximum principal stress and maximum shear stress at this point.*

Solution.

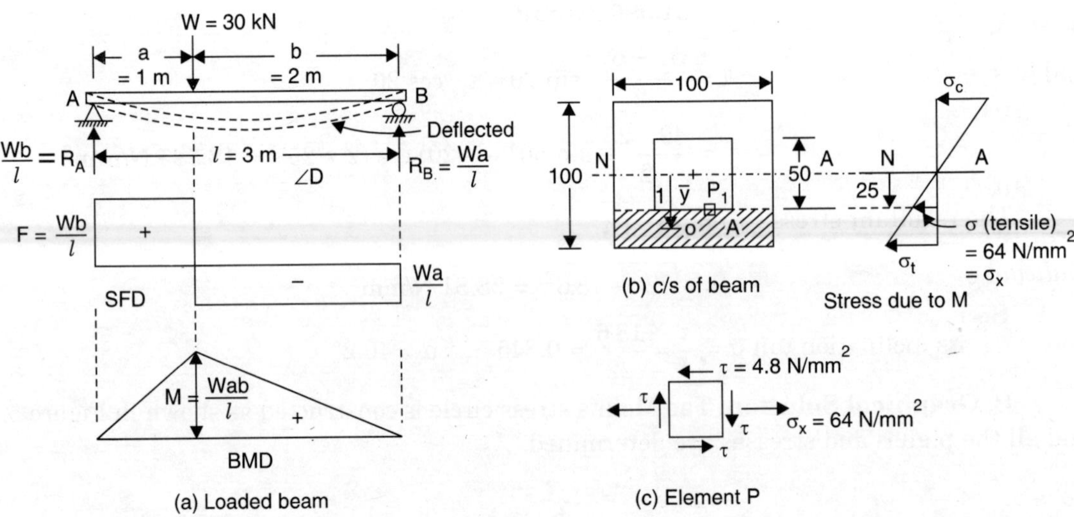

Figure 5.5.

Sheare stress at 1-1, see Figure 5.5

$$\tau = \frac{FA\bar{y}}{Ib}$$

where, $$F = \frac{Wb}{l} = 30 \times 10^3 \times \frac{2}{3} = 2 \times 10^4 \text{ N}$$

and $$I = \frac{1}{12}(100 \times 100^3 - 50 \times 50^3) = 7.81 \times 10^6 \text{ mm}^4$$

$$\therefore \quad \tau = \frac{(2 \times 10^4)(100 \times 25)(25 + 12.5)}{(7.81 \times 10^6) \times 50} = \mathbf{4.8\ N/mm^2}$$

Bending stress at 1-1

$$\frac{M}{I} = \frac{\sigma}{y}, \text{ where } M = \frac{Wab}{l} = (30 \times 10^3)\, 1 \times \frac{2}{3} = 2 \times 10^4 \text{ N}\cdot\text{m or } 2 \times 10^7 \text{ N}\cdot\text{mm}$$

$$\sigma = M \times \frac{Y}{I} = (2 \times 10^7) \times \frac{25}{(7.8 \times 10^6)} = \mathbf{64\ N/mm^2} = \sigma_x$$

The stresses on the element at P are shown in Figure 5.5 (c).

Max. principal stress from Eqn. (5.5), putting $\sigma_y = 0$, $\sigma_x = 64$ N/mm^2 (tensile)

$$\sigma_1 = \frac{\sigma_x}{2} + \sqrt{\left(\frac{\sigma_x}{2}\right)^2 + \tau_{xy}^{\ 2}}$$

$$= \frac{64}{2} + \sqrt{\left(\frac{64}{2}\right)^2 + 4.8^2} = 32 + 32.36$$

$$\sigma_1 = \mathbf{64.36\ N/mm^2}$$

The max. shear stress acts on planes at 45° to the principal planes, and

$$\tau_{max} = \mathbf{32.36\ N/mm^2}$$

Example 5.3. *A tubular shaft is subjected to combined bending and torsion. An element on the surface experiences a tensile bending stress of 40 N/mm^2 and a shear stress of 50 N/mm^2 as shown in Figure 5.6.*

Determine:

(*i*) *Normal and shear stresses on a plane inclined at 60° to the generator, as shown in Figure 5.6, and their resultant.*

(*ii*) *The maximum principal stress and shear stress.*

(*iii*) *The positions of the planes on which there is no normal stress.*

Indicate the stresses and planes in a Mohr's circle.

Solution. From Eqns. (5.1) and (5.2), putting $\sigma_y = 0$, $\theta = 180° - 60° = 120°$

$$\text{Normal stress } (\sigma) = \frac{\sigma_x + \sigma_y}{2} + \frac{\sigma_y - \sigma_x}{2}\cos 2\theta + \tau_{xy}\sin 2\theta$$

$$= \frac{40}{2} + \frac{-40}{2}\cos 240° + 50 \sin 240°$$

$$= \sigma = -13.3 \text{ N/mm}^2$$

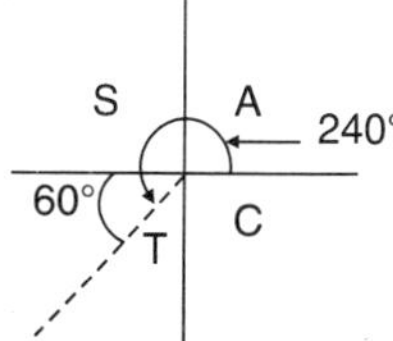

$$\left[\begin{array}{l} \because \cos 240° = -\cos 60° = -\dfrac{1}{2} \\ \sin 240° = -\sin 60° = -0.866 \end{array}\right]$$

$$\text{Shear stress } (\tau) = \frac{\sigma_y - \sigma_x}{2} \sin 2\theta - \tau_{xy} \cos 2\theta$$

$$= \frac{-40}{2} \sin 420° - 50 \cos 240° = \mathbf{42.32\ N/mm^2}$$

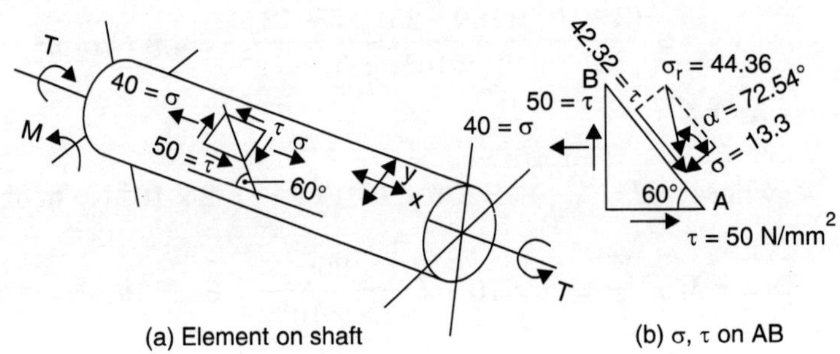

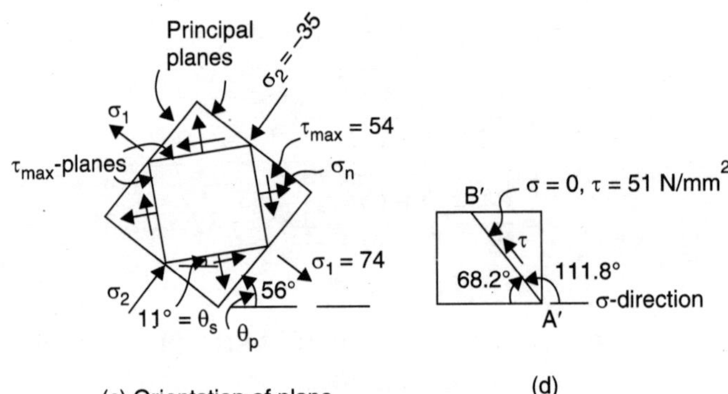

Figure 5.6.

$$\text{Resultant stress } (\sigma_r) = \sqrt{\sigma^2 + \tau^2} = \sqrt{(-13.3)^2 + 42.32^2} = \mathbf{44.36\ N/mm^2}$$

Principal stress from Eqn. (5.5), putting $\sigma_y = 0$,

$$\sigma_1 = \frac{\sigma_x}{2} + \sqrt{\left(\frac{\sigma_x}{2}\right)^2 + \tau_{xy}^{\ 2}}$$

$$= \frac{40}{2} + \sqrt{\left(\frac{40}{2}\right)^2 + 50^2} = 20 + 53.85$$

$$\sigma_1 = \mathbf{73.85\ N/mm^2}$$

$$\tau_{max} = \mathbf{53.85\ N/mm^2}$$

(*iii*) The planes on which the normal stress is zero are given by

$$\sigma = \sigma_x \sin^2\theta + \sigma_y \cos^2\theta + 2\,\tau_{xy} \sin\theta\cos\theta = 0$$

Dividing throughout by $\cos^2\theta$, we get

$$\sigma = \sigma_x \tan^2\theta + 2\tau_{xy} \tan\theta + \sigma_y = 0 \qquad ...(5.10)$$

or,

$$0 = 40 \tan^2\theta + 2 \times 50 \tan\theta$$

$$\therefore \qquad \tan\theta = \frac{-100}{40} = -2.5$$

$$\theta = 180° - 68.2° = 111.8°, \text{ see Figure 5.6 } (d).$$

S A 111.8°
68.2°
T C
tan 68.2° = 2.5

All the stresses and planes are shown in the Mohr's circle as shown in Figure 5.7.

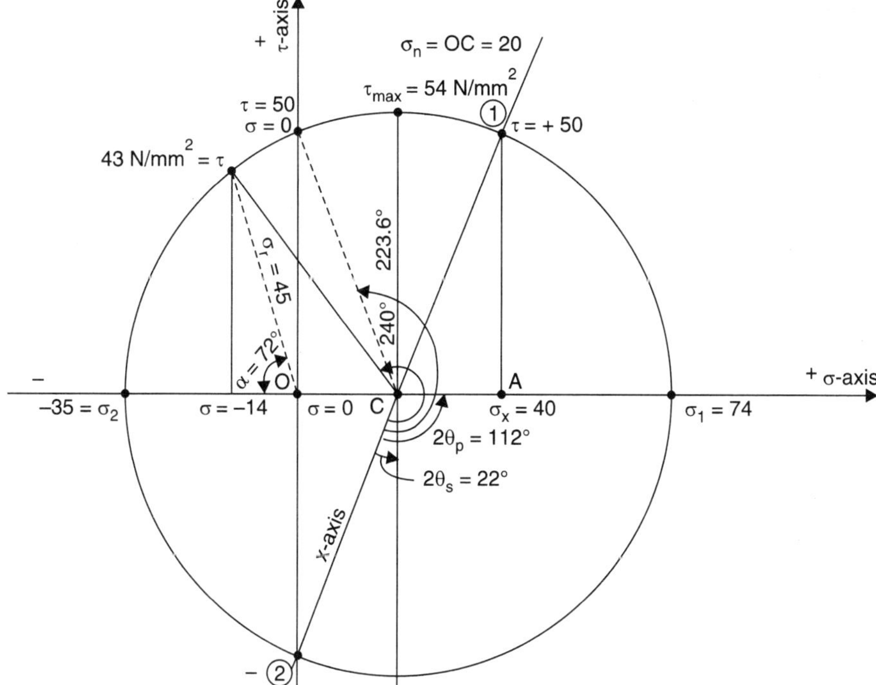

Origin-O; draw x & y-axes; measure OA = σ_x = 40, σ_y = 0 = O. Bisect OA to get centre C. At A, measure vertical = +50 to get (1) ; with C-(1) radius draw Mohr's circle and obtain (2). Note :O-(2) = –50 C-(2) gives x-axis and meaure all angles from C-(2) counter clockwise which give 2θ to different planes.

Figure 5.7. Mohr's circle

Note. In a general case, Eqn. (5.10) is solved as a quadratic in tan θ and two values of θ are obtained which give angles to the planes of zero normal stress, measured counter clockwise from the positive x-direction.

Example 5.4. (*a*) *A square element is subjected to equal tensile and compressive stresses in the two mutually perpendicular directions. Show that the diagonal planes are subjected to pure shear of the same magnitude.*

(*b*) *Also show that the linear strain in the direction of normal stresses is half the shear strain of the diagonal planes, and hence obtain the relation between E and G.*

Solution. (*a*) Consider the square element *ABCD* subjected to stresses as shown and let the normal and tangential stresses on the diagonal be σ_n and τ, see Figure 5.8 (*b*).

Resolving forces along the diagonal *DB*

$$\tau\sqrt{2} - (\sigma \cdot 1)\sin 45° - (\sigma \cdot 1)\sin 45° = 0$$

$$\tau\sqrt{2} = \sigma\sqrt{2}, \quad \tau = \sigma \qquad \ldots(i)$$

Resolving forces normal to the diagonal DB

$$\sigma_n \cdot \sqrt{2} + (\sigma \cdot 1) \cos 45° - (\sigma \cdot 1) \cos 45° = 0$$

$$\sigma_n \sqrt{2} = 0 \quad , \quad \boldsymbol{\sigma_n = 0}$$

i.e., the diagonal planes are subjected to pure shear.

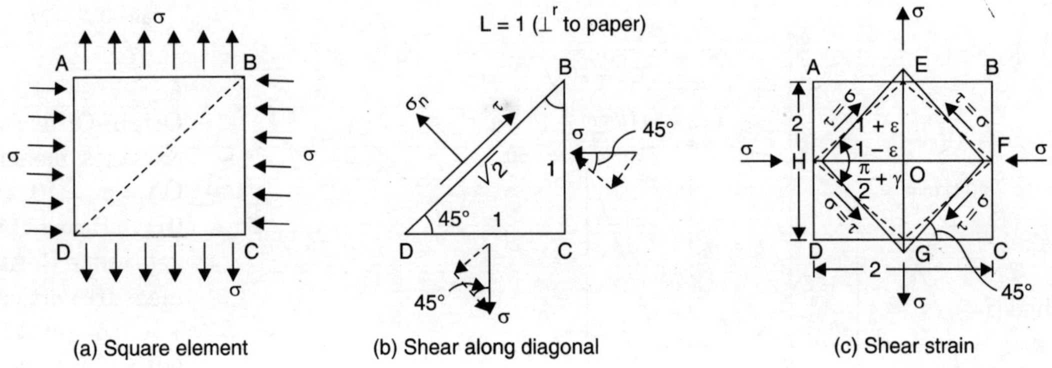

(a) Square element (b) Shear along diagonal (c) Shear strain

Figure 5.8.

(*b*) Consider the square element with sides 2 units (of unstrained length), for convenience,

Strain along EG: $\varepsilon = \dfrac{\sigma}{E} - \nu\left(\dfrac{-\sigma}{E}\right) = \dfrac{\sigma}{E}(1+\nu)$ ...(*i*)

Strain along $HF = \dfrac{-\sigma}{E} - \nu\dfrac{\sigma}{E} = -\dfrac{\sigma}{E}(1+\nu) = -\varepsilon$ ...(*ii*)

Strained length of $EO = 1 + \varepsilon$

Strained length of $HO = 1 - \varepsilon$

Due to the shear strain (= γ radian), the right angle EHG will increase to '$\dfrac{\pi}{2} + \gamma$',

$$\angle EHO = \frac{1}{2}\,\angle EHG = \frac{\pi}{4} + \frac{\gamma}{2}$$

$$\tan \angle EHO = -\frac{EO}{HO}$$

$$\tan\left(\frac{\pi}{4} + \frac{\gamma}{2}\right) = \frac{1+\varepsilon}{1-\varepsilon}$$

Expanding the above, $\dfrac{1+\varepsilon}{1-\varepsilon} = \dfrac{\tan\dfrac{\pi}{4} + \tan\dfrac{\gamma}{2}}{1 - \tan\dfrac{\pi}{4}\cdot\tan\dfrac{\gamma}{2}}$

$\because \quad \tan\dfrac{\pi}{4} = 1, \ \tan\dfrac{\gamma}{2} \approx \dfrac{\gamma}{2}$, for small angles,

$$\therefore \quad \frac{1+\varepsilon}{1-\varepsilon} = \frac{1+\dfrac{\gamma}{2}}{1-\dfrac{\gamma}{2}}$$

$$\varepsilon = \frac{\gamma}{2} \qquad ...(iii)$$

i.e., the linear strain in the direction of normal stresses is half the shear strain of the diagonal planes.

From (*ii*) and (*iii*), $\varepsilon = \frac{\gamma}{2} = \frac{\sigma}{E}(1 + \nu)$ ...(*iv*)

Rigidity modulus $(G) = \frac{\tau}{\gamma} = \frac{\sigma}{\gamma}$ from [a(*i*)]

Substituting for γ in (*iv*), $\frac{\sigma}{2G} = \frac{\sigma}{E}(1 + \nu)$

$$\therefore \quad \boldsymbol{E = 2G(1 + \nu)} \qquad ...(5.11)$$

Theories of Failure (or Yielding). In a simple tension test, when the elastic limit is reached (i.e., the material starts yielding), other quantities such as shear stress, strain energy, also attain definite values, and any one of these may be the deciding factor in the physical cause of failure. The problem of designing a pressure vessel, rotating disc, or some component containing two or three principal stress system so that the material remains elastic (i.e., no yielding) is rather more complex. It is necessary to establish experimentally the yield point in simple tension which corresponds to the onset of failure in the complex system.

A number of theoretical criteria have been proposed over the past century and are outlined below. It is now well established that for ductile metals, exhibiting yielding and subsequent plastic deformation, the shear strain energy theory correlates best with material behaviour, while for the brittle material, it is the maximum stress theory.

Principal Theories of Failure. Let σ_y = yield stress of the material in uniaxial (simple) tension

$\sigma_1, \sigma_2, \sigma_3$ = the principal stresses in any complex system $(\sigma_1 > \sigma_2 > \sigma_3)$

Then, allowable (permissible or safe) stress in simple tension with a factor of safety of '*n*'

$$\sigma_a = \frac{\sigma_y}{n} \qquad ...(5.11a)$$

The allowable shear stress in simple tension

$$\tau_a = \frac{\sigma_a}{2} \qquad ...(5.11b)$$

1. Maximum Principal Stress Theory (due to Rankine). According to this theory, the material fails when the maximum principal stress in the complex system becomes equal to the permissible stress in simple tension i.e.,

$$\sigma_1 = \sigma_a \qquad ...(5.12)$$

2. Maximum Principal Strain Theory (due to St. Venant). The material fails when the maximum principal strain in the complex stress system, becomes equal to the permissible strain in simple tension i.e.,

$$\varepsilon_1 = \varepsilon_a$$

$$\frac{1}{E}[\sigma_1 - \nu(\sigma_2 + \sigma_3) = \frac{\sigma_a}{E}$$

or $$\sigma_1 - \nu(\sigma_2 + \sigma_3) = \sigma_a \qquad ...(5.13)$$

3. Maximum Shear Stress or Stress Difference Theory (due to Guest and Tresca). The material fails when the maximum shear stress in the complex system becomes equal to the permissible shear stress in simple tension i.e.,

$$\tau_{max} = \tau_a$$

$$\frac{\sigma_1 - \sigma_3}{2} = \frac{\sigma_a}{2}, \quad \text{from Eqn. (5.11}b\text{)}$$

$$\sigma_1 - \sigma_3 = \sigma_a \qquad ...(5.14)$$

4. Strain Energy Theory (due to Haigh). The material fails when the strain energy per unit volume in the complex stress system becomes equal to that in simple tension i.e.,

$$\frac{1}{2E}[\sigma_1^2 + \sigma_2^2 + \sigma_3^2 - 2\nu(\sigma_1\sigma_2 + \sigma_2\sigma_3 + \sigma_3\sigma_1)] = \frac{\sigma_a^2}{2E}$$

or $$\sigma_1^2 + \sigma_2^2 + \sigma_3^2 - 2\nu(\sigma_1\sigma_2 + \sigma_2\sigma_3 + \sigma_3\sigma_1) = \sigma_a^2 \qquad ...(5.15)$$

5. Shear or Distortion Strain Energy Theory (due to Von Mises and Henky). When the material fails the shear strain energy in the complex system per unit volume becomes equal to that in simple tension (σ_a, 0, 0)

$$\frac{1}{12G}[(\sigma_1 - \sigma_2)^2 + (\sigma_2 - \sigma_3)^2 + (\sigma_3 - \sigma_1)^2 = \frac{2\sigma_a^2}{12G}$$

or $$(\sigma_1 - \sigma_2)^2 + (\sigma_2 - \sigma_3)^2 + (\sigma_3 - \sigma_1)^2 = 2\sigma_a^2 \qquad ...(5.16)$$

Example 5.5. *The principal stresses at a point in an elastic material are 50, 40 and –30 N/mm². If the material yields at 100 N/mm², calculate the factor of safety according to the five different theories.*

[*Take,* $E = 2 \times 10^5$ *N/mm²*,* $\nu = 0.3$]

Solution. 1. Maximum principal stress theory [From Eqn. (5.12)].

$$\sigma_1 = \sigma_a = 50 \text{ N/mm}^2$$

$$\therefore \quad \text{F.S.} = n = \frac{\sigma_y}{\sigma_a} = \frac{100}{50} = \mathbf{2}$$

2. Maximum principal strain theory [From Eqn. (5.13)]

$$\sigma_a = \sigma_1 - \nu(\sigma_2 + \sigma_3)$$

$$\Rightarrow \quad \sigma_a = 50 - 0.3(40 - 30) = 47 \text{ N/mm}^2$$

$$\therefore \quad \text{F.S.} = n = \frac{\sigma_y}{\sigma_a} = \frac{100}{47} = \mathbf{2.13}$$

3. Maximum shear stress theory [Eqn. (5.14)]

$$\tau_{max} = \tau_a$$

$$\Rightarrow \quad \frac{\sigma_1 - \sigma_3}{2} = \frac{\sigma_a}{2}$$

$$\Rightarrow \quad \sigma_a = \sigma_1 - \sigma_3 = 50 - (-30) = 80 \text{ N/mm}^2$$

$$\therefore \quad \text{F.S.} = n = \frac{\sigma_y}{\sigma_a} = \frac{100}{80} = \mathbf{1.25}$$

4. Strain energy theory [From Eqn. (5.15)]

$$\sigma_1^2 + \sigma_2^2 + \sigma_3^2 - 2\nu(\sigma_1\sigma_2 + \sigma_2\sigma_3 + \sigma_3\sigma_1) = \sigma_a^2$$

$$50^2 + 40^2 + (-30)^2 - 2 \times 0.3\,[(50 \times 40) + (40 \times -30) + (-30 \times 50)] = 5400 = \sigma_a^2$$

$$\sigma_a = \sqrt{5400} = 73.5 \text{ N/mm}^2$$

$$\text{F.S.} = n = \frac{\sigma_y}{\sigma_a} = \frac{100}{73.5} = \mathbf{1.36}$$

5. Shear (Distortional) strain energy theory [From Eqn. (5.16)]

$$(\sigma_1 - \sigma_2)^2 + (\sigma_2 - \sigma_3)^2 + (\sigma_3 - \sigma_1)^2 = 2\sigma_a^2$$

$$(50 - 40)^2 + (40 - (-30)^2) + (-30 - 50)^2 = 11{,}400 = 2\sigma_a^2$$

$$\Rightarrow \qquad \sigma_a = \sqrt{11{,}400/2} = 75.5 \text{ N/mm}^2$$

$$\therefore \qquad \text{F.S.} = n = \frac{\sigma_y}{\sigma_a} = \frac{100}{75.5} = \mathbf{1.32}$$

Example 5.6. *A M.S. bolt is subjected to an axial pull of 10 kN and transverse shear force of 5 kN. Taking the elastic limit in tension as 240 N/mm², ν = 0.3, and a factor of safety i.e., n = 3, estimate the diameter of the bolt required if the criterion of failure is (a) maximum principal stress, (b) maximum shear stress, (c) maximum strain energy and (d) shear strain energy.*

Solution. In simple tension

$$\sigma_a = \frac{\sigma_y}{n} = \frac{240}{3} = 80 \text{ N/mm}^2$$

At a point in the bolt

$$\sigma_x = \frac{P}{A} = \frac{10}{\pi d^2/4} = \frac{40}{\pi d^2} \text{ kN/mm}^2$$

$$\sigma_y = 0$$

and

$$\tau_{xy} = \frac{F}{A} = \frac{5}{\pi d^2/4} = \frac{20}{\pi d^2} \text{ kN/mm}^2$$

(*a*) Maximum principal stress

$$\sigma_{1,2} = \frac{\sigma_x + \sigma_y}{2} \pm \sqrt{\left(\frac{\sigma_x - \sigma_y}{2}\right)^2 + \tau_{xy}^2}, \quad \text{where} \quad \sigma_y = 0$$

$$= \frac{40}{2\pi d^2} \pm \sqrt{\left(\frac{40}{2\pi d^2}\right)^2 + \left(\frac{20}{\pi d^2}\right)^2}$$

$$\sigma_{1,2} = \frac{20}{\pi d^2}(1 \pm \sqrt{2})$$

From Eqn. (5.12) : $\sigma_1 = \sigma_a$

$$\frac{20}{\pi d^2}(1 + \sqrt{2}) = \frac{80}{1000}, \quad d = \mathbf{13.86 \text{ mm}}$$

(*b*) Maximum shear stress

$$\tau_{max} = \sqrt{\left(\frac{\sigma_x - \sigma_y}{2}\right)^2 + \tau_{xy}^{\ 2}}\,, \quad \text{where} \quad \sigma_y = 0$$

$$= \sqrt{\left(\frac{40}{2\pi d^2}\right)^2 + \left(\frac{20}{\pi d^2}\right)^2} = \frac{20\sqrt{2}}{\pi d^2} \text{ kN/mm}^2$$

and $$\tau_a = \frac{\sigma_a}{2} = \frac{80}{2} = 40 \text{ N/mm}^2$$

But, $$\tau_{max} = \tau_a$$

$\therefore$ $$\frac{20\sqrt{2}}{\pi d^2} = \frac{40}{1000}, \quad d = \mathbf{15\ mm}$$

(*c*) Maximum or total strain energy [By Eqn. 5.15]

$$\sigma_1^{\ 2} + \sigma_2^{\ 2} + \sigma_3^{\ 2} - 2\nu\,(\sigma_1\sigma_2 + \sigma_2\sigma_3 + \sigma_3\sigma_1) = \sigma_a^{\ 2}$$

But, $\sigma_1 = \dfrac{20}{\pi d^2}(1 + \sqrt{2}), \quad \sigma_2 = \dfrac{20}{\pi d^2}(1 - \sqrt{2}), \quad \sigma_3 = 0$

$$\therefore \quad \left(\frac{20}{\pi d^2}\right)^2 [(1+\sqrt{2})^2 + (1-\sqrt{2})^2 - 2 \times 0.3\,(1+\sqrt{2})(1-\sqrt{2})] = \left(\frac{80}{1000}\right)^2$$

d = **14.3 mm**

(*d*) Shear strain energy

Since, $\sigma_1 = \dfrac{20}{\pi d^2}(1+\sqrt{2}), \quad \sigma_2 = \dfrac{20}{\pi d^2}(1-\sqrt{2}), \quad \sigma_3 = 0$

From Eqn. (5.16) : $(\sigma_1 - \sigma_2)^2 + (\sigma_2 - \sigma_3)^2 + (\sigma_3 - \sigma_1)^2 = 2\sigma_a^{\ 2}$

$$\therefore \quad \left(\frac{20}{\pi d^2}\right)^2 [(2\sqrt{2})^2 + (1-\sqrt{2})^2 + (1+\sqrt{2})^2 + (1+\sqrt{2})^2] = 2\left(\frac{80}{1000}\right)^2$$

d = **14.5 mm**

PROBLEMS

5.1. (*a*) At a point in an elastic material, the stresses on two mutually perpendicular planes are 80 MN/m² tension and 40 MN/m² compression, the shear stress being, 20 MN/m². Find the stress components and resultant stress on a plane at 60° to the plane on which the tensile stress acts.

(*b*) Determine the principal stresses, and the maximum shear stress.

(*c*) Determine the plane of zero normal stress. What is the normal stress on the planes of maximum shear ?

(*d*) Verify all the above by constructing the Mohr's stress circle.

(*e*) Sketch an oriented element showing the principal planes and the associated stresses.

5.2. On a certain two-dimensional element bending stresses of 200 N/mm² and shear stresses of 100 N/mm² act. Determine :

(*i*) the normal and shear stresses on a plane at 30° to the bending stress direction and their resultant.

(*ii*) principal stresses and their directions

(*iii*) maximum shear stresses and their directions.

Indicate by constructing the Mohr's stress circle.

[136.6, 136.6, 241.4, 45° ; 241.4, – 41.4, 67.5°, 157.5, 141.4 N/mm^2, 22.5°, 112.5°]

5.3. At a point in an elastic material, the stresses on two mutually perpendicular planes are 80 MN/m^2 tension and 60 MN/m^2 compression. The maximum principal stress in the material is to be limited to 100 MN/m^2 tension. To what shearing stress may the material be subjected to on the given planes and what will then be the maximum shear stress at the point ?

5.4. In a beam, at a point near about the support, the bending stress is 80 MN/m^2 and also the shear stress, as shown in Figure P. 5.4. If the maximum principal stress is limited to 90° MN/m^2, what is the greatest shearing stress that can be applied on the given planes ? What is then the minimum principal stress and its direction ?

If another point at the same level is considered nearer to the centre of the span, state with reason whether the principal stress will increase or decrease.

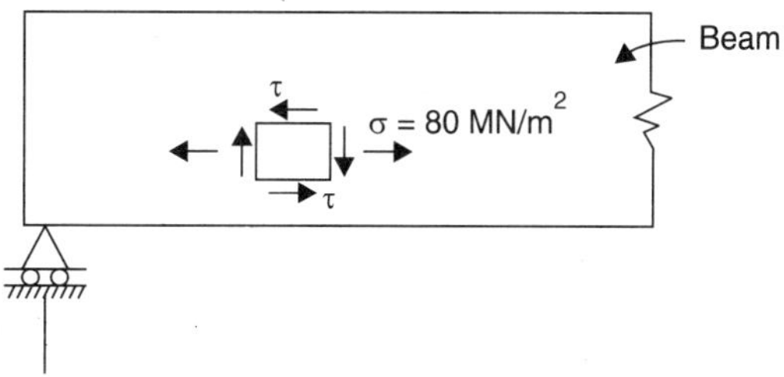

Figure P. 5.4

5.5. At a point in a strained material, the principal stresses are 100 N/mm^2 tensile and 40 N/mm^2 compressive.

Determine the normal, tangential and resultant stresses on a plane through the point at 30° to the major principal plane, using Mohr's circle of stress.

5.6. (*a*) A square element is subjected to a state of simple shear of intesity, τ. Show that the principal stresses are +ve and –ve and act on diagonal planes. Work from first principles.

(*b*) Also show that the linear strain in the direction of principal stresses is half the shear strain and hence, obtain the relation between *E* and *G*.

5.7. Draw Mohr's circle for the following cases and mark on them, the principal stresses and the maximum shear stress :

(*a*) $\sigma_x = 0$, $\sigma_y = 80$ N/mm^2, $\tau_{xy} = 40$ N/mm^2

(*b*) $\sigma_x = 100$ N/mm^2, $\sigma_y = 0$, $\tau_{xy} = -40$ N/mm^2

(*c*) $\sigma_x = \sigma_y = 100$ N/mm^2, $\tau_{xy} = 0$

(*d*) $\tau_{xy} = -50$ N/mm^2, $\sigma_x = \sigma_y = 0$.

5.8. A solid circular shaft 100 mm dia is subjected to a bending moment of 12.5 kN · m and a twisting moment. The yield strengths in tension and compression are respectively 270 N/mm^2 and 335 N/mm^2 and Poisson's ratio = 0.3. Determine the value of twisting moment required to produce yielding based on (*a*) maximum stress theory, (*b*) maximum shear theory, (*c*) maximum strain theory, and (*d*) distortion energy theory.

6 Deflection of Beams

Deflection (y, δ) of a loaded beam at any section x, is its deviation from its original position to its deflected form, measured from the neutral (or centroidal) axis. The deflected form of the neutral axis (N.A) is called the **elastic curve** or **deflection curve**.

Equation to the Elastic Curve. If R is the radius of the N.A. of the deflected beam, see Figure 6.1, flexure equation is

$$\frac{M}{I} = \frac{E}{R}, \quad \text{or} \quad \frac{M}{EI} = \frac{1}{R} \qquad \text{...(6.1)}$$

and the radius of curvature in terms of the coordinates x and y is given by

$$\frac{1}{R} = \left(\pm\frac{d^2y}{dx^2}\right) \times \frac{1}{\left[1+\left(\frac{dy}{dx}\right)^2\right]^{3/2}} = \frac{y''}{(1+y'^2)^{3/2}} \qquad \text{...(6.2)}$$

Since, the slope $\frac{dy}{dx}$ $(= y')$ is very small, its square is negligibly small and

$$\frac{1}{R} = \frac{d^2y}{dx^2} = y'' \qquad \text{...(6.3)}$$

From Eqns. (6.1) and (6.3),

$$\frac{M}{EI} = \frac{1}{R} = \frac{d^2y}{dx^2}$$

or

$$\boxed{\mathbf{EI\frac{d^2y}{dx^2} = M = M_x}} \qquad \text{...(6.4)}$$

where M_x is the B.M. at the section expressed as a function of x. Equation (6.4) when integrated once gives slope (dy/dx) at the section, and twice, the deflection y. The constants of integration can be evaluated from the support conditions:

at supports : $y = 0$; at built in or fixed ends : $\dfrac{dy}{dx} = 0$;

for maximum deflection : $\dfrac{dy}{dx} = 0.$

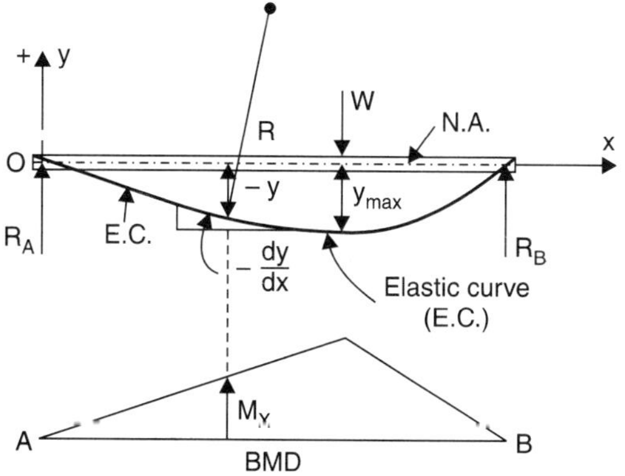

Figure 6.1. Equation to the elastic curve

The product '*EI*' is called the **flexural rigidity** of the beam.

Differentiating Eqn. (6.4),

$$EI\ \frac{d^3y}{dx^3} = \frac{dM}{dx} = F, \quad \text{i.e., S.F. at the section} \qquad ...(6.5)$$

Again differentiating, $EI\ \dfrac{d^4y}{dx^3} = \dfrac{dF}{dx} = -\,w,$ i.e., *udl* ...(6.6)

which are of use in some cases.

The methods of solution are:

(*a*) Macaulay's method or double integration method,

(*b*) Moment-area method,

(*c*) Conjugate beam method,

(*d*) Strain energy method.

The first two methods are discussed here.

(*a*) Slope and Deflection by Macaulay's Method

Example 6.1. *For loaded beam in Figure 6.2, find the slope and deflection at C and D. [Take E = 2 × 10⁵ N/mm², I = 4.4 × 10⁸ mm⁴]*

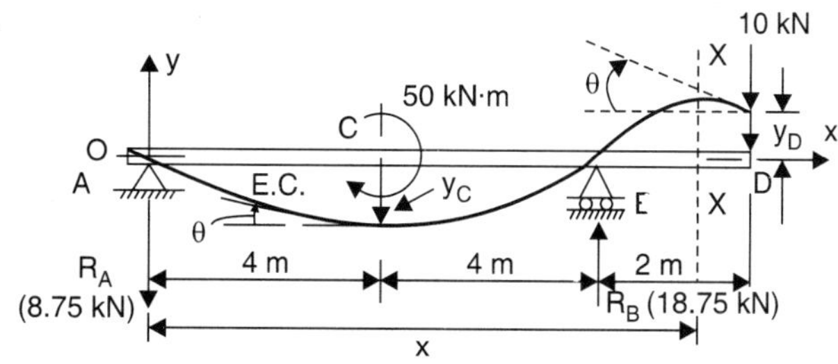

Figure 6.2.

Solution. $M_A = 0 : R_B \times 8 = 10 \times 10 + 50 = 150$ kN

$$R_B = \frac{150}{8} = 18.75 \text{ kN} \uparrow, \quad R_A + R_B = 10$$

$$R_A = 10 - 18.75 = -8.75 \text{ kN or } 8.75 \text{ kN} \downarrow$$

Take origin at A and section X-X between B and D

$$M_X = -8.75\,x + 50\,(x-4)^0 + 18.75\,(x-8) = EI\,\frac{d^2y}{dx^2}$$

(Convention)

Integrating, $EI\,\dfrac{dy}{dx} = -8.75\,\dfrac{x^2}{2} + 50\,(x-4) + 18.75\,\dfrac{(x-8)^2}{2} + c_1$ ...(*i*)

(Convention)

Again Integrating, $EI\,y = -\dfrac{8.75}{2}\cdot\dfrac{x^3}{3} + 50\,\dfrac{(x-4)^2}{2} + \dfrac{18.75}{2}\,\dfrac{(x-8)^3}{3} + c_1x + c_2$ ...(*ii*)

(Convention)

Putting $x = 0$, $y = 0$ in (*ii*), $c_2 = 0$, negative quantities in brackets not considered.

When $x = 8$, $y = 0$ in (*ii*), then

$$0 = -\frac{8.75}{6} \times 8^3 + 25 \times 4^3 + 8c_1 = 0, \; c_1 = -106.67$$

Putting $x = 4$ in (*ii*), $EI\,Y_C = -1.46 \times 4^3 - 106.67 \times 4 = -520.12$ kN·m³

$$y_C = \frac{-520.12\,(10^3 \times 10^9)\,\text{N}\cdot\text{mm}^3}{(2\times10^5)(4.4\times10^8)\,\text{N}\cdot\text{mm}^2} = -5.9 \text{ mm, i.e., } 5.9 \text{ mm} \downarrow$$

Putting $x = 4$ in (*i*), $EI\left(\dfrac{dy}{dx}\right)_C = -4.375 \times 4^2 - 106.67 = -176.67$ kN·m²

or $$\theta_C = \left(\frac{dy}{dx}\right)_C = \frac{-176.67\,(10^3\times10^6)\,\text{N}\cdot\text{mm}^2}{(2\times10^5)(4.4\times10^8)\,\text{N}\cdot\text{mm}^2} = -2 \text{ radian,}$$

i.e., 2 rad. clockwise.

Putting $x = 10$ in (*ii*),

$$EI\,y_D = -1.46 \times 10^3 + 25 \times 6^3 + 3.125 \times 2^3 - 106.67 \times 10 = 2898.3 \text{ k N}\cdot\text{m}^3$$

or $$y_D = \frac{2898.3 \times 10^{12}}{(2\times10^5)(4.4\times10^8)} = 32.93 \text{ mm} \uparrow \text{(as assumed).}$$

Putting $x = 10$ in (*i*),

$$EI\left(\frac{dy}{dx}\right)_D = -4.375 \times 10^2 + 50 \times 6 + 9.375 \times 2^2 - 106.67 = -206.67 \text{ kN}\cdot\text{m}^2$$

$$\theta_D = \left(\frac{dy}{dx}\right)_D = \frac{-206.67\,(10^3\times10^6)}{(2\times10^5)(4.4\times10^8)} = -2.35 \text{ radian,}$$

i.e., 2.35 rad. clockwise

Example 6.2. *Determine the deflection at the free end of the beam in Figure 6.3. Flexural rigidity of the beam is 30 MN·m².*

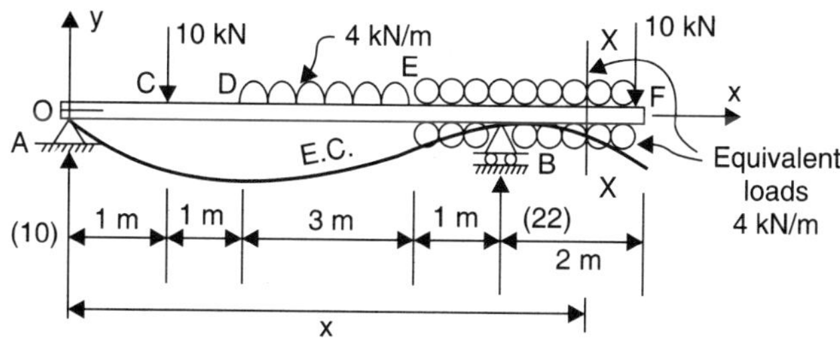

Figure 6.3.

Solution. $R_A + R_B = 10 + (4 \times 3) + 10 = 32$ kN

$$\sum M_A = 0: \quad R_B \times 6 = 10 \times 8 + (4 \times 3)\,3.5 + 10 \times 1 = 132 \text{ kN}$$

$$R_B = \frac{132}{6} = 22 \text{ kN} \uparrow, \quad R_A = 32 - 22 = 10 \text{ kN} \uparrow$$

The *udl* is extended upto *F* and of the same magnitude upward between *E* and *F* introduced, in this method. The section *X-X* is taken between *B* and *F*.

$$EI\frac{d^2y}{dx^2} = M_X = 10x - 10(x-1) + \frac{4(x-2)^2}{2} - \frac{4(x-5)^2}{2} - 22\,(x-6)$$

$$EI\frac{dy}{dx} = 10\frac{x^2}{2} - \frac{10(x-1)^2}{2} + 2\frac{(x-2)^3}{3} - 2\frac{(x-5)^3}{3} - 22\frac{(x-6)^2}{2} + c_1 \qquad ...(i)$$

$$EI\,y = 5\frac{x^3}{3} - 5\frac{(x-1)^3}{3} + \frac{2}{3}\frac{(x-2)^4}{4} - \frac{2}{3}\frac{(x-5)^4}{4} - 11\frac{(x-6)^3}{3} + c_1x + c_2 \qquad ...(ii)$$

When $x = 0$, $y = 0$ in (*ii*), $c_2 = 0$

When $x = 6$, $y = 0$, in (*ii*),

$$0 = \frac{5}{3}\times 6^3 - \frac{5}{3}\times 5^3 + \frac{1}{6}\times 4^4 - \frac{1}{6}\times 1 + c_1 \times 6, c_1 = -32.36$$

Putting $x = 8$, $c_1 = -32.36$, $c_2 = 0$ in (*ii*),

$$EI\,y_F = \frac{5}{3}\times 8^3 - \frac{5}{3}\times 7^3 + \frac{1}{6}\times 6^4 - \frac{1}{6}\times 3^4 - \frac{11}{3}\times 2^3 - 32.36 \times 8 = 203.29 \text{ kN·m}^3$$

$$y_F = \frac{203.29 \times 10^{12} \text{ N}\cdot\text{mm}^2}{30\,(10^6 \times 10^6)\text{ N}\cdot\text{mm}^2} = \mathbf{6.71 \text{ mm}}$$

Example 6.3. (*a*) *In Figure 6.4, a prop is given at the free end of the cantilever at the same level as the fixed end. Determine the prop reaction P, and bending moment anywhere on the cantilever.*

(*b*) *What would have been the deflection of the free end if there were no prop? [Take flexural rigidity = 50 MN·m²].*

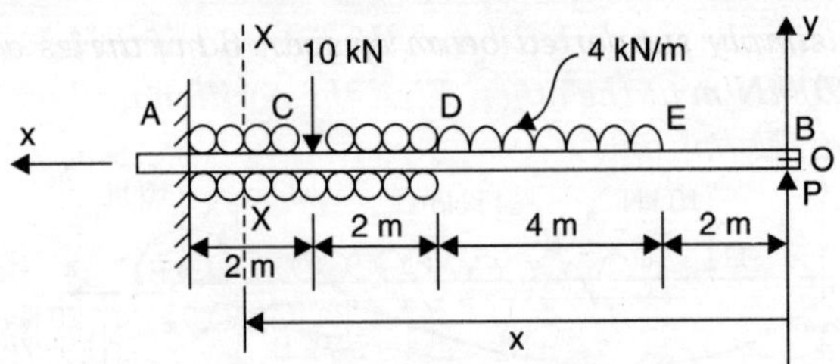

Figure 6.4. Propped cantilever

Solution. (*a*) $EI \dfrac{d^2y}{dx^2} = M_X = Px - 4\dfrac{(x-2)^2}{2} + \dfrac{4(x-6)^2}{2} - 10(x-8)$ ...(*i*)

$$EI \frac{dy}{dx} = P\frac{x^2}{2} - 2\frac{(x-2)^3}{3} + 2\frac{(x-6)^3}{3} - 10(x-8) + c_1 \quad ...(ii)$$

$$EI\, y = \frac{P}{2}\frac{x^3}{3} - \frac{2}{3}\frac{(x-2)^4}{4} + \frac{2}{3}\frac{(x-6)^4}{4} - 10\frac{(x-8)^2}{2} + c_1x + c_2 \quad ...(iii)$$

When $x = 0, y = 0, \quad c_2 = 0$

Putting $x = 10, y = 0, c_2 = 0$ in (*iii*), $0 = \dfrac{P}{6}.10^3 - \dfrac{1}{6}.8^4 + \dfrac{1}{6}.4^4 - 5.2^2 + 10c_1$ ...(*iv*)

$\Rightarrow$ $166.67\,P - 660 + 10c_1 = 0$ ...(*v*)

Putting $x = 10, \dfrac{dy}{dx} = 0$ in (*ii*), $0 = \dfrac{P}{2}.10^2 - \dfrac{2}{3}.8^3 + \dfrac{2}{3}.4^3 - 10\times 2 + c_1$

$\Rightarrow$ $50\,P - 318.66 + c_1 = 0$ ...(*vi*)

Solving (*v*) & (*vi*), $P = 7.58$ kN, $c_1 = -60.34$

Max. B.M. occurs at a value of x where S.F. = 0 :

$\therefore$ $P\uparrow = 4(x-2), \quad x = \dfrac{P}{4} + 2 = 3.9$ m from the prop.

$$M_{max} = 7.58 \times 3.9 - \frac{4(3.9-2)^2}{2} = \mathbf{22.34\ kN{\cdot}m}$$

Note. R_A, M_A and y at *C, D* and *E* can easily be determined.

(*b*) If there were no prop, $P = 0$:

at $x = 10$, $\dfrac{dy}{dx} = 0$, from (*vi*), $c_1 = 318.66$

at $x = 10, y = 0$: from (*iii*) and (*v*), $-660 + 10c_1 + c_2 = 0$

$\therefore$ $c_2 = -2526.6$

At $x = 0$, from (*iii*), $EI\, y_B = c_2 = -2526.6$ kN· m^3

$$\therefore \quad y_B = \frac{-2526.6\times 10^{12}}{50(10^6\times 10^6)} = 50.53 \text{ mm} \downarrow$$

Example 6.4. *A simply supported beam of span 6 m carries a uniformly varying load from zero at one end to 20 kN/m at the other. Find the position and magnitude of the maximum deflection. [Take E = 200 GN/m² and I = 10⁸ mm⁴.]*

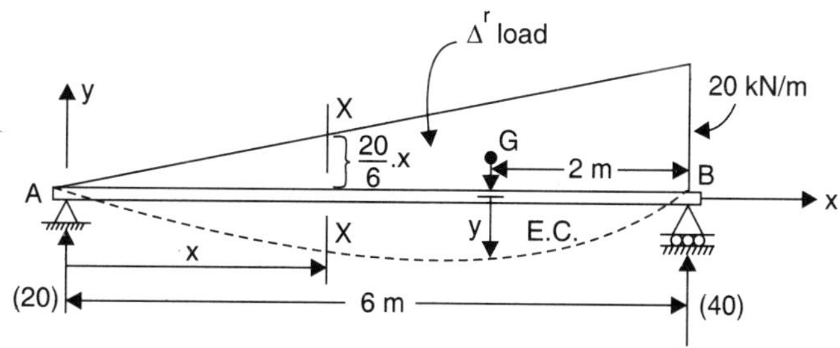

Figure 6.5.

Solution. $$R_A + R_B = \frac{0+20}{2} \times 6 = 60 \text{ kN}$$

$\sum M_A = 0:$ $R_B \times 6 = 60 \times 4,\quad R_B = 40 \text{ kN},\ R_A = 60 - 40 = 20 \text{ kN}$

$$EI = \frac{d^2y}{dx^2} = M_X = \frac{1}{2}\left(0 + \frac{20}{6}\right)x \cdot \frac{x}{3} = \frac{5}{9}\,x^3 \qquad ...(i)$$

$$EI\,\frac{dy}{dx} = \frac{5}{9}\cdot\frac{x^4}{4} + c_1 \qquad ...(ii)$$

$$EI\,y = \frac{5}{36}\cdot\frac{x^5}{5} + c_1x + c_2 \qquad ...(iii)$$

When $x = 0$, $y = 0$; from (*iii*), $c_2 = 0$

At $x = 6$ m, $y = 0$; $c_2 = 0$, from (*iii*), $0 = \frac{1}{36}\cdot 6^5 + c_1 \times 6,\quad c_1 = -36$

For max. deflection $\frac{dy}{dx} = 0:$ $\quad 0 = \frac{5}{36}x^4 - 36,\quad x = 4.012$ m

$$EI\,y_{max} = \frac{1}{36}(4.012)^5 - 36\,(4.012) = -115.57 \text{ kN}\cdot\text{m}^3$$

$$\therefore \quad y_{max} = \frac{-115.57 \times 10^{12}\ \text{N}\cdot\text{mm}^3}{(200 \times 10^3)\,10^8\ \text{N}\cdot\text{mm}^2} = \mathbf{5.78\ mm} \downarrow$$

(*b*) Moment-area Method

Mohr's Theorems:

Theorem I. The angle between the tangents at two points *A* and *B* on the elastic curve is equal to the area of the B.M. diagram between these points divided by *EI* (usually a constant), see Figure 6.6, as

$$\theta_{BA} = \frac{\sum M_{AB}}{EI} \qquad ...(6.7)$$

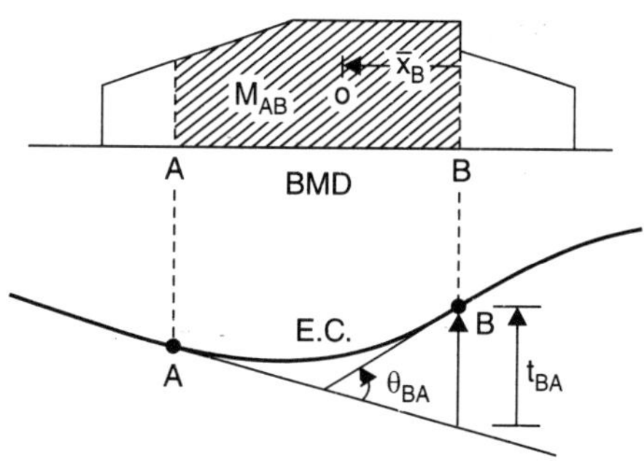

Figure 6.6. Moment-area method

Theorem II. The displacement of B on the elastic curve from the tangent at A is equal to the sum of the moments of the areas of the B.M. diagram between A and B, taken about B, divided by EI (usually a constant).

$$t_{BA} = \frac{1}{EI}\sum M_{AB}\cdot \bar{x}_B \quad ...(6.8)$$

Example 6.5. *Find the central deflection for simply supported beam (a) with concentrated load at mid-span, (b) with udl.*

Solution. (a) simply supported beam with concentrated load at mid-span, *Fig. 6.7*

Slope at support

$$\theta_{BC} = \frac{1}{EI} M_{CB}$$

$$= \frac{1}{EI}\left[\frac{1}{2}\cdot\frac{L}{2}\cdot\frac{WL}{4}\right] = \frac{WL^2}{16EI} \quad ...(6.9)$$

Central deflection

$$y_C = t_{BC} = \frac{1}{EI} M_{CB}\cdot \bar{x}_B$$

$$= \frac{1}{EI}\cdot\frac{WL^2}{16}\cdot\frac{L}{3} = \mathbf{\frac{WL^2}{48EI}} \quad ...(6.10)$$

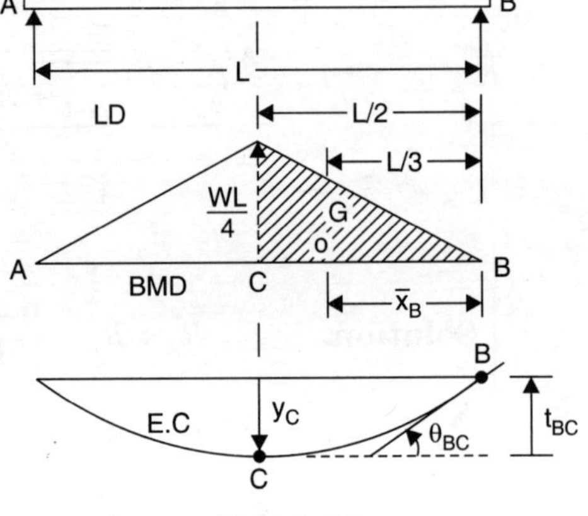

Figure 6.7.

(b) Simply supported beam with udl, *Fig. 6.8*

Slope at support $\theta_{BC} = \frac{1}{EI} M_{CB}$

$$= \frac{1}{EI}\left[\frac{2}{3}\cdot\frac{wL^2}{8}\cdot\frac{L}{2}\right]$$

$$= \frac{wL^3}{24EI} \quad ...(6.11)$$

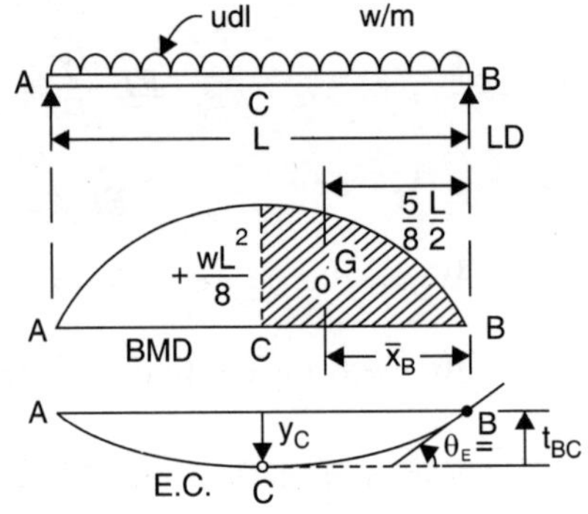

Figure 6.8.

Central deflection $(y_C) = t_{BC} = \frac{1}{EI} M_{CB}\cdot \bar{x}_B$

$$= \frac{1}{EI}\cdot\frac{wL^3}{24}\cdot\frac{5L}{16} = \mathbf{\frac{5wL^4}{84EI}} \quad ...(6.12)$$

Example 6.6. *Find the deflection at the free end of a cantilever : (a) with concentrated load at the free end, (b) with udl over the whole length, Fig. 6.9.*

Solution. (a) Cantilever with concentrated load at the free end, *Fig. 6.9(a):*

$$\text{Slope at the free end } \theta_{BA} = \frac{1}{EI} M_{AB} = \frac{1}{EI}\left[\frac{1}{2}.L{\cdot}WL\right] = \frac{WL^2}{2EI} \quad ...(6.13)$$

Deflection at the free end

$$y_B = t_{BA} = \frac{1}{EI} M_{AB}\cdot \bar{x}_B$$

$$= \frac{1}{EI} \cdot \frac{WL^2}{2} \cdot \frac{2L}{3} = \frac{\mathbf{WL^3}}{\mathbf{3EI}} \qquad ...(6.14)$$

(b) Cantilever with udl over the whole length, *Fig. 6.9(b)*:

Slope at the free end $\theta_B = \frac{1}{EI} M_{AB} = \frac{1}{EI}\left[\frac{1}{3}\frac{wL^2}{2} . L\right] = \frac{wL^3}{6EI}$ ...(6.15)

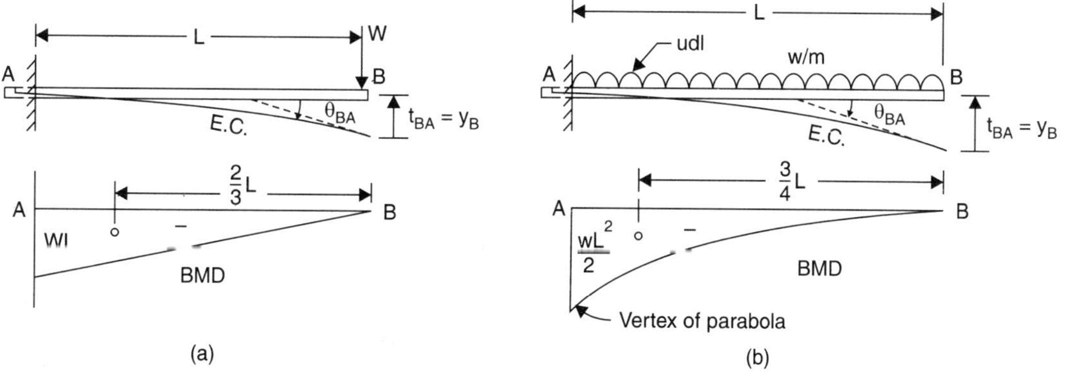

Figure 6.9.

Deflection at the free end $y_B = t_{BA} = \frac{1}{EI} M_{AB} \cdot \bar{x}_B$

$$= \frac{1}{EI} \cdot \frac{wL^3}{6} \cdot \frac{3L}{4} = \frac{\mathbf{wL^4}}{\mathbf{8EI}} \qquad ...(6.16)$$

Example 6.7. *Find the deflection at C in Figure 6.10.*

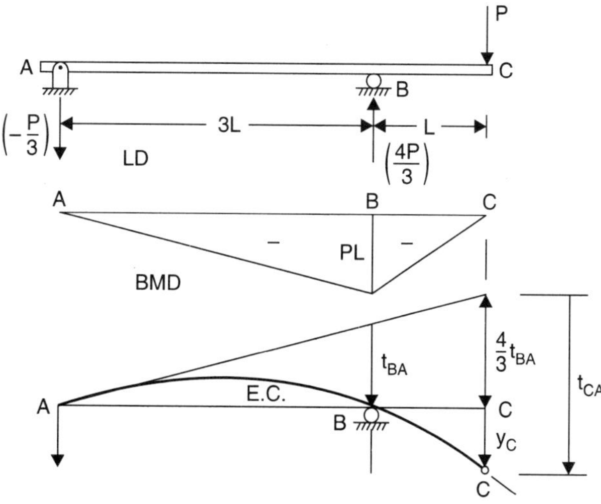

Figure 6.10.

Solution. The BMD and the Elastic Curve (E.C) are shown in *Fig. 6.10*.

Taking moments of BMD about C, $EI\, t_{CA} = \sum M_{AC} \cdot \bar{x}_C$

$$= \left(\frac{1}{2} \cdot 3L \cdot PL\right)(L + L) + \left(\frac{1}{2} \cdot L \cdot PL\right)\frac{2}{3} L$$

$$= 3PL^3 + \frac{PL^3}{3} = \frac{10}{3} PL^3 \; (-)$$

Taking moments of BMD between A & B, about B,

$$EI\, t_{BA} = M_{AB} \cdot \bar{x}_B = \left(\frac{1}{2} \cdot 3L \cdot PL\right) L = \frac{3}{2} PL^3 \; (-)$$

Deflection $\quad y_C = t_{CA} - \frac{4}{3} t_{BA} = \frac{1}{EI}\left[\frac{10}{3} PL^3 - \frac{4}{3} \cdot \frac{3}{2} PL^3\right] = \mathbf{\frac{4}{3} \frac{PL^3}{EI}} \downarrow$

Example 6.8. *Find the deflections at C and D of the beam in Figure 6.11. Given $E = 2 \times 10^5$ N/mm².*

Solution. Reactions : $R_A + R_B = 5 \times 8 + 10 = 50$ kN

$\Sigma M_A = 0$: $R_B \times 8 = 10 \times 10 + (5 \times 8) \times 4 = 260$, $R_B = \frac{260}{8}$ 32.5 kN ↑

$$R_A = 50 - 32.5 = 17.5 \text{ kN} \uparrow$$

$$EI\, t_{CA} = \sum M_{AC} \cdot \bar{x}_C = \left(\frac{1}{2} \times 4 \times 70\right)\frac{4}{3} - \frac{1}{3}(40 \times 4)\frac{4}{4} = 133.34 \text{ kN} \cdot \text{m}^3$$

$$EI\, t_{BA} = \sum M_{AB} \cdot \bar{x}_B = \left(\frac{1}{2} \times 8 \times 140\right)\frac{8}{3} - \frac{1}{3}(160 \times 8)\frac{8}{4} = 640 \text{ kN} \cdot \text{m}^3$$

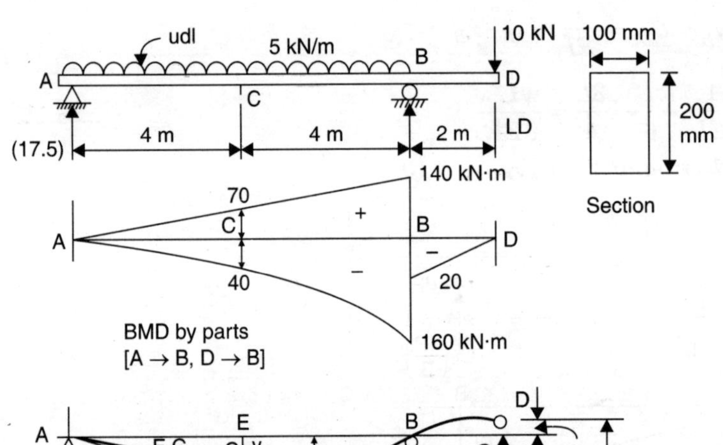

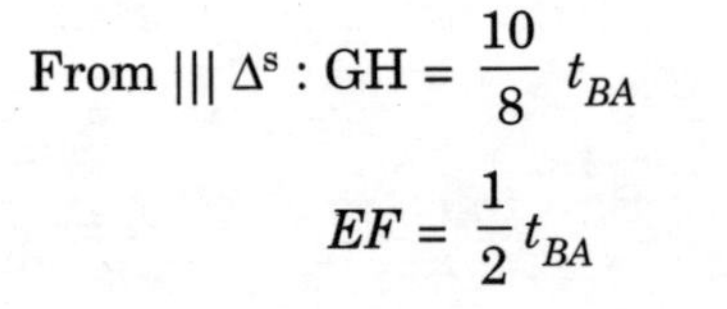

$$I = \frac{bd^3}{12} = \frac{100 \times 200^3}{12} = \frac{2}{3} \times 10^8 \text{ mm}^4$$

$$EI = 2 \times 10^5 \frac{\text{N}}{\text{mm}^2} \times \frac{2}{3} \times 10^8 \text{ mm}^4 = \frac{4}{3} \times 10^{13} \text{ N} \cdot \text{mm}^2$$

From ||| Δs : $GH = \frac{10}{8} t_{BA}$

$$EF = \frac{1}{2} t_{BA}$$

Figure 6.11. Deflections by area-moments

$$EI\, t_{DA} = \sum M_{AD} \cdot \bar{x}_D = \left(\frac{1}{2} \times 8 \times 140\right)\left(\frac{8}{3} + 2\right) - \frac{1}{3}(160 \times 8)\left(\frac{8}{4} + 2\right) - \left(\frac{1}{2} \times 2 \times 20\right)\frac{2}{3} \times 2$$

$$= 880 \text{ kN} \cdot \text{m}^3$$

Deflection at D, $\quad y_D = t_{DA} - \frac{10}{8} t_{BA} = \frac{1}{EI}(880 - 800) = \frac{80(10^3 \times 10^9)}{\frac{4}{3} \times 10^{13}} = \mathbf{6\ mm}$

Deflection at C, $\quad y_C = \frac{t_{BA}}{2} - t_{CA} = \frac{1}{EI}\left(\frac{640}{2} - 133.34\right) = \frac{186.66 \times 10^{12}}{\frac{4}{3} \times 10^{13}} = \mathbf{14\ mm}$

Example 6.9. *Determine the prop reaction in Figure 6.12. Also determine the fixing moment and reaction.*

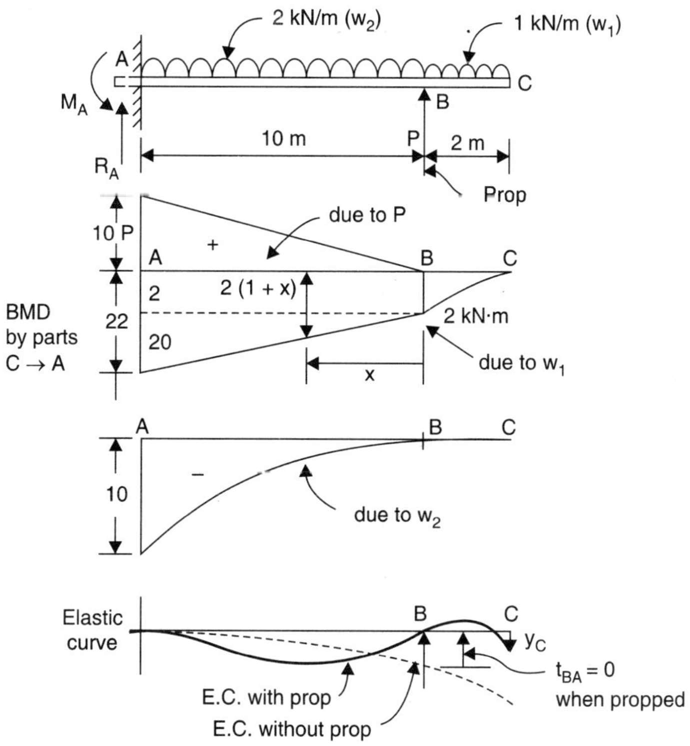

Figure 6.12. Propped cantilever

Solution. The prop is at the same level as the fixed end.

$$\therefore \quad EI\, t_{BA} = 0 = \sum M_{BA} \cdot \bar{x}_B$$

$$0 = \left(\frac{1}{2} \cdot 10P \cdot 10\right)\frac{2}{3} \cdot 10 - (2.10)5 - \left(\frac{1}{2} \cdot 20 \cdot 10\right)\frac{2}{3} \cdot 10 - \frac{1}{3}(100 \times 10)\frac{3}{4} \times 10$$

$$\frac{1000P}{3} = 3266.67, \qquad P = 9.8 \text{ kN}$$

$$M_A = 10P - 22 - 100 = 10 \times 9.8 - 122 = -24 \text{ kN·m}$$

$\therefore$ Fixing moment = **24 kN·m**

Reaction at A : $R_A + P \uparrow = \downarrow 2 \times 10 + 1 \times 2$

$\therefore \quad R_A = 22 - 9.8 = \mathbf{12.2\ kN} \uparrow$

Example 6.10. *Determine the deflection at the free end of the cantilever in Figure 6.13. Given E = 2 × 10⁵ N/mm², I = 4.8 × 10⁸ mm⁴.*

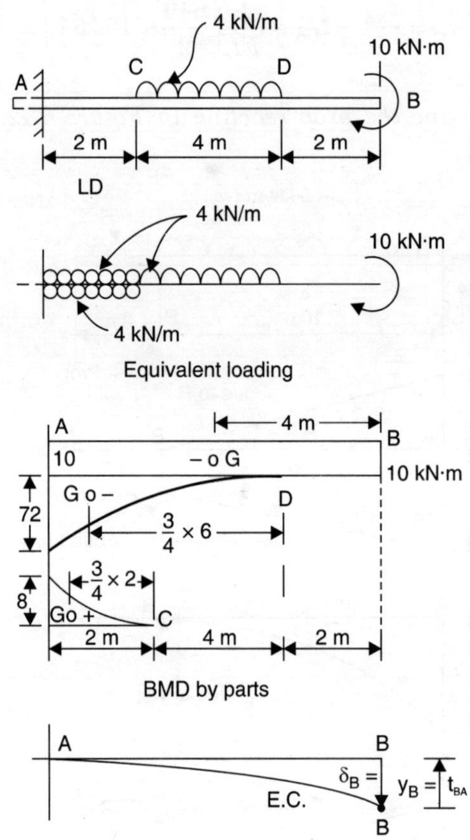

Figure 6.13. Deflection at the free end of cantilever

$$EI\, t_{BA} = \Sigma M_{AB} \cdot \bar{x}_B$$

$$= -(10.8)4 - \frac{1}{3}(72 \times 6)\left(\frac{3}{4} \times 6 + 2\right) + \frac{1}{3}(8 \times 2)\left(\frac{3}{4} \times 2 + 6\right)$$

$$EI\, y_B = -\ 1216\ \text{kN·m}^3$$

$$\therefore \quad y_B = \frac{-\ 1216\,(10^3 \times 10^9)}{(2 \times 10^5)(4.8 \times 10^8)} = -\ \mathbf{12.66\ mm}, \text{ i.e., } \downarrow$$

Example 6.11. *A timber beam 4 m long of rectangular cross-section is simply supported at its ends and is subjected to a concentrated load of 10 kN at mid-span. If the width of the beam is one-third the depth, determine its dimensions given the following permissible values:*

Bending stress = 8 N/mm²

Shear stress = 0.4 N/mm²

Deflection = 8 mm

Modulus of elasticity = 7000 N/mm²

Solution. See Figure 6.14.

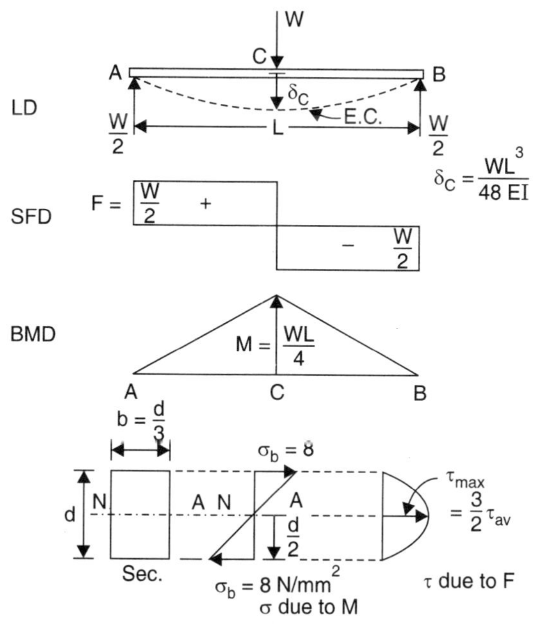

Figure 6.14. Design of timber beam

$$b = \frac{d}{3}, \quad I = \frac{bd^3}{12} = \frac{(d/3)\,d^3}{12} = \frac{d^4}{36}$$

(*i*) Beam section to resist bending : $\sigma_b = 8\ \text{N/mm}^2$

$$M = \frac{WL}{4} = \frac{(10 \times 10^3)\,4000}{4} = 10^7\ \text{N·mm}$$

Since, $$\frac{M}{I} = \frac{\sigma_b}{y}$$

$\therefore$ $$\frac{10^7}{d^4/36} = \frac{8}{d/2}, \quad d^3 = 22.5 \times 10^6, \quad d = 282\ \text{mm}$$

$$b = \frac{d}{3} = \frac{282}{3} = 94\ \text{mm}$$

$\therefore$ **94 × 282 mm** section is required.

(*ii*) Beam section to resist shear : $\tau_{max} = 0.4\ \text{N/mm}^2$

$$\tau_{max} = \frac{3}{2}\tau_{av}\ \text{, where } \tau_{av} = \frac{F}{A} = \frac{W/2}{A} = \frac{5000}{(d/3)\,d}$$

$$0.4 = \frac{3}{2} \times \frac{5000}{d^2/3}, \quad d^2 = 5.625 \times 10^4, \quad d = 237\ \text{mm}$$

$$b = \frac{d}{3} = \frac{237}{3} = 79\ \text{mm}$$

$\therefore$ **79 × 237 mm** section is required.

(*iii*) Beam section for the allowable deflection : $\delta_C = 8$ mm

$$\delta_C \text{ or } y_C = \frac{WL^3}{48\,EI}$$

$$8 = \frac{(10 \times 10^3)\,4000^3}{48 \times 7000 \times (d^4/36)}$$

$$\therefore \qquad d^4 = 8.57 \times 10^8, \quad d = 304 \text{ mm}, \; b = \frac{d}{3} = 102 \text{ mm}.$$

Section required is **102 × 304 mm** which is the largest for the three criteria and hence can be adopted.

Note. If $b = \frac{d}{4}$ is chosen the section required will be 82 × 327 mm, which is economical.

Example 6.12. *Calculate the maximum instantaneous stress induced in a simply supported beam 4 m long when a load of 7 kN is dropped at mid-span from a height of 0.4 m. The beam section is 100 × 200 mm and E = 2 × 10⁵ N/mm². What is the corresponding deflection ?*

Solution. Impact factor, see page 166—167,

$$I_f = \frac{\delta}{\delta_{St}} = \frac{\sigma}{\sigma_{St}} = 1 + \sqrt{1 + \frac{2h}{\delta_{St}}}$$

For a simply supported beam with static load P = 7 kN at mid-span,

$$\delta_{St} = \frac{PL^3}{48\,EI} = \frac{7000 \times (4 \times 1000)^3}{48\,(2 \times 10^5)\left(\dfrac{100 \times 200^3}{12}\right)} = 0.7 \text{ mm}$$

$$\therefore \qquad I_f = 1 + \sqrt{1 + \frac{2 \times 400}{0.7}} = 34.82$$

Static stress $$\sigma_{St} = \frac{M \times y}{I} = \frac{\dfrac{PL}{4} \times \dfrac{d}{2}}{(bd^3)/12} = 1.5 \times \frac{PL}{bd^2}$$

$$= 1.5 \times \frac{7000 \times 4000}{100 \times 200^2} = 10.5 \text{ N/mm}^2$$

Max. instantaneous stress = 10.5 × 34.82 = **365.5 N/mm²**

and the corresponding deflection δ = 0.7 × 34.82 = **24.4 mm**

PROBLEMS

6.1. A simply supported beam of 10 m span and EI = 100 N·m² is subjected to a concentrated load of 10 kN at a distance of 4 m from the left support. Calculate the deflection at mid-span.

6.2. A cantilever 2 m long is 150 mm wide and 200 mm deep. It carries a concentrated load of 50 kN at its free end. Find the slope and deflection at the free end. [Take $E = 2 \times 10^5$ N/mm²]

6.3. A beam of uniform cross-section and flexural rigidity 30 MN·m² is hinged at A and rests on support B, 6 m, from A. A clockwise moment of 100 kN·m. acts at C, 4 m apart from A. Determine the deflection at C, and also the maximum deflection and its position.

6.4. A uniform beam AB, 8 m long, rests symmetrically on support C and D, 4 m apart. A load of 40 kN is applied at each end A and B. Neglecting the weight of the beam, calculate the deflection at the ends A and B and at mid-span.
[Take E = 200 GN/m², $I = 50 \times 10^6$ mm⁴]

6.5. A simply supported beam 4 m long carries a *udl* of 10 kN/m for half the span from the left support. Determine the maximum deflection and its location. The beam has a rectangular section 50 × 100 mm and $E = 2 \times 10^5$ N/mm^2.

6.6. Find the slope and deflection at *B*, *C*, and *D* for the cantilever shown in Figure P. 6.6. [Take EI = 50 MN·m^2]

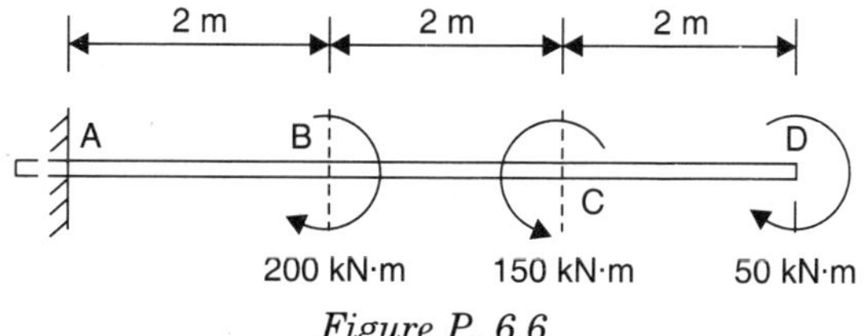

Figure P. 6.6

6.7. A cantilever 2 m long carries a load of 20 kN at its free end and 30 kN at mid-span. Calculate the slope and deflection at the free end. [Take $E = 2 \times 10^5$ N/mm^2, $I = 1.5 \times 10^8$ mm^4.]

6.8. Determine the prop reaction of the cantilever shown in Figure P. 6.8.

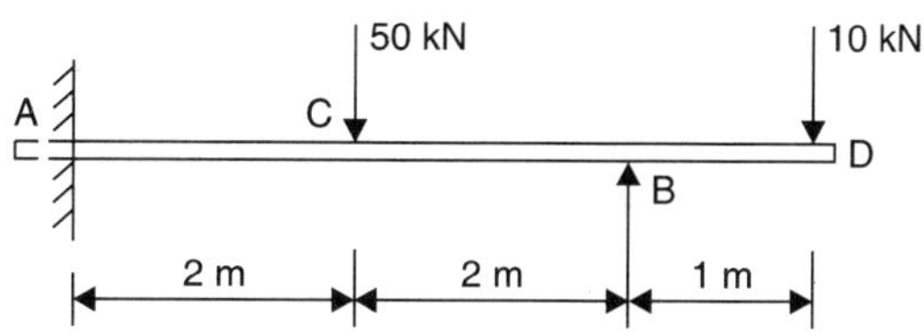

Figure P. 6.8

6.9. (*a*) Find the slope and deflection at the free end of the cantilever shown in Figure P. 6.9. [Take EI = 20 MN·m^2.]

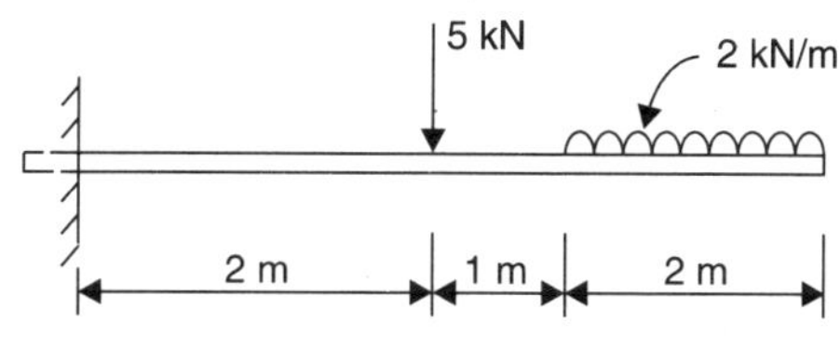

Figure P. 6.9

(*b*) If a prop is given at the free end such that it is brought to the level of the fixed end, what is the prop reaction ?

6.10. Determine the deflection at the free end *C*, of the beam shown in Figure P. 6.10. [Take EI = 30 MN/m^2.]

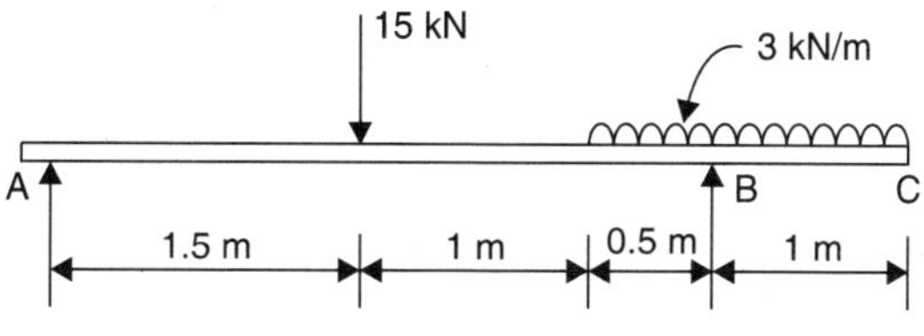

Figure P. 6.10

6.11. A simply supported beam 2 m long and 100 × 200 mm section is subjected to an impact load of 10 kN being dropped at mid-span from a height '*h*'. If for the material of the beam, the impact yield stress is 400 N/mm^2 and $E = 2 \times 10^5$ N/mm^2, determine the height of drop and the corresponding deflection.

7 Thin and Thick Cylinders

A. Thin Cylinder subjected to fluid pressure p, inside (see Figure 7.1) in a thin cylinder of inside diameter d, the circumferential or hoop stress σ_c, and longitudinal stresses σ_l, are assumed to be uniform across the thickness t, of the metal and $t < \frac{d}{20}$, suitable for low pressure vessels.

***(i)* Circumferential of hoop stress (σ_c).** If the efficiency of the longitudinal joint is η_l (as in the case of a boiler-riveted), see Figure 7.1 (*a*),

$$\uparrow p\,(d \cdot l) = \downarrow \sigma_c\,(2 \cdot l \cdot t)\,\eta_l$$

$$\therefore \qquad \sigma_c = \frac{pd}{2t\eta_l} \qquad \text{...(7.1)}$$

If there is no riveted joint (seamless pipe), $\boldsymbol{\sigma_c = \frac{pd}{2t}}$ ...(7.2)

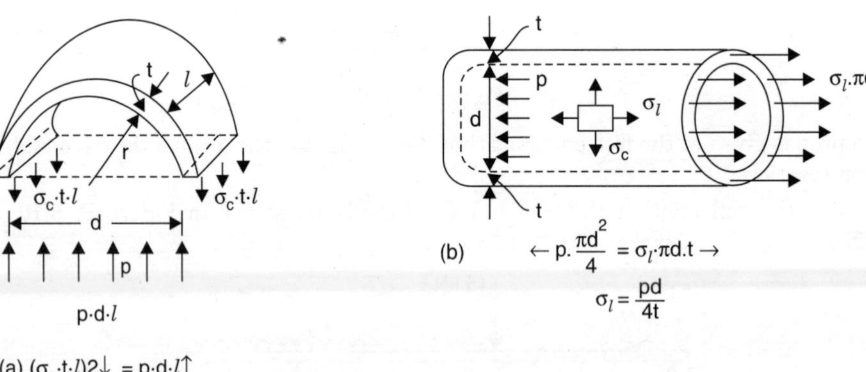

Figure 7.1. Thin cylinder

***(ii)* Longitudinal stress (σ_l).** If the efficiency of the circumferential joint is η_c, see Figure 7.1 (*b*),

$$\leftarrow p\,\frac{\pi d^2}{4} = \sigma_l\,(\pi d \cdot t)\,\eta_c \rightarrow$$

$$\therefore \qquad \sigma_l = \frac{p \cdot d}{4 \cdot t \cdot \eta_c} \qquad \text{...(7.3)}$$

If there is no riveted joint, $\boldsymbol{\sigma_l = \frac{pd}{4t}}$ ...(7.4)

Hence, for a seamless pipe, $\sigma_c = 2\,\sigma_l$, and the maximum shear stress

$$\tau_{\max} = \frac{\sigma_c - \sigma_l}{2} = \frac{\sigma_l}{2} = \frac{pd}{8t}, \qquad \text{...(7.5)}$$

B. Thin Spherical Shell under Internal Pressure p, See Figure 7.2. The circumferential or hoop stress σ, is given by

$$\leftarrow p\,\frac{\pi d^2}{4} = \sigma\,(\pi d \cdot t) \rightarrow$$

$$\therefore \qquad \sigma = \frac{p \cdot d}{4t} \qquad \text{...(7.6)}$$

and the principal stresses are equal in two directions.

Hoop strain $\varepsilon_c = \frac{\sigma}{E} - \nu\frac{\sigma}{E} = \frac{\sigma}{E}(1-\nu)$ ...(7.7)

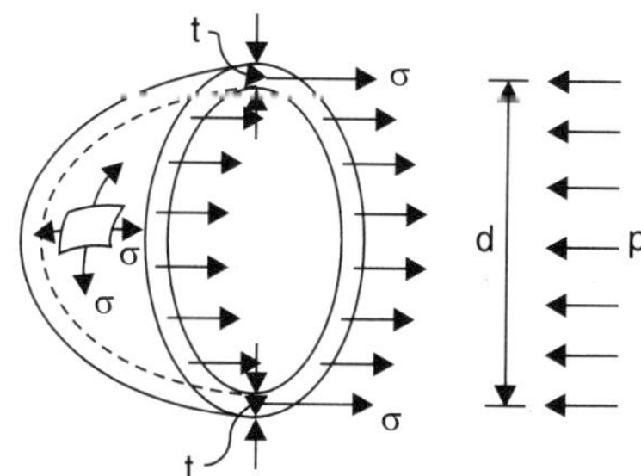

Figure 7.2. Thin spherical shell

It can be shown that volumetric strain = 3 × hoop strain

$$\varepsilon_v = 3\varepsilon_c \qquad \text{...(7.8)}$$

$$\varepsilon_v = \frac{\Delta V}{V}, \quad \Delta V = \varepsilon_v \times \frac{\pi d^3}{6} \qquad \text{...(7.9)}$$

When a cylindrical shell has hemispherical ends (see Figure 7.3), it is customary to assume the hoop strain for the two hemispherical ends same as the circumferential strain of the cylinder (for compatibility i.e., to avoid any distortion at the junction under pressure).

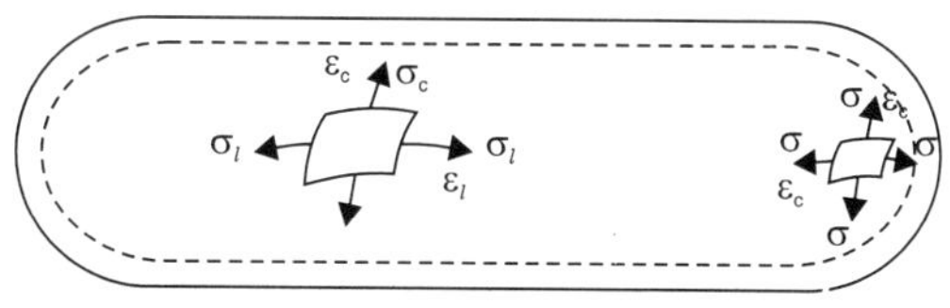

Figure 7.3. Boiler shell

Example 7.1. *A boiler is subjected to an internal pressure of 2 N/mm². The thickness of the boiler plate is 16 mm and the permissible tensile stress 120 N/mm². If the efficiency of the longitudinal and circumferential joints are 90% and 40%, respectively, what should be the maximum permissible diameter of the boiler ?*

Solution. (*i*) $\sigma_c = \frac{p \cdot d}{2t\,\eta_l}$

$$120 = \frac{2 \times d}{2 \times 16 \times 0.90}, \quad d = 1728 \text{ mm}$$

(*ii*) $$\sigma_l = \frac{p \cdot d}{4 \times t \cdot \eta_c}$$

$$120 = \frac{2 \times d}{4 \times 16 \times 0.40}, \quad d = 1536 \text{ mm}$$

∴ Max. permissible diameter

$$d = 1536 \text{ mm, or } \mathbf{1.536\ m}$$

Example 7.2. *A vertical cylindrical tank, open at top has I.D. = 2.5 m and 10 m high. The tank steel has an yield point of 220 N/mm². What should be the thickness of the plated steel with a factor of safety of 3, if the efficiency of the longitudinal joint is 75% ?*

Solution. Water pressure at the base of the tank

$$p = \rho gh = 1000 \text{ kg/m}^3 \times 9.81 \times 10 = 98{,}100 \text{ N/m}^2$$

or $$= 0.0981 \text{ N/mm}^2$$

$$\sigma_c = \frac{p \cdot d}{2t \cdot \eta_l}, \text{ but } \sigma_c = \frac{220}{3} = 73.33 \text{ N/mm}^2$$

$$73.3 = \frac{0.0981 \times 2500}{2t \times 0.75}, \quad t = 2.23 \text{ mm, say } \mathbf{3\ mm}$$

Example 7.3. *A shell 3.25 m long and 1 m diameter is subjected to an internal pressure of 1 MN/m², if the thickness of the shell is 10 mm, find the circumferential and longitudinal stresses. Find also the maximum shear stress and change in dimensions and volume of the shell. [Take E = 200 GN/m² and ν = 0.3.]*

Solution. (*i*) $$\sigma_c = \frac{p \cdot d}{2t} = \frac{1 \times 1000}{2 \times 10} = \mathbf{50\ N/mm^2}$$

(*ii*) $$\sigma_l = \frac{p \cdot d}{4t} = \frac{\sigma_c}{2} = \frac{50}{2} = \mathbf{25\ N/mm^2}$$

(*iii*) $$\tau_{max} = \frac{\sigma_c - \sigma_l}{2} = \frac{50 - 25}{2} = \mathbf{12.5\ N/mm^2}$$

(*iv*) $$\varepsilon_c = \frac{\sigma_c}{E} - \nu \frac{\sigma_l}{E} = \frac{1}{E}(50 - 0.3 \times 25) = \frac{42.5}{E}$$

Circumferential strain $(\varepsilon_c) = \frac{\Delta d}{d}$ = diametral strain

$$\Delta d = 1000 \times \frac{42.5}{200 \times 10^3} = \mathbf{0.2125\ mm:}$$

Note: $MN/m^2 \equiv N/mm^2$
$MN/m^2 = 10^6\ N/m^2$
$GN/m^2 = 10^9\ N/m^2$

(*v*) $$\varepsilon_l = \frac{\sigma_l}{E} - \nu \frac{\sigma_c}{E} = \frac{1}{E}(25 - 0.3 \times 50) = \frac{10}{E}$$

Longitudinal strain $\varepsilon_l = \frac{\Delta L}{L}$

∴ $$\Delta L = (3.25 \times 10^3) \frac{10}{200 \times 10^3} = \mathbf{0.1625\ mm}$$

(*vi*) Volumetric strain $\varepsilon_v = \dfrac{\dfrac{\pi}{4}(d+\delta d)^2 (L+\delta L) - \dfrac{\pi}{4} d^2 \cdot L}{\dfrac{\pi}{4} d^2 \cdot L}$

Neglecting powers and products of small quantities,

$$\varepsilon_v = 2\frac{\delta d}{d} + \frac{\delta L}{L} \qquad \therefore \boldsymbol{\varepsilon_v = 2\,\varepsilon_c + \varepsilon_l} \qquad \text{...(7.10)}$$

$$\therefore \qquad \varepsilon_v = 2\varepsilon_c + \varepsilon_l = \frac{1}{200 \times 10^3}(2 \times 42.5 + 10) = 4.75 \times 10^{-4}$$

$$\varepsilon_v = \frac{\Delta V}{V}, \quad \Delta V = (4.75 \times 10^{-4}) \frac{\pi}{4} \,.\, 1000^2 \times 3250 = \mathbf{1.23 \times 10^6\ mm^3}$$

Example 7.4. *A cylindrical shell 1 m long, 180 mm I.D., 8 mm thick is filled with water at atmospheric pressure. If an additional 20 × 10³ mm³ of water is pumped into the cylinder, find the pressure exerted by water on the wall of the cylinder. Find also the hoop stress induced.*

[Take E_{steel} = 200 GN/m², K_{water} = 2.1 GN/m², ν = 0.3]

Solution. Let the internal pressure be p N/mm².

(*i*) Increase in the capacity of the cylinder $\Delta V_1 = \varepsilon_v \times$ Volume

$$\sigma_c = \frac{pd}{2t} = \frac{p \times 180}{2 \times 8} = 11.25\,p$$

$$\sigma_l = \frac{\sigma_c}{2} = \frac{11.25p}{2} = 5.625\,p$$

$$\varepsilon_c = \frac{1}{E}(\sigma_c - \nu\sigma_l) = \frac{p}{E}(11.25 - 0.3 \times 5.625) = 9.5135\,\frac{p}{E}$$

$$\varepsilon_l = \frac{1}{E}(\sigma_l - \nu\sigma_c) = \frac{p}{E}(5.625 - 0.3 \times 11.25) = 2.25\,\frac{p}{E}$$

$$\varepsilon_v = 2\,\varepsilon_c + \varepsilon_l = \frac{p}{E}(19.027 + 2.25) = 21.277\,\frac{p}{E}$$

$$\varepsilon_v = \frac{\Delta V}{V}, \quad \Delta V_1 = 21.277\,\frac{p}{E}\left[\frac{\pi}{4}180^2 \times 1000\right] = 541.4 \times 10^6\,\frac{p}{E}$$

(*ii*) Decrease in the volume of water originally in

$$\Delta V_2 = \frac{p}{K} \times \text{Volume}$$

$$\Delta V_2 = \frac{p}{K}\left[\frac{\pi}{4} \,.\, 180^2 \cdot 1000\right] = 25.4 \times 10^6\,\frac{p}{K}$$

Additional volume of water pumped in at atmospheric pressure = $\Delta V_1 + \Delta V_2$

$$20 \times 10^3\ \text{mm}^3 = \frac{p}{200 \times 10^3} \times 541.4 \times 10^6 + 25.4 \times 10^6 \times \frac{p}{2.1 \times 10^3}$$

$$20{,}000 = 2707p + 12{,}100p$$

$$\therefore \qquad p = \frac{20{,}000}{14{,}807} = \mathbf{1.35\ N/mm^2}$$

Hoop stress $\sigma_c = \frac{pd}{2t} = \frac{1.35 \times 180}{2 \times 8} = \mathbf{15.19\ N/mm^2}$

C. Thick Cylinder: The wall of the cylinder is thick $\left(t > \frac{d}{20}\right)$ and the radial and hoop stresses across the wall vary from inside to outside the cylinder. The cylinder may be subjected to appreciable pressure both inside and outside, see Figure 7.4.

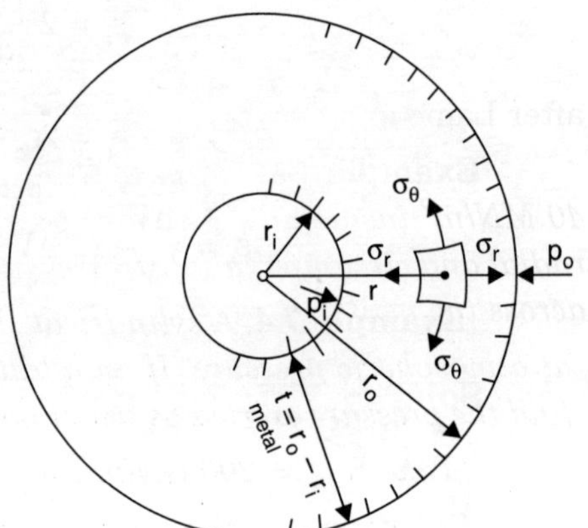

Figure 7.4. Thick cylinder

From the theory of elasticity,

Radial stress $\sigma_r = a - \frac{b}{r^2}$ ...(7.10*a*)

Circumferential or hoop stress $\sigma_\theta = a + \frac{b}{r^2}$ ...(7.11)

On an element at a radial distance *r*, from the centre of the cylinder; *a* and *b* are constants and can be obtained from the known boundary conditions.

Lame's Equations

Case I. When the thick cylinder is subjected to internal pressure p_i, only.

Using the boundary conditions in Eqn. (7.10):

At $r = r_i$, $\sigma_r = -p_i$: $-p_i = a - \frac{b}{r_i^2}$ ...(*i*)

At $r = r_o$, $\sigma_r = 0$: $0 = a - \frac{b}{r_o^2}$ ...(*ii*)

Solution of (*i*) and (*ii*) yields

$$a = \frac{p_i r_i^2}{r_o^2 - r_i^2}, \qquad b = \frac{p_i r_i^2 r_o^2}{r_o^2 - r_i^2}$$

Substituting for *a* and *b* in Eqns. (7.10) and (7.11),

$$\left.\begin{aligned} \boldsymbol{\sigma_r} &= \frac{\mathbf{p_i\, r_i^2}}{\mathbf{r_o^2 - r_i^2}} - \frac{\mathbf{p_i\, r_i^2 r_o^2}}{\mathbf{r^2(r_o^2 - r_i^2)}} \\ \boldsymbol{\sigma_\theta} &= \frac{\mathbf{p_i\, r_i^2}}{\mathbf{r_o^2 - r_i^2}} + \frac{\mathbf{p_i\, r_i^2 r_o^2}}{\mathbf{r^2(r_o^2 - r_i^2)}} \end{aligned}\right\} \textbf{Lame's equations} \qquad ...(7.12)$$

Case II. When the thick cylinder is subjected to both internal pressure p_i, and external pressure p_o.

Using the boundary conditions in Eqn. (7.10)

At $r = r_i$, $\sigma_r = -p_i$: $-p_i = a - \frac{b}{r_i^2}$ ...(*iii*)

At $r = r_o$, $\sigma_r = -p_o$: $-p_o = a - \frac{b}{r_o^2}$ ...(*iv*)

Solution of (*iii*) and (*iv*) yields *a* and *b* which when substituted in Eqns. (7.10) & (7.11) give the Lame's equations,

$$\left.\begin{aligned}\sigma_r &= \frac{p_i r_i^2 - p_o r_o^2}{r_o^2 - r_i^2} - \frac{(p_i - p_o) r_i^2 r_o^2}{r^2 (r_o^2 - r_i^2)} \\ \sigma_\theta &= \frac{p_i r_i^2 - p_o r_o^2}{r_o^2 - r_i^2} + \frac{(p_i - p_o) r_i^2 r_o^2}{r^2 (r_o^2 - r_i^2)}\end{aligned}\right\} \quad ...(7.13)$$

after Lame and Clapeyron in 1833.

Example 7.5. *A pipe of O.D. = 300 mm and 60 mm thick is subjected to fluid pressure of 40 MN/m² inside and 1 MN/m² outside. Calculate the maximum and minimum intensities of radial and circumferential stresses in the pipe section and plot their variation at 15 mm interval across the wall.*

Solution. $\sigma_r = a - \frac{b}{r^2}, \quad \sigma_\theta = a + \frac{b}{r^2}$

At $r = r_i = \frac{300 - 2 \times 60}{2} = 90$ mm, $\sigma_r = -40$ N/mm²

At $r = r_o = \frac{300}{2} = 150$ mm, $\sigma_r = -1$ N/mm²

$$\therefore \quad -40 = a - \frac{b}{90^2} \quad ...(i)$$

and

$$-1 = a - \frac{b}{150^2} \quad ...(ii)$$

Solving (*i*) and (*ii*), $a = 20.94$, $b = 4.94 \times 10^5$

$$\therefore \quad \sigma_r = 20.94 - \frac{4.94 \times 10^5}{r^2}$$

$$\sigma_\theta = 20.94 + \frac{4.94 \times 10^5}{r^2}$$

σ_r = max., when $r = r_o = 150$ mm : $\sigma_r = -1.00$ N/mm²

σ_θ = max., when $r = r_i = 90$ mm : $\sigma_\theta = 81.92$ N/mm²

Hoop stress is maximum at the inner boundary of the wall and should not exceed the permissible tensile stress of the material.

Stress variation across the wall, see Figure 7.5.

r (mm)	90	105	120	135	150
σ_r (N/mm²)	–40	–23.87	–13.37	–6.16	–1.00
σ_θ (N/mm²)	81.92	65.75	55.25	48.04	42.90

Note. The radial stress is compressive throughout.

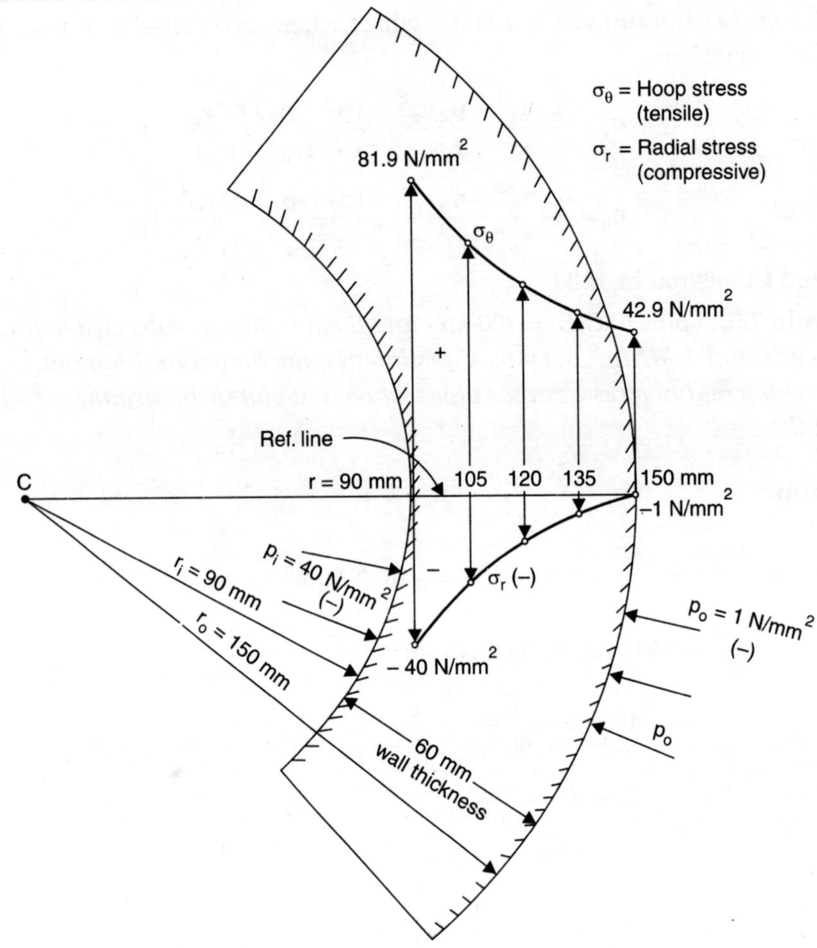

Figure 7.5. Variation of stress across the wall of thick cylinder

Example 7.6. *A thick cylinder of I.D. = 160 mm is subjected to an internal pressure of 40 N/mm². If the allowable stress for the material is 120 N/mm², find the thickness required.*

Solution.

$$\sigma_r = a - \frac{b}{r^2} : \quad -40 = a - \frac{b}{80^2} \qquad ...(i)$$

$$\sigma_\theta = a + \frac{b}{r^2} : \quad 120 = a + \frac{b}{80^2} \qquad ...(ii)$$

Solving (*i*) and (*ii*), $a = 40$, $b = 80^3$

At $r = r_o$, $\sigma_r = 0$ (no external pressure)

$$0 = 40 - \frac{80^3}{r_o^2}, \quad r_o = 113.14 \text{ mm}$$

Thickness of metal $\quad t = 113.14 - \dfrac{160}{2} = \mathbf{33.14\ mm}$

PROBLEMS

7.1. A cylindrical shell 2 m long, 600 mm diameter and 10 mm thick is subjected to an internal pressure of 2N/mm^2. Determine the change in dimensions of the shell and its volume. [Take, E = 200N/m^2, ν = 0.3.]

7.2. A steel boiler shell 2 m diameter is subjected to an internal pressure of 1.2 N/mm^2. If the allowable tensile stress for steel is 100 N/mm^2 and efficiency of the longitudinal joint is 85%, determine the thickness of the shell.

7.3. A spherical pressure vessel is 1 m diameter is subjected to an internal pressure of 1.5 N/mm^2. If the permissible stress for the shell material is 60 N/mm^2 and efficiency of the joint is 80%, determine the wall thickness.

7.4. A cylindrical shell 3 m long, 1 m diameter, and 12 mm thick is subjected to an internal pressure of 1.5 N/mm^2. Determine the maximum shear stress induced and changes in the dimensions of the shell. [Take, E = 200 GN/m^2, ν = 0.3.]

7.5. A copper tube 60 mm I.D., 1.5 m long and 1.5 mm thick has closed ends and is filled with water under pressure. Neglecting the distortion of end plates, what pressure is reached when an additional volume of 35 m^3 of water is pumped into the tube.

[Take, E_{copper} = 100 GN/m^2 and ν = 0.3, K_{water} = 2.1 GN/m^2 .]

7.6. A steel cylinder 2 m long, 1 m diameter and 10 mm thick is filled with water under atmospheric pressure. More water is pumped in until the pressure is raised to 2 N/mm^2. On relieving the pressure, the water let out measured 3 × 10^6 mm^3. Assuming the ends to remain flat, estimate the bulk modulus of water.

[Take, E_{steel} = 200 GN/m^2, ν = 0.3.]

7.7. A spherical shell of copper 300 mm diameter and 1.6 mm thick is filled with water under atmospheric pressure. What pressure is reached when 25 × 10 mm^3 of water is pumped in ?

[Take, E_c = 100 GN/m^2, ν = 0.286, K_{water} = 2.1 GN/m^2.]

7.8. A boiler drum consists of a cylindrical portion 2 m long, 1 m dia, and 20 mm thick, and the two ends are hemispherical. It is initially filled with water under atmospheric pressure. Additional water is pumped in and the pressure is raised to 6 N/mm^2. Assuming the hoop strain for the hemispherical ends same as the circumferential strain for the cylindrical portion, determine the additional water pumped in. [Take, E = 200 GN/m^2, ν = 0.3, K_{water} = 2.1 GN/m^2]

[**Hint :** Additional volume of water at atmospheric pressure

$$= (2\varepsilon_c + \varepsilon_l)\,\frac{\pi d^2}{4}\cdot L + 3\,\varepsilon_c\cdot\frac{\pi d^3}{6} + \frac{p}{k}\left[\frac{\pi d^2}{4}\cdot L + \frac{\pi d^3}{6}\right].\Bigg]$$

7.9. The maximum stress permitted in a thick cylinder of radii 200 mm and 300 mm is 20 N/mm^2. If the external pressure is 5 N/mm^2, what internal pressure can be applied ?

Plot curves showing the variation of hoop and radial stresses across the wall.

7.10. A thick walled tube has I.D. = 120 mm and O.D. = 200 mm. Determine the maximum and minimum hoop stresses due to external pressure of 40 N/mm^2.

7.11. A pipe of 400 mm I.D. and 100 mm thick contains a fluid at a pressure of 8 N/mm^2. Find the maximum and minimum hoop stress across the section.

7.12. What should be the thickness of metal of a thick cylinder of I.D. = 500 mm to withstand a fluid pressure of 10 N/mm^2, assuming a permissible stress of 90 N/mm^2 for the material ?

7.13. A thick cylinder has I.D. = 200 mm. If the internal pressure is 10 N/mm^2 and the maximum permissible stress in the cylinder is 16 N/mm^2, what is the minimum thickness required ? If the internal pressure is to be increased to 15 N/mm^2, what external pressure has to be applied ?

8 Columns and Struts

Any compression member may be called a **strut** but it usually means a 'long column' which fails by buckling before the limiting compressive stress is reached. A short column (usually vertical) when subjected to axial compression fails when the limiting compressive stress (yield stress) is exceeded. For the purpose of this classification, the term **slenderness ratio** is defined as

$$\frac{\text{Length of member}}{\text{Least radius of gyration}} = \frac{l}{k} \qquad \text{...(8.1)}$$

is used, and the columns are classified as follows:

Short columns	$\frac{l}{k} < 40$	l – small
Intermediate columns	$40 \le \frac{l}{k} \le 80$	$l - 30d$
Long columns	$\frac{l}{k} > 80$	$l > 30d$

The columns buckle about the axis of least moment of inertia I, and the least radiation of gyration k, is used in obtaining the slenderness ratio as

$$I = Ak^2 \quad \text{or} \quad k = \sqrt{\frac{I}{A}} \qquad \text{...(8.2)}$$

Euler's Formula for Long Columns. The least value of the axial load which will cause the column (strut) to buckle is called the crippling (or the buckling or critical) load. The crippling load for long columns (of length L, under axial load), **with both ends hinged**, is given by

$$\text{Eulers crippling load } P_{cr} = \frac{\pi^2 EI}{L^2} \qquad \text{...(8.3)}$$

Effective Length or Equivalent Length (l). It is that length of the column which gives the same critical buckling loads as that of a column with both ends hinged (which is taken as a standard case), and depends on the end conditions, see Figure 8.1. The strength of a column to resist buckling depends on the **slenderness ratio** and the **end conditions**.

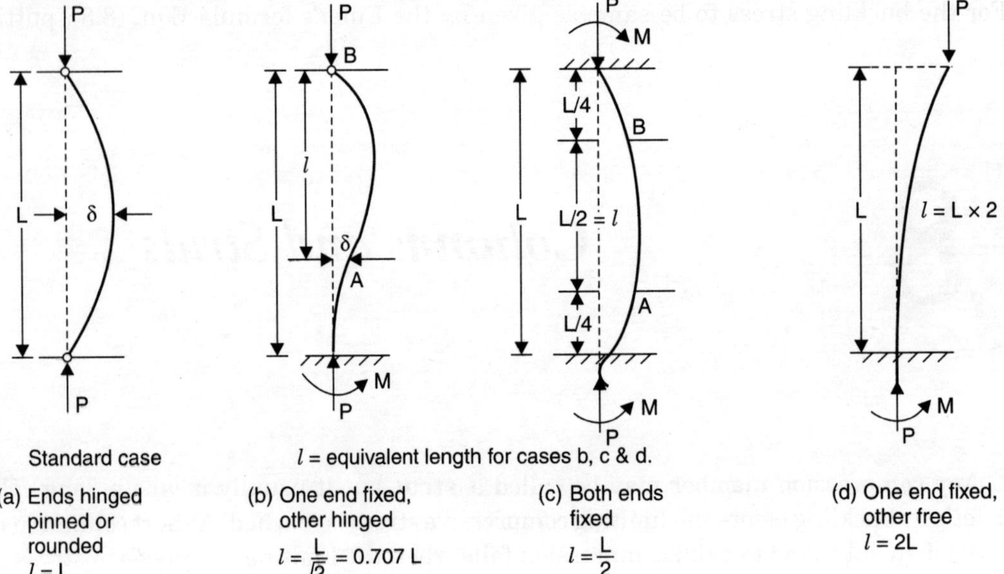

Figure 8.1. Equivalent length for different end conditions

Rankine-Gordon Formula for Intermediate Columns. The crippling load of columns (short to long) for both ends hinged (or pinned) is given by

$$P = \frac{\sigma_c \cdot A}{1 + a\left(\frac{L}{k}\right)^2} \qquad ...(8.4)$$

where σ_c = stress at yield point in compression,

A = area of cross-section of the column,

L = length of the column, equivalent length (l) should be taken for end conditions other than hinged,

a = a constant.

The values of the constants 'σ_c' and 'a' depend on the material of the column, as follows:

Material	*σ_c (N/mm²)*	*a*
M.S.	330	1/7500
C.I.	560	1/1600
W.I.	255	1/9000
Timber	50	1/750

Equation (8.4) can be written for intermediate columns as

$$\sigma_c = \frac{P}{A} + \frac{Pal^2}{Ak^2} = \text{direct stress + buckling stress (Euler)} \qquad ...(8.5)$$

For the buckling stress to be same as given by the Euler's formula Eqn. (8.3), putting $I = Ak^2$

$$P = \frac{\sigma_c \cdot I}{al^2} = \frac{\pi^2 \cdot E \cdot I}{l^2}, \quad \therefore \quad a = \frac{\sigma_c}{\pi^2 \cdot E} \qquad \text{....(8.6)}$$

The length of a column which gives the same buckling load both by Euler and Rankine formulae can be obtained as

$$P = \frac{\pi^2 \cdot E \cdot I}{l^2} = \frac{\sigma_c \cdot A}{1 + a\left(\frac{l}{k}\right)^2}$$

$$\pi^2 EI \left(1 + \frac{al^2}{k^2}\right) = \sigma_c Al^2$$

Putting $I = Ak^2$, $l^2 (\sigma_c A - \pi^2 EAa) = \pi^2 EAk^2$

$$\therefore \qquad l^2 = \frac{\pi^2 Ek^2}{\sigma_c - \pi^2 Ea} \qquad \text{....(8.7)}$$

in which both direct and buckling stresses are taken into account and 'l' is the equivalent length from which the actual length of the column can be found for the given end conditions.

Example 8.1. *A M.S. column consists of two channels ISJC 200 × 136 N/m per channel placed at 100 mm back to back and two plates 250 × 10 mm, one riveted to each flange. The column is 6 m high and its ends firmly built in. The properties of each channel are:*

Area = 1717 mm², $I_{xx} = 11.61 \times 10^6$ mm⁴,

$I_{yy} = 84.2 \times 10^4$ mm⁴, $C_{yy} = 19.7$ mm

Determine : (a) the safe load that the column can carry with a factor of safety of 3.5,(i) by Euler's formula, and (ii) by Rankine's formula. (b) the length of the column for which the crippling load will be the same by both the formulae.

For M.S., $E = 200$ kN/mm², $\sigma_c = 330$ N/mm², $a = \frac{1}{7500}$.

Solution. (*a*) (*i*) Euler's crippling load

$$P_{cr} = \frac{\pi^2 \cdot E \cdot I}{L^2}$$

To find I of the built up column:

$$I_{xx} = 2\left[11.61 \times 10^6 + \left\{\frac{250 \times 10^3}{12} + (250 \times 10)\,105^2\right\}\right]$$

$$= 78.38 \times 10^6 \text{ mm}^4$$

$$I_{yy} = 2\left[\{84.2 \times 10^4 + 1717\,(19.7 + 50)^2\} + \frac{10 \times 250^3}{12}\right]$$

$$= 44.4 \times 10^6 \text{ mm}^4$$

Least value of $I = I_{yy} = 44.4 \times 10^6$ mm^4.

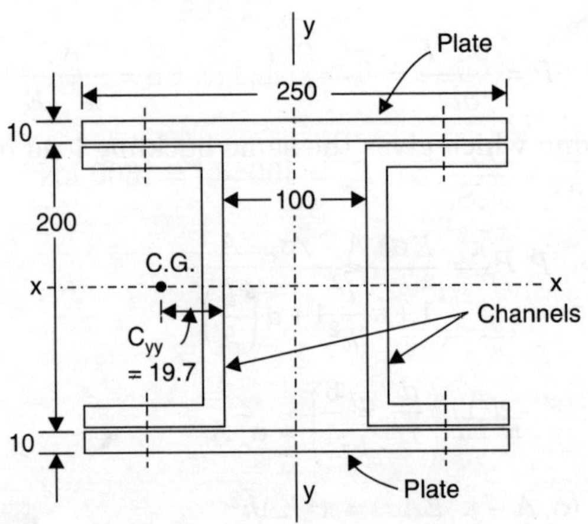

Figure 8.2. Built-up column

For firmly built in or fixed ends, $l = \dfrac{L}{2} = \dfrac{6}{2} = 3$m or 3000 mm

$$P_{cr} = \frac{\pi^2 (200 \times 10^3)(44.4 \times 10^6)}{3000^2} = 97.38 \times 10^5 \text{ N} \quad \text{or} \quad 9738 \text{ kN}$$

Safe load $\quad P = \dfrac{P_{cr}}{\text{F.S.}} = \dfrac{9738}{3.5} = \mathbf{2782\ kN}$

(*ii*) Rankine's formula

$$P_{cr} = \frac{\sigma_c \cdot A}{1 + a\dfrac{l^2}{k^2}}, \qquad k^2 = \frac{I}{A} = \frac{44.4 \times 10^6}{2(1717 + 2500)} = 5264.4 \text{ mm}^2$$

$$= \frac{330 \times 8434}{1 + \dfrac{1}{7500} \times \dfrac{3000^2}{5264.4}} = 22.66 \times 10^5 \text{ N or } 2266 \text{ kN}$$

Safe load $\quad P = \dfrac{P_{cr}}{\text{F.S.}} = \dfrac{2266}{3.5} = \mathbf{647.5\ kN}$

(*b*) The length of the column to give the same crippling load by both the formulae is given by Eqn. (8.7)

$$l^2 = \frac{\pi^2 Ek^2}{\sigma_c - \pi^2 Ea} = \frac{\pi^2 (200 \times 10^3) 5264}{330 - \pi^2 (200 \times 10^3) \dfrac{1}{7500}}$$

$$= 1.55 \times 10^8 \text{ mm}^2 \text{ or } 155 \text{ m}^2$$

$$l = 12.47 = \frac{L}{2} \text{ (for both ends fixed)}$$

$\therefore$ Length of column (L) = **25 m**

Example 8.2. *Determine the size of a hollow C.I. cylindrical column 5m long with its ends firmly built in to carry on axial load of 400 kN. Take the ratio of diameters as 3 : 4 and a factor of safety of 4 in the Rankine's formula. For C.I,* $\sigma_c = 560\ N/mm^2$*,* $a = \frac{1}{1600}$.

Solution. Safe load $= \frac{P_{cr}}{\text{F.S.}}$, $\quad P_{cr} = 400 \times 4 = 1600$ kN

Rankine's formula $\quad P_{cr} = \dfrac{\sigma_c A}{1 + a\dfrac{l^2}{k^2}}$ ...(*i*)

$$A = \frac{\pi}{4}(D^2 - d^2), \quad \frac{d}{D} = \frac{3}{4}$$

$$= \frac{\pi}{4} D^2 \left(1 - \frac{9}{16}\right) = 0.3436\ D^2$$

$$k^2 = \frac{I}{A} = \frac{\frac{\pi}{64}(D^4 - d^2)}{\frac{\pi}{4}(D^2 - d^2)} = \frac{1}{16}(D^2 + d^2) = \frac{D^2}{16}\left(1 + \frac{9}{16}\right)$$

$$= 0.0978\ D^2$$

For fixed ends $\quad l = \dfrac{L}{2} = \dfrac{5000}{2} = 2500$ mm

Substituting the above values in (*i*), we get

$$1600 \times 10^3 = \frac{560(0.3436\,D^2)}{1 + \dfrac{1}{1600} \times \dfrac{2500^2}{0.0976\,D^2}}$$

$$16 \times 10^5 = \frac{192.422\ D^2}{1 + 40000/D^2}$$

$$16 \times 10^5\ D^2 + 64 \times 10^9 = 192.422\ D^4 \qquad \text{...}(ii)$$

Putting $D^2 = x$, in (*ii*), $x^2 - 8315.144\ x - 3.326 \times 10^8 = 0$

$$\Rightarrow \qquad x = \frac{8315.144 + \sqrt{8315.144^2 - 4 \times 1 \times (-3.326 \times 10^8)}}{2 \times 1}$$

$\Rightarrow \qquad x = 22862.8 = D^2, \quad D =$ **151.2 mm**

$\therefore \qquad d = \dfrac{3}{4} D = \dfrac{3}{4} \times 151.2 =$ **113.4 mm**

Example 8.3. *Compare the crippling loads given by Euler and Rankine formulae of a hollow M.S. cylindrical column 2.5 m long of O.D. = 40 mm, I.D. = 30 mm, axially loaded through pinned ends. For what height of the column, does the Euler's formula cease to apply ?*

Take $E = 200\ kN/mm^2, \sigma_c = 330\ N/mm^2, a = \frac{1}{7500}$.

Solution. $I = \frac{\pi}{64}(40^4 - 35^4) = 5.2 \times 10^4 \text{ mm}^4$

$$A = \frac{\pi}{64}(40^2 - 35^2) = 294.52 \text{ mm}^2$$

$$k^2 = \frac{I}{A} = 176.56 \text{ mm}^2$$

For pinned ends $l = L = 2500$ mm

$$\text{Euler} \quad P_{cr} = \frac{\pi^2 EI}{L^2} = \frac{\pi^2 (200 \times 10^3)(5.2 \times 10^4)}{2500^2}$$

$$= 16.4 \times 10^3 \text{ N or } \mathbf{16.4\ kN}$$

and $$\text{Rankine's} \quad P_{cr} = \frac{\sigma_c \cdot A}{1 + a \cdot \frac{l^2}{k^2}} = \frac{320 \times 294.52}{1 + \frac{1}{7500} \times \frac{2500^2}{176.56}} = 17 \times 10^3 \text{ N} \quad \text{or} \quad \mathbf{17\ kN}$$

$$\therefore \quad \frac{P_{cr}\text{ (Euler)}}{P_{cr}\text{ (Rankine)}} = \frac{16.4}{17} = 0.965 \simeq 1$$

$$\text{Euler's} \quad P_{cr} = \frac{\pi^2 EI}{l^2} = \frac{\pi^2 E(Ak^2)}{l^2} = \frac{\pi^2 EA}{(l/k)^2}$$

$$\left(\frac{l}{k}\right)^2 = \frac{\pi^2 E}{P_{cr}/A} = \frac{\pi^2 E}{\sigma_c} = \frac{\pi^2 (200 \times 10^3)}{330} = 5981.6$$

$$\therefore \quad \frac{l}{k} = 77.34$$

Euler's formula ceases to apply when the slenderness ratio becomes less than 77.34 and the length

$$l < 77.34\sqrt{176.56}$$

$$< 1027.66 \text{ mm or } 1.027 \text{ m}$$

or $$\frac{l}{D} = \frac{1027.66}{40} < 25.7, \quad \text{say } l < \mathbf{26\ D}$$

Long Columns Under Eccentric Loading. For a column under a load P, with eccentricity e, from the symmetrical axis xx, (see Fig. 8.3), then the maximum and minimum stresses developed are given by

$$\sigma_{y_{1/2}} = \frac{P}{A} \pm \frac{M}{z}, \quad M = Pe \sec\frac{\alpha l}{2}, \text{ where } \alpha = \sqrt{P/EI}$$

For σ_{max} at y_1, $z = \frac{I_{xx}}{y_1}$, $I_{xx} = A \times k_b^2$, $I = A \times k^2$

$$\therefore \quad \sigma_{max} = \frac{P}{A} + \frac{Pe\, y_1 \sec\frac{l}{2}}{A \times k_b^2}$$

$$\sigma_{max} = \frac{P}{A}\left[1 + \frac{ey_1}{k_b^2}\sec\frac{l}{2k}\sqrt{\frac{P}{EA}}\right] \qquad \text{...(8.8)}$$

which is the **Secant Formula** for the maximum compressive stress in the column

where l = equivalent length,

k_b = radius of gyration about the axis of bending ($\sqrt{I_{xx}/A}$),

k = least radius of gyration i.e., about the buckling axis ($\sqrt{I_{yy}/A}$).

Note : If in addition to eccentric load P, there is an axial load P_o, then

$$\sigma_{y_{1/2}} = \frac{(P_o + P)}{A} \pm \frac{M}{z}, \quad M = P \cdot e \sec \frac{\alpha l}{2}$$

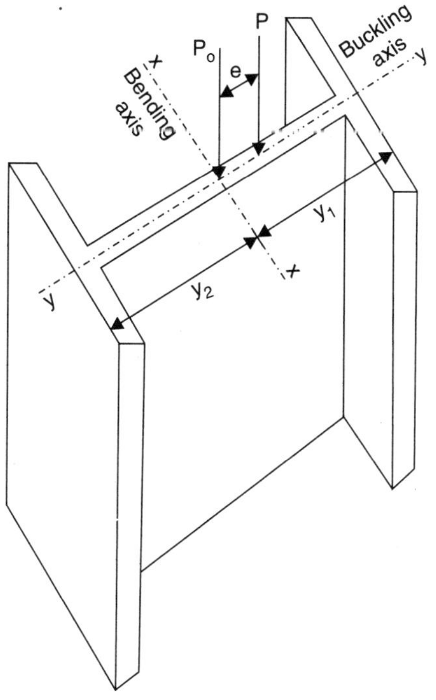

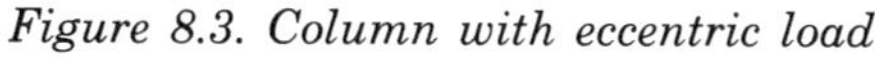
Figure 8.3. Column with eccentric load

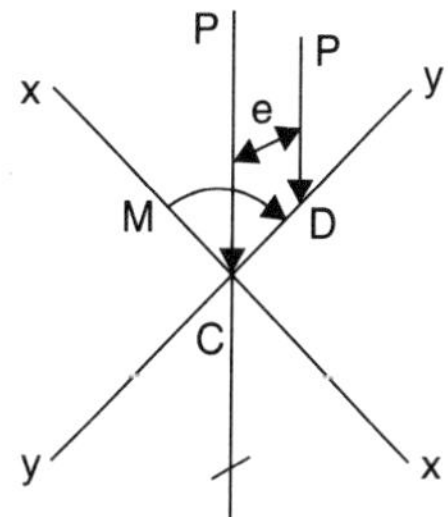

Figure 8.3a

Note: P at D

|||

P at C and M (= $P.e$) *i.e.*, a force ≡ same force (moved) and a couple

$M = P.e \sec \frac{\alpha l}{2}$, for long olumns.

Example 8.4. *A steel hollow cylindrical column with O.D. = 100 mm and I.D. = 90 mm is 3 m long and hinged at both ends. A load of 60 kN is applied at an eccentricity of 10 mm from its axis. Determine the maximum and minimum compressive stresses developed. What is the maximum eccentricity so that no tension is induced.* $E = 2 \times 10^5$ N/mm^2

Solution. $\sigma = \frac{P}{A} \pm \frac{M}{z}$, where $M = P \cdot e \sec \frac{\alpha l}{2}$, $\alpha = \sqrt{\frac{P}{EI}}$

$$A = \frac{\pi}{4}(120^2 - 90^2) = 4948 \text{ mm}^4$$

$$I = \frac{\pi}{64}(120^4 - 90^4) = 69.58 \times 10^5 \text{ mm}^4 = I_{xx} = I_{yy}$$

$$k^2 = \frac{I_{yy}}{R} = 1406.25 \text{ mm}^2$$

$$z = \frac{I_{xx}}{y_1} = \frac{69.58 \times 10^5}{90/2} = 1.546 \times 10^5 \text{ mm}^3$$

$$\alpha = \sqrt{\frac{P}{E \cdot I_{yy}}} = \sqrt{\frac{60 \times 10^3}{(2 \times 10^5)(69.58 \times 10^5)}} = 2.076 \times 10^{-4}$$

$$\sec \frac{\alpha l}{2} = \sec \frac{(2.076 \times 10^{-4})3000}{2} = 1.051 \qquad [\because \ l = 3000 \text{ mm}]$$

$$\sigma_{y_{1/2}} = \frac{60 \times 10^3}{4968} \pm \frac{(60 \times 10^3)10 \times 1.051}{1.546 \times 10^5} = 12.126 \pm 3.88$$

$$\sigma_{max} = \mathbf{16.006 \ N/mm^2}, \quad \sigma_{min} = \mathbf{8.246 \ N/mm^2}$$

For no tension, $\frac{P}{A} = \frac{M}{z}$, i.e., $\frac{P}{A} = \frac{P \cdot e \times 1.051}{z}$, See Fig. 8.3(*a*)

$$e = \frac{z}{1.051 A} = \frac{1.546 \times 10^5}{1.051 \times 4948} = \mathbf{29.72 \ mm}$$

Example 8.5. (*a*) *A steel stanchion is built up of three ISMB 225 as shown in Figure 8.4. The length is 10m and is used as a strut with both ends rigidly fixed. Find the safe load it can carry with a factor of safety of 6.*

(*b*) *What safe load can be applied if its line of action on A-A axis is 50 mm away from the centroid of the section and the maximum stress developed ?*

[For steel, $a = \frac{1}{7500}$, $\sigma_c = 3300 \ N/mm^2$, $E = 200 \ kN/mm^2$.*]*

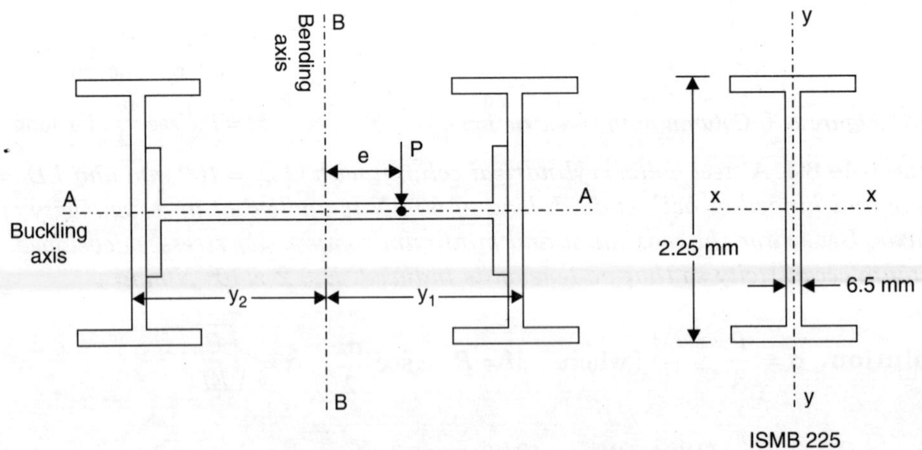

Figure 8.4. Built-up section

Solution. From steel tables for each ISMB 225

$$A = 3972 \text{ mm}^2, I_{xx} = 34.42 \times 10^6 \text{ mm}^4, I_{yy} = 2.18 \times 10^6 \text{ mm}^4.$$

For the built-up section

$$A = 3972 \times 3 = 11{,}916 \text{ mm}^2$$

$$I_{AA} = 2I_{xx} + I_{yy} = 2(34.42 \times 10^6) + 2.18 \times 10^6 = 71.04 \times 10^6 \text{ mm}^4.$$

$$I_{BB} = 34.42 \times 10^6 + 2\left[2.18 \times 10^6 + 3972\left(\frac{6.5}{2} + \frac{225}{2}\right)^2\right]$$

$$= 145.22 \times 10^6 \text{ mm}^4$$

$$k^2 = \frac{I_{\text{least}}}{A} = \frac{71.04 \times 10^6}{11{,}916} = 5961.73 \text{ mm}^2$$

$$k_b^2 = \frac{I_{\text{bending}}}{A} = \frac{145.22 \times 10^6}{11{,}916} = 12{,}186.9 \text{ mm}^2$$

$$k_b = \sqrt{12{,}186.9} = 110.394 \text{ mm}$$

$$k = \sqrt{5961.73} = 77.21 \text{ mm}$$

For both ends rigidly fixed

$$l = \frac{L}{2} = \frac{10}{2} = 5\text{m, or } 5000 \text{ mm}$$

(*a*) Rankine's $P_{cr} = \dfrac{\sigma_c A}{1 + a\dfrac{l^2}{k^2}}$

$$= \frac{330 \times 11{,}916}{1 + \dfrac{1}{7500} \times \dfrac{5000^2}{5961.73}} = 25.22 \times 10^5 \text{ N or } 2522 \text{ kN}$$

$$\text{Safe load} = \frac{P_{cr}}{\text{F.S.}} = \frac{2522}{6} = \mathbf{420.35\ kN}$$

(*b*) Due to eccentric loading, $e = 50$ mm (on *A-A* axis)

$$\sigma = \frac{P}{A} \pm \frac{M}{z} \quad \text{...}(i)$$

where $M = P \cdot e$

and $z = \dfrac{I_{BB}}{y_1} = \dfrac{Ak_b^2}{y_1}$

Putting the value of M and z in (*i*), then

$$\sigma = \frac{P}{A} \pm \frac{Pe \cdot y_1}{Ak_b^2}$$

$$\sigma = \frac{P}{A}\left(1 \pm \frac{ey_1}{k_b^2}\right) = (\text{Rankine's } P_{cr}) = \frac{\sigma_c}{1 + a\dfrac{l^2}{k^2}}$$

$$\therefore \quad P = \frac{\sigma_c \cdot A}{1 + a \cdot \dfrac{l}{k^2}} \cdot \frac{1}{1 + \dfrac{ey_1}{k_b^2}} = \frac{P_{cr}}{1 + \dfrac{ey_1}{k_b^2}}$$

i.e., Safe eccentric load $= \dfrac{\text{safe central load}}{1 + \dfrac{e \cdot y_1}{k_b^2}}$ ...(8.9)

$$\therefore \quad P_{safe} = \frac{420.35}{1 + \dfrac{50\left(\dfrac{225}{2} + 6.5\right)}{12{,}186.9}} = \frac{420.35}{1.488} = \mathbf{282.5\ kN}$$

$$\sigma_{max} = \frac{P}{A}\left(1 + \frac{e \cdot y_1}{k_b^2}\right) = \frac{282.5 \times 10^3}{11{,}916} \times 1.488 = \mathbf{35.28\ N/mm^2}$$

Alternatively, from the Secant Formula, [Eq. (8.8)]

$$\sigma_{max} = \frac{P}{A}\left[1 + \frac{ey_1}{k_b^2} \sec \frac{l}{2k}\sqrt{\frac{P}{E \cdot A}}\right]$$

$$= \frac{282.5 \times 10^3}{11{,}916}\left[1 + \frac{50\left(\dfrac{225}{2} + 6.5\right)}{12{,}186.9} \sec \frac{5000}{2 \times 77.21}\sqrt{\frac{282{,}500}{(2 \times 10^5) \times 11{,}916}}\right]$$

$$\therefore \quad = \frac{282.5 \times 10^3}{11{,}916}[1 + 0.488] = \mathbf{35.28\ N/mm^2}$$

PROBLEMS

8.1. A steel column consists of two channels ISLC 350 × 380.6 N/m per channel placed at 200 mm back to back. The plates 300 mm × 10 mm are riveted one to each flange. The column is 6m long fixed at one end and another end hinged. Determine the safe load the column can carry using Rankine's formula and a factor of safety of 3.5.

[For steel σ_c = 330 N/mm², $a = \dfrac{1}{7500}$

Properties of each channel : Area = 4947 mm², I_{xx} = 93.13 × 10⁶ mm⁴

I_{yy} = 3.95 × 10⁶ mm⁴, C_{yy} = 24.1 mm.]

8.2. (*a*) From some tests on steel struts with ends fixed in position but free in direction, the following results were obtained:

Trial No.	1	2
Slenderness ratio	70	170
Average stress at failure (MN/m²)	200	69

Assuming that the Rankine's formula holds good, find the two constants in the formula.

(*b*) If a steel bar of rectangular section 60 mm × 20 mm and of length 1.25 m is used as a strut with both ends fixed in position and direction, find the safe load using the constants obtained in (*a*) above, and employing a factor of safety of 4.

8.3. Determine the crippling load for a strut of *T*-section of flange width 150 mm and overall depth 120 mm and thickness of both flange and stem 20 mm. The strut is 4 m long and hinged at both ends. [Take E = 200 GN/m^2]

8.4. A M.S. stanchion is built up of one 200 mm (overall depth) × 140 mm (wide) flange R.S.J. with one 160 mm × 10 mm plate riveted to each flange. For what length of the column will both the Rankine and Euler formulae give the same crippling load. The column is fixed at one end and the other end is hinged.

[For steel $E = 200$ GN/m^2, $\sigma_c = 330$ N/mm^2, $a = \dfrac{1}{7500}$.

Properties of the R.S.J. : Area = 3671 mm^2, $I_{xx} = 26.25 \times 10^6$ mm^4,

$I_{yy} = 3.29 \times 10^6$ mm^4, $r_{xx} = 84.6$ mm, $r_{yy} = 29.9$ mm]

8.5. A built up *I*-section has an overall depth of 1100 mm, 300 mm wide flanges 50 mm thick and web 20 mm thick. It is used as a beam simply supported at ends and deflects by 10 mm at mid-span when subjected to a *udl* of 40 kN/m. Estimate the safe load when the section is used as a column with both ends fixed. Assume a factor of safety of 4 and use Euler's formula. [Take E = 200 GN/m^2]

$$\left[\textbf{Hint: } \text{For a SSB with } udl\ \delta_C = \frac{5}{384}\frac{wL^4}{EI_{xx}}\right]$$

[Take least I for Euler P_{cr}] (**Ans.** 2320 kN)

8.6. A hollow C.I. column, with fixed ends, has to carry an axial load of 1 MN. If the column is 5m long and O.D. = 250 mm, what thickness of metal is required ?

Use Rankine's formula, taking $a = \dfrac{1}{1600}$ and a working stress of 80 MN/m^2.

[**Hint:** For σ = 80 MN/m^2, use P = 1 MN in the Rankine's formula] (**Ans.** 29.5 mm)

8.7. A tubular steel strut has O.D. = 60 mm and I.D. = 45 mm. It is 2.5 m long and has hinged ends. The load is applied parallel to the axis at an eccentricity of '*e*'. What should be the maximum value of '*e*' for a load of 70% of the Euler value, if the yield stress is 310 N/mm^2 and E = 200 GN/m^2.

$$\left[\textbf{Hint: } 310 = \frac{P}{A} + \frac{M}{z}, \text{ where } M = Pe \sec\frac{\alpha l}{2}\right]$$

8.8. An alloy bar 1 m long and 15 mm × 5 mm in section was loaded axially as a column till it buckled. Assuming the Euler's formula for pinned ends to apply, estimate the maximum central deflection before the material attains, its yields point at 300 N/mm^2.
Take E = 80 GN/m^2.

$$\left[\textbf{Hint: } 300 = \frac{P}{A} + \frac{M}{z}, M = p\delta, P = \text{Euler } P_{cr}\right]$$

[Take least I for P_{cr} and z] (**Ans.** 156 mm)

8.9. A short length of tube 25 mm I.D. and 32 mm O.D. failed in compression at a load of 70 kN. When a 2.4 m length of the same tube was tested as a strut with fixed ends, the failing load was 25 kN. Assuming that σ_c in the Rankine's formula is given by the first test, find the value of the constant a in the same formula. Hence, estimate the crippling load for a piece of the same tube 1.5 m long when both ends are hinged.

QUESTIONS AND PROBLEMS

Questions

A. OBJECTIVE–I

I. Say True or False; if false, give the correct statement:

(*a*) Moment of inertia of any section is maximum about the axis passing through its C.G.

(*b*) The polar moment of inertia of a circular section is twice the moment of inertia about an axis passing through its C.G.

(*c*) In a shaft subjected to torsion, the angle of twist will increase as the length of the shaft increases.

(*d*) Buckling is due to axial load while bending is due to transverse load.

(*e*) Short columns fail by buckling while long columns fail by crushing.

(*f*) In a shaft subjected to torsion, the shear stress is maximum at the centre.

(*g*) In a loaded beam, the shear stress is maximum at the extreme fibres while the flexure stress is maximum at the neutral axis.

(*h*) The strength of a beam is proportional to its section modulus.

(*i*) In a cantilever with *udl* over the entire length, the tensile stresses are induced at the top, while in a simply supported beam with *udl* over the entire span (or concentrated load at mid-span), the tensile stresses are induced at the bottom.

(*j*) In a loaded beam, the neutral axis passes through its centroid.

(*k*) On a principal plane, the shear stress is maximum.

(*l*) The planes of maximum shear are at 45° to the principal planes and the normal stress on the maximum shear planes is the average of the major and minor principal stresses.

(*m*) The maximum shear is equal to the radius of the Mohr's stress circle and is equal to half the difference of the principal stresses.

(*n*) The normal stress on the planes of maximum shear is zero.

(*o*) A thin cylinder is subjected to the longitudinal stress, only if the ends are closed.

(*p*) The crippling load is inversely proportional to the square of the slenderness ratio.

(*q*) The maximum stress or deflection due to a suddenly applied (shock) load is twice that caused by gradually applied (static) load.

(*r*) For a simply supported beam with a concentrated load, the maximum deflection occurs under the load.

(*s*) For a simply supported beam with *udl* over the entire span, the maximum deflection occurs at mid-span.

(*t*) For a cantilever with a *udl* over the entire length, the maximum deflection occurs at the free end.

(*u*) A column is a vertical member subjected to compression, while a strut is a compression member in any direction. (**False:** *a, c, e, f, g, k, n,*)

II. Rewrite after filling the blanks:

(*a*) The rod of a ceiling fan is subjected to .... stress.

(*b*) In a reveted lap joint, the revets are subjected to .... stress.

(*c*) The strain energy stored in a body upto the elastic limit is known as ....

(*d*) The proof resilience under tension is .... of that under shear loading.

(*e*) The circumferential stress is known as .... stress.

(*f*) In a thin cylindrical shell the $\frac{d}{t}$ ratio is ≥ ....

(*g*) Longitudinal stress is .... of the hoop stress in seamless cylinderical shells, and such shells are designed for .... stress.

(*h*) In a beam, a roller is capable of giving .... reaction, the hinged (or pinned) end is capable of giving .... and a fixed end is capable of giving .... as well as ....

(*i*) A tie rod is a .... member and a column (strut) is a .... member.

(*j*) Euler's formula is used for .... columns while the Rankine-Gordon formula is used for .... columns (short, long, medium).

(*k*) A close coiled helical spring subjected to direct axial load, mainly develops .... stress.

III. Choose the correct answer:

(*a*) Maximum bending moment occurs in a beam where

(*i*) S.F. = 0, (*ii*) S.F. = maximum

(*iii*) S.F. = minimum, (*iv*) S.F. = changes sign.

(*b*) At the point of contraflexure

(*i*) B.M. = 0, (*ii*) B.M. = maximum

(*iii*) B.M. = minimum (*iv*) B.M. changes sign.

(*c*) Change in temperature induces stresses in a structural member if

(*i*) both ends are hinged

(*ii*) both ends are fixed

(*iii*) one end fixed and the other is on roller

(*iv*) one end fixed and the other free.

(*d*) If a load is applied to an elastic body, then the stress developed is

(*i*) equal to the strain (*ii*) less than strain

(*iii*) greater than strain (*iv*) proportional to the strain.

(**Ans.** *a – i ; b – iv, i ;* c *– ii, d – iv*)

IV. Match the following:

(*a*) Strain (*i*) mm^3

(*b*) Young's modulus of elasticity (*ii*) mm

(*c*) Polar moment of inertia (*iii*) N/mm

(*d*) Section modulus (*iv*) mm^{-1}

(*e*) Torque	(*v*) N·mm/mm^3
(*f*) Stiffness of a spring	(*vi*) no dimensions
(*g*) Slenderness ratio	(*vii*) mm^4
(*h*) Radius of gyration	(*viii*) N/mm^2
(*i*) The constant of a column'α'	(*ix*) No dimensions
(*j*) Resilience	(*x*) kN·m.

(**Ans.** *a-vi*, *b-viii*, *c-vii*, *d-i*, *e-x*, *f-iii*, *g-ix*, *h-ii*, *i-iv*, *j-v*)

B. OBJECTIVE – II

(*a*) Say True or False; if false, give the correct statement:

(*i*) The torque-twist plot is similar to load-extension plot, but the curve does not droop.

(*ii*) Brinell test is suitable for extremely hard materials and finished products.

(*iii*) In a torsion test yielding of the specimen commences in the core.

(*iv*) The tensile strength of material directly varies as the hardness.

(*v*) Percentage elongation is a measure of ductility.

(*vi*) In impact test the notch of the specimen is on the tension side.

(*vii*) In compound bar subjected to axial load the stresses in the bars are dissimilar and proportional to their elastic moduli.

(*viii*) In a compound bar subjected to a rise in temperature, the stresses in the bars are similar and directly proportional to their areas.

(*ix*) A hollow shaft is always stronger than a solid shaft of the same weight, material and length.

(*x*) Poisson's ratio $\mu = \frac{1}{m}$, where m is always greater than 2, and usually lies between 2 and 4.

(*xi*) Brittle coatings are thicker than photoelastic coatings.

(*xii*) Isoclinics give the locus of all points having the same direction of principal stresses and they are independent of applied load.

(*xiii*) A fringe (dark or isochromatic) represents the locus of all points having the same maximum shear stress (or principal stress difference).

(*xiv*) In a crossed plane polariscope, if the polariser-analyser combination is rotated, the isochromatics move.

(*xv*) The photographs of dark-and bright-field fringe patterns will enable to determine the fring order to an accuracy of half wavelength, at the point of interest.

(*xvi*) The maximum shear stress is half the difference of the principal stresses and acts on planes at 45° to the principal planes, forming complimentary shear stress.

[False: (*ii*), (*iii*), (*v*), (*vii*), (*viii*), (*xi*), (*xiv*)]

(*b*) Match the following:

1. Group I

A	**B**
(*i*) C.I.	(*a*) Isotropic and ductile
(*ii*) Wood	(*b*) Isotropic and brittle
(*iii*) M.S.	(*c*) Orthotropic
(*iv*) Aluminium	(*d*) Alloy
(*v*) Brass	(*e*) Isotropic and highly ductile

2. Group II

A	**B**
(*i*) Bending test on timber beam	(*a*) Modulus of rigidity
(*ii*) Hardness test	(*b*) Toughness
(*iii*) Torsion test	(*c*) Ductility
(*iv*) Impact test	(*d*) Resistance to surface indentation
(*v*) Tension test	(*e*) Endurance limit
(*vi*) Fatigue test	(*f*) M.O.R.

3. Group III

A	**B**
(*i*) Compression test on C.I.	(*a*) Cup and cone fracture
(*ii*) Impact test on M.S.	(*b*) Star shaped
(*iii*) Torsion test on M.S.	(*c*) Splitting of fibres
(*iv*) Tension test on aluminium	(*d*) Fracture along 45° to the axis
(*v*) Bending test on wood along grains	(*e*) Local plastic flow

4. Group IV

A	**B**
(*i*) BHN	(*a*) Railway axles
(*ii*) VPN	(*b*) Springs
(*iii*) Rockwell superficial hardness	(*c*) Cup and cone-fibrous, dull-gray or silky
(*iv*) Rebound hardness	(*d*) Irregular-bright, crystalline or granular
(*v*) Impact test	(*e*) Razor blades
(*vi*) Fatigue test	(*f*) Raw stock
(*vii*) High resilience	(*g*) Polished and finished hard materials
(*viii*) Ductile fracture	(*h*) Ship's hull, pressure vessels, crane hooks
(*ix*) Brittle fracture	(*i*) Difficult profiles; portable

5. Group V

A	**B**
(*i*) Creep	(*a*) Dimensionless number
(*ii*) RHN	(*b*) N·mm/mm^3

(*iii*) BHN	(*c*) N/mm^2
(*iv*) Spring stiffness	(*d*) kg$_f$/mm^2
(*v*) Bulk modulus	(*e*) N/mm
(*vi*) Impact strength	(*f*) 1% strain in 100,000 hr.
(*vii*) Resilience	(*g*) Joules.

ANSWERS

Group:

I. (*i*) (*b*), (*ii*) (*c*), (*iii*) (*b*), (*iv*) (*e*), (*v*) (*d*),

II. (*i*) (*f*), (*ii*) (*d*), (*iii*) (*a*), (*iv*) (*b*), (*v*) (*c*), (*vi*) (*e*) ;

III. (*i*) (*d*), (*ii*) (*b*), (*iii*) (*e*), (*iv*) (*a*), (*v*) (*c*) ;

IV. (*i*) (*f*), (*ii*) (*g*), (*iii*) (*e*), (*iv*) (*i*), (*v*) (*h*), (*vi*) (*a*), (*vii*) (*b*), (*viii*) (*c*) (*ix*) (*d*) ;

V. (*i*) (*f*), (*ii*) (*a*), (*iii*) (*d*), (*iv*) (*e*), (*v*) (*c*), (*vi*) (*g*), (*vii*) (*b*).

(*c*) Choose the correct answer:

(*i*) A point of contraflexure in a loaded beam is one at which

(*a*) the load is zero
(*b*) the shear force is zero
(*c*) the bending moment is zero
(*d*) the deflection is zero
(*e*) the slope is zero.

(*ii*) A beam of uniform strength is one in which

(*a*) area of cross-section is uniform
(*b*) modulus of section is uniform
(*c*) the moment of resistance is uniform
(*d*) extreme fibre stress is uniform at all the sections.

(*iii*) The major principal plane in a stressed material is the one in which

(*a*) the normal stress is zero and the shear stress is maximum
(*b*) the normal stress is maximum and the shear stress is zero
(*c*) both the normal and shear stresses are maximum
(*d*) both the normal and shear stresses are zero.

(*iv*) When tensile and shear stresses act on a plane in the stressed material, the principal stresses are

(*a*) of the same sign
(*b*) of different signs
(*c*) zero in magnitude
(*d*) one is maximum, the other is zero.

(*v*) Direct shear stress is conducted to determine

(*a*) neutral axis
(*b*) modulus of rigidity

(*c*) ultimate shear strength

(*d*) modulus of elasticity.

[Ans. (*i*) – *e*, (*ii*) – *d*, (*iii*) – *b*, (*iv*) – *d*, (*v*) – *c*]

C. SHORT QUESTIONS (VIVA)

1. Distinguish among: axial stress, shear stress, and volumetric strain.

2. Explain: Poisson's ratio and Bulk modulus.

3. Distinguish between elastic limit and proportional limit; up to which limit Hook's law applies ?

4. Draw a typical stress-strain curve for

(*a*) a ductile material

(*b*) a brittle material

Indicate the salient points on the curve. How does a true stress-strain diagram differs from (*a*) above ?

5. What are elastic constants? Define each of them. Is there any relation between them ? If so write down.

6. Explain: (*a*) St. Venant's principle

(*b*) Stress concentration

7. Distinguish among:

(*i*) Strain energy, (*ii*) Resilience, (*iii*) Toughness.

8. Distinguish between Fatigue and Creep.

9. Explain 'fracture' with reference to ductile and brittle materials.

10. What is meant by 'beam of uniform strength' ? Can a carriage spring be called like that ? Give reasons.

11. Explain: (*a*) Torsional rigidity

(*b*) Stiffness of a spring.

12. Differentiate between:

(*a*) Strain and deformation

(*b*) Strain gauge and extensometer.

13. In a static bending test why two-point symmetrical loading is preferred ?

14. Why in a tension test the load drops after reaching a maximum value ?

15. Why short specimens are used in compression test ?

16. How do you select load in a Brinell hardness test ?

17. What is the necessity of applying a minor load in Rockwell hardness test ?

18. Distinguish between 'Indentation hardness' and 'Rebound hardness'; where each is employed ?

19. Distinguish between an open-coil and closed coil helical spring.

20. Why modulus of rigidity is not determined by direct shear test ? By which other test is it determined ?

21. Why a notch is provided in the specimen for Impact Test ?

22. Explain: (*a*) Proof stress, (*b*) Endurance limit.

23. What is the necessity of standardisation of tests, conducted in the laboratory ?

24. List the tests conducted on a ductile material in the laboratory.

25. What mechanical properties qualify the selection of material for:

(*a*) Beam (*b*) A cable

(*c*) Actuating tool (*d*) Moving part of a high speed mechinery

(*e*) Spring steel

26. What are the factors which affect the indentation hardness ?

27. Why the, load-deflection plot' for leaf springs makes a loop ? How the width of loop is affected by the use of lubricants ?

28. Explain 'yield point', what does it indicate ? Does it exist for brittle materials like C.I. ?

29. Explain the terms 'gauge length' and 'notch sensitivity'.

30. Indicate by a neat sketch the general form of a two-dimensional stress system at a point in an elastic body.

[**Hint:** See *Figure A-1*]

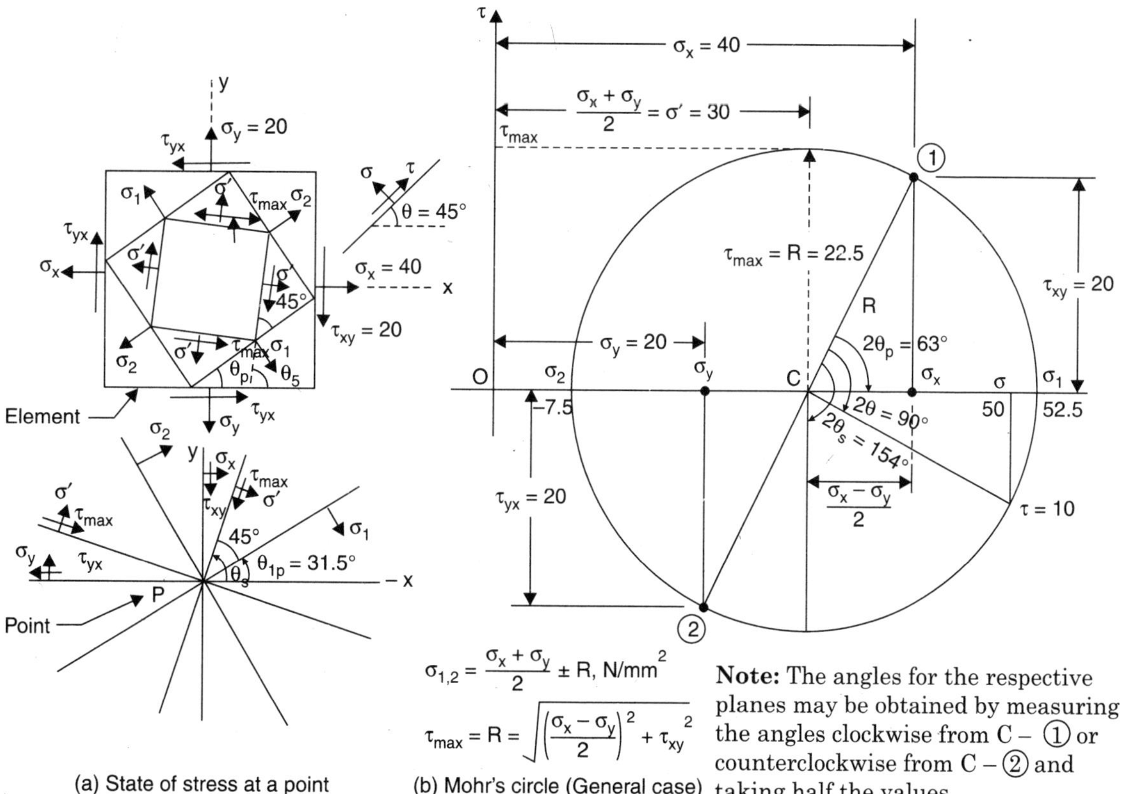

(a) State of stress at a point (b) Mohr's circle (General case)

Figure A-1 General two-dimensional state of stress at a point

31. Explain with the help of neat sketches

(*a*) principal planes and principal stresses

(*b*) planes of maximum shear and maximum shear stress at a point in a stressed material.

32. State naming the investigators:

(*a*) Maximum principal stress theory of failure

(*b*) Maximum principal strain theory of failure

Does failure mean rupture ?

33. Give examples of single and double shear.

34. What is nondestructive testing ?

Explain : (*a*) Magnaflux method

(*b*) Ultrasonic method

(*c*) Schmidt rebound hammer

35. (*a*) Explain 'bend tests' and 'cupping tests'.

(*b*) How would you conduct a tension test on a pipe ?

36. (*a*) Explain 'slip rings' 'load cells' and 'Torque bars'

(*b*) What is a 'Strain gauge transducer' ?

37. What is dynamic strain and how you would measure it ? Explain with sketches.

38. What is a 'rosette' ? When it becomes necessary ? Give two common rosette arrangements ?

39. Explain 'model fringe value' and 'material fringe value' in photoelasticity and how you would obtain these values ?

40. Explain: (*a*) Separation techniques

(*b*) Freezing technique in photoelasticity.

D. CONCEPTUAL PROBLEMS

1. A circular bar of length 0.6 m is 20 mm dia in the middle one third of its length and 40 mm dia. on the two end portions. If the bar is subjected to an axial pull of 100 kN, find the stresses produced in the bar and the total elongation. [Take E = 80000 N/mm^2.]

Hint : $\sigma = \dfrac{P}{A}$, $\Delta L = \sum \dfrac{\sigma}{E} \times L_1$ **[Ans.** 318.3, 79.6 N/mm^2 ; 1.194 mm]

2. A 10 mm steel rod 1 m long is inserted inside an aluminium tube 20 mm external dia, 2 mm thick and of the same length, and rigidly connected at their ends such that they are coaxial. If $E_{St} = 2 \times 10^5$ N/mm^2 $= 3E_{Al}$, $\alpha_{St} = 11 \times 10^{-6}$/°C, $\alpha_{Al} = 2\alpha_{St}$, Calculate the stresses in the two materials when they are subjected to

(*i*) axial load of 20 kN

(*ii*) Temperature increase by 50°C

Hint: (*i*) *Figure A-2* (*i*)

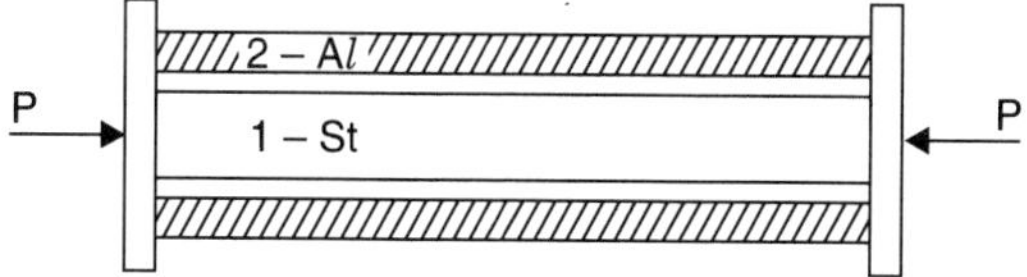

(i) Compound bar with axial load

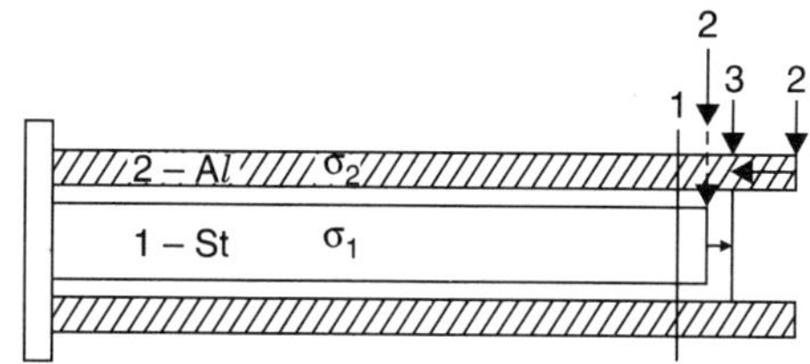

1. Initial position
2. Positions if free
3. Final position
   (ends being rigidly connected)

(ii) Compound bar subjected to temperature rise

Figure A-2

Strain is same : $\frac{\sigma_1}{E_1} = \frac{\sigma_2}{E_2}$

Load P is shared : $\sigma_1 A_1 + \sigma_2 A_2 = P$

Solve for σ_1 and σ_2 which are similar stresses proportional to the moduli.

(*ii*) *Figure A-2* (*ii*)

$$\sigma_1 A_1 = \sigma_2 A_2$$

$$\varepsilon_{1T} + \varepsilon_{1t} = \varepsilon_{2T} - \varepsilon_{2c}$$

$$\varepsilon_{1t} - \varepsilon_{2c} = \varepsilon_{2T} - \varepsilon_{1T} = \alpha T$$

where T = temperature, t = tension, c = compression (due to)

$$\frac{\sigma_1}{E_1} + \frac{\sigma_2}{E_2} = 50\,(\alpha_2 - \alpha_1) \qquad \text{...}(ii)$$

Solve for σ_1 and σ_2 from (*i*) and (*ii*); σ_1 and σ_2 are dissimilar stresses inversely proportional to the area.

Note : The length of the compound bar does not come in calculations.

(**Ans.** 172.06, 57.35 ; 35.7, –24.8, N/mm^2)

3. A steel rail is 30 m long and is at a temperature of 24°C. Estimate the elongation when the temperature increases to 44°C. Calculate the thermal stress developed in the rail if

(*i*) no expansion joint is provided

(*ii*) a 6 mm gap is provided for expansion

If the stress developed is not to exceed 20 N/mm², what is the gap left between the rails ? [Take E_{St} = 2 × 10⁵ N/mm², α_{St} = 11 × 10⁻⁶/°C]

Hint : $\Delta L = L\,\alpha\,\Delta T$ (**Ans.** 6.6 mm ; 44, 4, N/mm² ; 3.6 mm)

4. The rails of a railway track are welded at 30°C. Estimate the thermal stress developed in summer when the temperature recorded at noon is 50°

$$E_{St} = 2 \times 10^5 \text{ N/mm}^2,\ \sigma_{St} = 11 \times 10^{-6}/°\text{C}.$$

Hint : Expansion is prevented ; $\sigma_T = \alpha\,\Delta T E$ (**Ans.** – 44 N/mm²)

5. A steel block of *x, y, z* dimensions 200 × 300 × 100 mm is subjected to 200 N/mm² tensile stress in *x*-direction, 150 N/mm² compressive stress in *y*-direction, and 100 N/mm² compressive stress in *z*-direction. If the poisson's ratio is $\frac{1}{3}$ and E_{St} = 2 × 10⁵ N/mm², calculate the new dimensions of the block, the volumetric strain and resilience.

Hint :

$$\varepsilon_x = \frac{\sigma_x}{E} + \frac{\sigma_y}{mE} + \frac{\sigma_z}{mE}, \qquad \Delta x = \varepsilon_x \cdot X$$

$$\varepsilon_y = \frac{-\sigma_y}{E} - \frac{\sigma_x}{mE} + \frac{\sigma_z}{mE}, \qquad \Delta y = \varepsilon_y \cdot Y$$

$$\varepsilon_z = \frac{-\sigma_z}{E} - \frac{\sigma_x}{mE} + \frac{\sigma_y}{mE}, \qquad \Delta z = \varepsilon_z \cdot Z$$

$$\text{Poisson's ratio } \mu = \frac{1}{m}$$

Vol. strain = Algebraic sum of the three principal strains

$$= \frac{(\sigma_1 + \sigma_2 + \sigma_3)(1 - 2\mu)}{E}$$

Resilience U

$$= \frac{1}{2E}\,[\sigma_1^2 + \sigma_2^2 + \sigma_3^2 - 2\mu\,(\sigma_1\sigma_2 + \sigma_2\sigma_3 + \sigma_3\sigma_1)] \text{ per unit volume}$$

(**Ans.** 100.2833, 299.725, 99.14167 mm, –8.333 × 10⁻⁵ ; 0.2395 N·mm/mm³)

6. If a steel cube of side 100 mm is subjected to a stress of 200 N/mm² tensile along each of the principal axes, determine the new dimensions of the cube the volumetric strain and the bulk modulus.

$$E_{St} = 2 \times 10^5 \text{ N/mm}^2, \mu = \frac{1}{3}$$

Hint : Vol. strain = 3 × principal strains

(**Ans.** 100.1 mm^3, 3 × 10^{-3}, 6.67 × 10^4 N/mm^2)

7. Work out the problem no. 6 if the stress is 200 N/mm^2 compressive.

(**Ans.** 99.9 mm^3, 6.67 × 10^4 N/mm^2)

8. Determine the value of μ and E for the material given

$$G = 5 \times 10^4 \text{ N/mm}^2, K = 8 \times 10^4 \text{ N/mm}^2$$

(**Ans.** 0.24, 1.24 × 10^5 N/mm^2)

9. A 3-element rectangular rosette strain gauge gave the following strains at a point in a two-dimensional stress system. 0° : 400 μ m/m, 45° : 200 μm/m, 90° : – 100μ m/m. Determine the principal strains and stresses

(*i*) analytically

(*ii*) by construction of Mohr's strain circle

Assume E = 2.07 × 10^5 N/mm^2, μ = 0.3

(**Ans.** 405, –105, μm/m ; 85, 3.7, N/mm^2 ; θ = 5° 40′ to x-y axes)

10. At a certain point in an elastic material, normal stresses (= σ) act on two mutually perpendicular planes, tensile on one plane and compressive on another plane. Show that the element with planes at 45° to the above planes is in pure shear of the same magnitude (σ) and the linear strain in the direction of normal stresses is half the shear strain of the mutually perpendicular shear planes ; hence obtain the relation between E and G.

Hint : Consider an element of 2 units (unstrained length), *Figure A-3* $\tan\left(\frac{\pi}{4} + \frac{\gamma}{2}\right) = \frac{1+\varepsilon}{1-\varepsilon}$

$$\left[(\textbf{Ans. } \varepsilon = \frac{\gamma}{2}, E = 2G\,(1+\nu)\right]$$

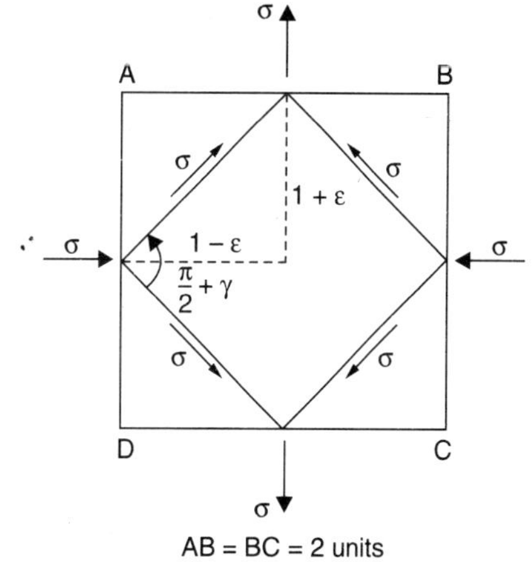

Figure A-3

11. On a certain two-dimensional element bending stresses of 200 N/mm^2 and shear stress of 100 N/mm^2 act. Determine:

(*i*) the normal and shear stresses on a plane at 30° to the bending stress direction.

(*ii*) principal stresses and their direction

(*iii*) maximum shear stresses and their direction.

Construct Mohr's circle for (*ii*) and (*iii*). Indicate in a neat sketch the planes as determined above for (*ii*) and (*iii*).

Hint : (*i*) In *Figure A-4*, on a θ° (= 30°) plane

$$\sigma = \frac{\sigma_x + \sigma_y}{2} - \frac{\sigma_x - \sigma_y}{2} \cos 2\theta + \tau_{xy} \sin 2\theta$$

$$\tau = \frac{\sigma_x - \sigma_y}{2} \sin 2\theta + \tau_{xy} \cos 2\theta$$

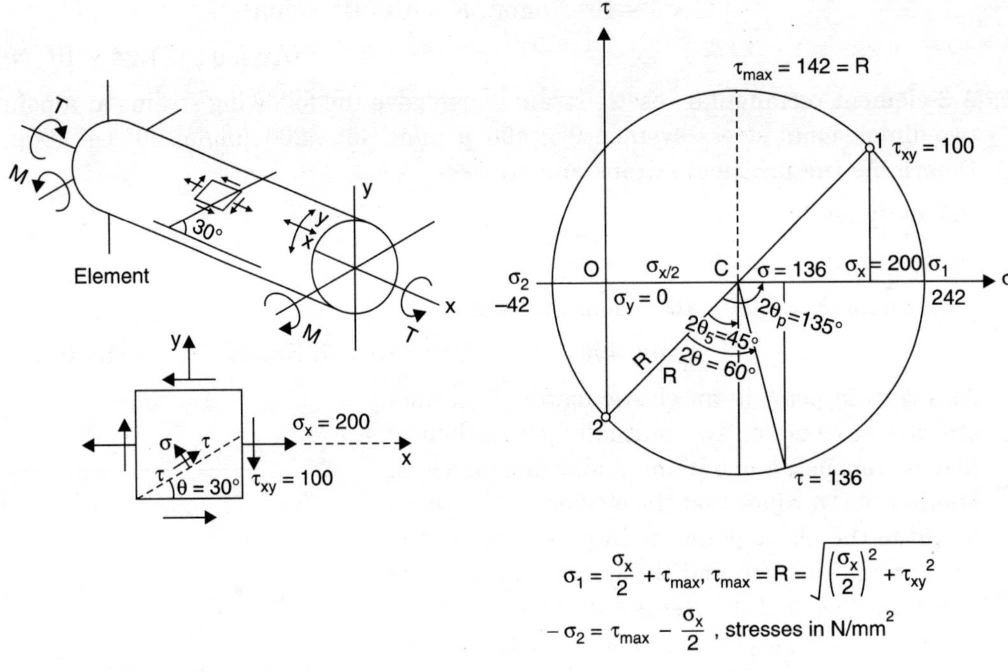

(a) Shaft under combined bending and torsion

(b) Mohr's circle (Problem 11)

Figure A-4

(*ii*) Principal stresses

$$\sigma_{1,2} = \frac{\sigma_x + \sigma_y}{2} \pm \sqrt{\left(\frac{\sigma_x - \sigma_y}{2}\right)^2 + \tau_{xy}^2}$$

$$\tan 2\theta_p = \frac{2\tau_{xy}}{\sigma_x - \sigma_y}$$

(*iii*) $$\tau_{max} = \sqrt{\left(\frac{\sigma_x - \sigma_y}{2}\right)^2 + \tau_{xy}^2} = \frac{\sigma_x - \sigma_y}{2}$$

at 45° to the principal planes ;

Mohr's circle for the problem [see Fig. A–4(*b*)]

(**Ans.** 136.6, 136.6 ; 241.4, – 41.4, 141.4 ; N/mm^2 ; 67.5°, 157.5°, 22.5°, 112.5°)

12. Design the diameter of a solid shaft to carry simultaneously a torque of 1200 N·m and a bending moment of 800 N·m if the maximum shear stress is not to exceed 70 N/mm^2 and the maximum normal stress is not to exceed 110 N/mm^2.

Hint. Equivalent torque $T_e = \sqrt{M^2 + T^2}$

Equivalent moment $M_e = \frac{1}{2}(M + T_e)$

Find '*D*' from : $\tau = \frac{16T_e}{\pi D^3}$, $\sigma = \frac{32\,M_e}{\pi\,D^3}$

Required '*D*' = larger of the two values. [Ref (10)]

(**Ans.** 47.16 mm).

13. (*a*) What are the assumptions in simple bending theory ?

(*b*) A timber beam 4 m long of rectangular cross-section is simply supported at its ends and is subjected to a central concentrated load of 10 kN. Determine the dimensions of the beam given the following :

Width of beam	=	$\frac{1}{4}$ depth
Design bending stress	=	8 N/mm^2
Design shear stress	=	0.4 N/mm^2
Allowable deflection	=	8 mm
Modulus of elasticity	=	7000 N/mm^2

Hint : $d = 4b$; find b for M and SF; b = largest of the three.

(**Ans.** 81.74 × 326.96 mm, from permissible deflection)

14. A simply supported beam 1.5 m long and 50 × 200 mm cross-section, is subjected to an impact load at mid-span by a weight of 200 N dropped from a height '*h*'. If for material of the beam the impact yield stress = 400 N/mm^2 and $E = 2 \times 10^5$ N/mm^2, determine the height of drop.

Hint : Max. bending stress due to impact loading at mid-span given by

$\sigma = \sqrt{\frac{6\,mgh\,Ey^2}{LI}}$; for a rectangular beam σ^2 = 18 Wh E/AL,] (Ref. 10]

(**Ans.** 3.333 m)

15. A hollow steel shaft of external dia 80 mm and internal dia 40 mm is to transmit 200 kW at 500 rpm. It is subjected to an end thrust of 30 kN. To what bending moment the shaft may be subjected to if the maximum stress in the material is not to exceed 80 N/mm^2 ? What is then the maximum stress developed ?

Hint : In the limiting case both end thrust and bending stress are compressive and the maximum principal stress (compressive)

$$80 = \frac{\sigma}{2} + \sqrt{\left(\frac{\sigma}{2}\right)^2 + \tau^2_{xy}}, \quad \sigma = \frac{P}{A} + \frac{MY}{I}, \quad \tau = \frac{Tr}{J}, \text{Power} = T\omega$$

[**Ans.** 2.45 kN·m, 50.4 N/mm^2].

16. A solid steel shaft transmits 600 kW at 250 rpm with a maximum shear stress of 60 N/mm^2. What is the shaft diameter ? Determine the diameters of a hollow shaft ($d = 0.6\ D$) of the same material and length, to transmit the same power at the same speed and stress ?

Compare their stiffness and the percentage saving of material if the solid shaft is replaced by the hollow shaft.

Hint : (*i*) Solid shaft (D_1) : $P = Tw, \quad \tau = \dfrac{16T}{\pi D_1^3}$

(*ii*) Hollow shaft $\left(k = \dfrac{d}{D} = 0.6\right); \quad \tau = \dfrac{16T}{\pi D^3 (1 - k^4)}$

(*iii*) % Saving of metal $= \dfrac{D_1^2 - (D^2 - d^2)}{D_1^2} \times 100 \quad (= 29.2\%)$

[**Ans.** $D_1 = 125$, $D = 131$, $d = 78.4$ mm]

(*iv*) Stiffness $k = \dfrac{T}{\theta} = \dfrac{GJ}{L}$ [**Ans.** $k_s/k_h = 0.95$]

17. In a delta-rosette, the strains measured at a point in a two dimensional stress system are 0° : 300 μm/m, 60° : – 400 μm/m, 120° : 100 μm/m. Taking $E = 2 \times 10^5$ N/mm^2 and $\nu = 0.3$, determine the principal stresses, the maximum shear stress and their directions

(*i*) analytically,

(*ii*) by construction of Mohr's strain circle.

[**Ans.** 63, – 63, 63 N/mm^2, $\theta_{1p} = 21°$, $\theta_{1s} = 66°$]

Appendix - A

IS AND BS SPECIFICATIONS

The following ISS Booklets can be obtained from: The Director, Bureau of Indian Standards, Manak Bhavan, 9-Bhadur Shah Zafar Marg, New Delhi- 110002. India.

Test / Material	Specification
1. Tension Test	BS 18
Mild steel rod	IS : 1608 – 1972
Steel wire	IS : 1521 – 1960, IS : 1633 – 1960 Parts I and II
Aluminium and aluminium alloy tubes and wires	IS : 2657 – 1964, IS : 2658 – 1964
Copper and Copper alloys	IS : 2654 – 1964
Copper and Copper alloy tubes	IS : 2655 – 1964
Copper and Copper alloy wires	IS : 2656 – 1964
Steel tubes	IS : 1894 – 1962
Steel strips and sheets	IS :1666 (Part I) – 1960
	IS : 1663 (Part II) – 1962
2. Torsion testing of steel wire	IS : 1717 – 1971
3. Izod Impact Test for steel	IS : 1598 – 1960
	and – 1959 } BS 131
4. Charpy Impact Test for steel (U-notch)	IS : 1499 – 1977
5. Beam Impact (V-notch) on steel	IS : 1757 – 1961
6. Brinell Hardness Test	BS 240
Light metals and their alloys	IS : 1790 – 1961
Steel	IS : 1500 – 1968
Copper and copper alloys	IS : 3054 – 1965
Grey cast iron	IS : 1789 – 1961
7. Vickers Hardness Test	BS 427
Light metals and their alloys	IS : 1810 – 1961
Copper and copper alloys	IS : 2866 – 1965
Steel	IS : 1501 – 1968

No.	Test	IS Code	BS Code
8.	Rockwell Hardness Test for steel	IS : 1586 – 1968 IS : 3454 – 1967	BS 891
9.	Hardness Conversion Tables for metals	IS : 4258 – 1967	
10.	Fatigue Testing of Metals	IS : 5619 – 1970	
11.	Fatigue – rotating bar	IS : 5075 – 1969	
12.	Creep testing for steel at elevated temperatures	IS : 3407 – 1965 IS : 3408 – 1965 IS : 3409 – 1965	BS 3500
13.	Simple Bend Testing of Steel sheet and strip less than 3 mm thick	IS : 1692 – 1960	BS 1639
14.	Reverse Bend Test for steel and strip less than 3 mm thick	IS : 1403 – 1959	BS 1639
15.	Bend Test for steel products other than sheet, strip and wire	IS : 1599 – 1960	
16.	Bend Test on steel tubes	IS : 2329 – 1963	
17.	Flanging Test on steel tubes	IS : 2330 – 1963	
18.	Modification of Erichsen Cupping Test for steel sheet and strip	IS : 1756 – 1961	
19.	Directions of testing timber specimens	IS : 1708 – 1969	
20.	Nondestructive Testing using ultrasonic Pulse Echo	IS : 3664 – 1966	
21.	Test on Helical Springs	IS : 7906 – 1976	

Appendix – B

AVERAGE PHYSICAL PROPERTIES OF COMMON MATERIALS

SI. No.	*Material*	*Density (ρ, kg/m^3)*	*Coefficient of linear expansion (α, µm/m°C)*	*E (N/mm^2)*	*G (N/mm^2)*	*Ultimate strength, (N/mm^2)*			*Proportional limit (N/mm^2)*		*Elongation %*	*Poisson's ratio, v*
						Tension	*Compression*	*Shear*	*Tension*	*Shear*		
1.	Mild steel	7850	12	200,000	80,000	480	*	350	280	180	25	0.3
2.	Cast iron	7200	11	96,000	40,000	210	500	300	250	150	8	0.27
3.	Brass, Rolled (70% Cu, 30% Zn)	8500	18	100,000	40,000	350	*	300	170	110	8	0.35
4.	Bronze, Cast (90% Cu, 10% Sn)	8800	17	100,000	40,000	230	390	–	140	–	10	–
5.	Copper	8900	21	120,000	40,000	380	*	–	260	160	45	0.35
6.	Duralumin (Aluminium alloy)	2700	23	71,000	30,000	390	*	220	220	150	18	0.34
7.	Wrought iron	7700	12.1	190,000	70,000	350	*	240	210	130	35	–
8.	Monel metal (70% Ni, 30% Cu)	8800	14	180,000	70,000	550			400	–	20	–
9.	Nickel-Chrome steel			206,000	82,000	1650			1100		12	
10.	Timber			12,000	1000	70			40			

*≈ Yield point (slightly greater than the proportional limit in tension)

REFERENCES

1. Davis Troxell Wiskocil, The Testing and Inspection of Engineering Materials, 3rd Edn., ISE, McGraw-Hill Book Company Inc. Kogakusha Company, Ltd. Tokyo, 1964.
2. Durelli, A.J., Applied Stress Analysis, Prentice Hall of India, Pvt. Ltd., New Delhi, 1970.
3. Frought, M.M., Photoelasticity, Vols I and II, John Wiley & Sons, New York, 1941.
4. Gourd, L.M., An Introduction to Engineering Materials, Oxford and IBH Publishing Co., Calcutta – 700 016, 1982.
5. Hetenyi, M., Handbook of Experimental Stress Analysis, John Wiley & Sons, New York, 1950.
6. Holes, K.A., Experimental Strength of Materials (A Laboratory Manual), The English Universities Press Ltd., 102 Newgate Street, London E.C. 1, 1962.
7. Jessop, H.T. and Harris, F.C., Photoelasticity – Principle and Methods, Dover Publications, New York, 1960.
8. Mc. Gonnagle, Warren J., Nondestructive Testing, McGraw-Hill Book Company, Inc., New York, 1961.
9. Perry C.C. and Lissner H.R., The Strain Gage Primer, 2nd Edn., McGraw-Hill Book Company, New York, 1962.
10. Pytel Andrew and Singer Ferdinand L., Strength of Materials, 4th Edn., Harper and Row Publishers, New York, 1987.
11. Ryder, G.H., Strength of Materials, 3rd Edn., in SI Units, ELBS and Macmillan, 1969.
12. Srinivasan, N.K. and Ramakrishnan, S.S., The Science of Engineering Materials, Oxford and IBH Publishing Co., Calcutta – 700 016, 1983.
13. Srinath, L.S., et al., Experimental Stress Analysis, Tata McGraw Hill Publishing Company Ltd., New Delhi, 1984.
14. Suryanarayana, A.V.K:, Testing of Metallic Materials, 2nd Edn., Prentice Hall of India, Pvt. Ltd., New Delhi – 110 001, 1985.
15. Timoshenko, S.P. and Goodier, J.N., ISE, 3rd Edn., 21st Printing, McGraw-Hill Book Co., New Delhi., 1985.
16. TTTI Chandigarh, Programmed Text in Strength of Materials (Stresses and Strains), Tata McGraw-Hill Publishing Co. Ltd., New Delhi, 1972.
17. Warnock, F.V. and Benham, P.P., Mechanics of Solids and Strength of Materials, Sir Issac Pitman & Sons Ltd., London, 1965.
18. Gere, J.M. and Timoshenko, S.P., Mechanics of Materials, 2nd Edn., CBS Publishers and Distributors, Delhi — 110 032, 1986.
19. Nash, W.A., Theory and Problems of Strength of Materials, 2nd Edn., – ASE, SCHAUM's Outline Series, McGraw-Hill International Book Company, Singapore, 1983.
20. Cheng, Fa-Bwa, Applied Strength of Materials, Macmillan Publishing Company, New York, 1986.
21. Peterson, R.E., Stress Concentration Factors, John Wiley & Sons, New York, 1974.

EQUIPMENT MANUFACTURERS

1. Avery India Limited, 28/2, Waterloo Street, Kolkata.
2. Applied Research and Engg. Pvt. Ltd. (ARE) – 12 – inch diffused Light Research Polariscope, R.S. No. 356 A, Plot No. 41, Vishrambag, Sangli – 416 415, Maharashtra.
3. Industrial Engg. Instruments (IEI), IEICOS – Strain Gauge Tool Kit, 203, III Phase, Peenya Industrial Area, Peenya – 562 140 ;

   Also Bangalore Suburb – 560 058, Grams : 'IEIONIX'.
4. Toshniwal Instruments and Engg. Co., 10-A, Shivaji Marg, New Delhi – 110 015. (For Digital Ultrasonic Concrete Tester-(DUCT).
5. Magnaflux Corporation, Chicago, U.S.A. (For Magnaflux Machine).
6. SCHMIDT (Swiss Made) Rebound Hammer (Concrete Test Hammer) : Types N and NR, PROEQ SA, Riesbachstrasse 57 CH-8034, Zurich, Switzerland.
7. The Shore Instrument and Mfg. Company Inc. Jamaica, N.Y., U.S.A. (For the shore Scleroscope).

Index

L

M

N

O

P

Q

R